Advances in High-speed Rail Technology

"Advances in High-speed Rail Technology" presents the latest and most cutting-edge theories, techniques, and methodologies in the multidisciplinary field of high-speed railways, focusing on advances and findings from China. This series includes monographs, professional books and edited volumes from dedicated conferences and workshops. All volumes are authored or edited by established experts in their fields and undergo rigorous peer review, based on the editors' preview and selection and refereeing by independent experts. The intended audience includes researchers, engineers, industrial practitioners, graduate students, and professionals. Topics of interest in "Advances in High-speed Rail Technology" include, but are not limited to: Infrastructure, Mobile Equipment, Communication & Signal, Traction Power Supply, Operation Organization, etc.

Xiubin Zhang · Muhammad Mansoor Khan ·
Yeerjiang Halimu

Principles of Intelligent Rail Transit

上海交通大學出版社
SHANGHAI JIAO TONG UNIVERSITY PRESS

Springer

Xiubin Zhang
School of Electronic Information
and Electrical Engineering
Shanghai Jiao Tong University
Shanghai, China

Muhammad Mansoor Khan
School of Electronic Information
and Electrical Engineering
Shanghai Jiao Tong University
Shanghai, China

Yeerjiang Halimu
School of Computer and Information
Engineering
Xinjiang Agricultural University
Urumqi, Xinjiang, China

ISSN 2363-5010 ISSN 2363-5029 (electronic)
Advances in High-speed Rail Technology
ISBN 978-981-19-6071-0 ISBN 978-981-19-6072-7 (eBook)
https://doi.org/10.1007/978-981-19-6072-7

Jointly published with Shanghai Jiao Tong University Press
The print edition is not for sale in China (Mainland). Customers from China (Mainland) please order the print book from: Shanghai Jiao Tong University Press.

This Springer imprint is published by the registered company Springer Nature Singapore Pte Ltd.
The registered company address is: 152 Beach Road, #21-01/04 Gateway East, Singapore 189721, Singapore

Preface

After more than ten years, my colleagues and many students (Dr. Muhammad Mansoor Khan, Hao Wu, Junhao Ying, Zhen Mo, Changan Ji and many masters and bachelors) have done research work involving transmission and energy saving, network and communication, detection and control, sensing and signal processing, state recognition, image processing, mathematical models and intelligent algorithms. There are also practical technologies of artificial intelligence in rail transit system and so on. Some of them have also directly participated in the technical transformation, R&D or operation management of rail transit system equipment and achieved corresponding results, which have long prompted me to compile relevant achievements into a book. Only because I was busy, I put it behind me for a while. Now, my student Yeerjiang·Halimu is deeply studying intelligent theory and its algorithm, which has activated my thinking and work motivation.

Although I have published several works, I still want to summarize and sort out my achievements in the academic field, so as to encourage my students to continue their efforts and make some contributions to the improvement and perfection of relevant theories and technologies.

I hope the contents described in this book (textbook) can help college students, graduate students and other readers to understand and master the knowledge of Rail Transit Intelligent Technology. Because rail transit is a huge and complex system, involving a wide range of knowledge, many of which are worthy of separate research.

I am very grateful to Muhammad Mansoor Khan, Yeerjiang Halimu and other students for their efforts and help in the publication of this book! In return for your hard work, I specially write this preface.

Shanghai, China
February 2022

Prof. Xiubin Zhang

Contents

About the Authors

Xiubin Zhang (1946.9–) is Professor in the School of electronic information and electrical engineering, Shanghai Jiao Tong University.

Main research interests include information processing and intelligent control, image processing and pattern recognition, artificial intelligence, electrical engineering and automation, creative thinking and method theory, etc.

He has published ten academic monographs, such as the principle of automotive intelligent technology. And he has one software copyright of "intelligent visual perception technology". He also obtained more than 80 Chinese invention patents.

Muhammad Mansoor Khan (1966.9–) is Associate Professor in the School of electronic information and electrical engineering, Shanghai Jiao Tong University.

Main research interests include new energy, energy conservation and environmental protection, smart grid, FACTS, power electronics technology, intelligent control, etc.

He has co-authored three monographs with Professor Xiubin Zhang and obtained 20 Chinese invention patents.

Yeerjiang Halimu (1974.10–) is Teacher in the School of computer and information engineering, Xinjiang Agricultural University, and doctoral student majoring in control science and engineering, Jiangnan University.

Main research interests include machine learning, pattern recognition and intelligent systems, intelligent optimization algorithm research, etc.

He collaborated with Prof. Xiubin Zhang to publish a monograph and four papers.

Chapter 1
Brief History of Rail Transit Development

Take history as a mirror to know the rise and fall. China has experienced a very "sad" process in history. The brilliant scientific and technological achievements of the Chinese nation were "buried" in the "dust" of history by the fatuous and incompetent feudal rulers. In particular, the corrupt rulers of the Qing Dynasty were "unconscious" when the wave of modern western industrial revolution had already spread to the world. For a long time, the feudal dynasty stubbornly refused to change the backward and decadent feudal system, and also took a rejection attitude toward the advanced "Western learning" at that time. It called the scientific theory and technology of the west "in the ascendant" as "barbarian learning" and had no intention to "learn from others" to make up for the "shortage" of its own science and technology. Since Yongzheng's "hundred-year ban on education", the "introduction of Western learning to the East", which had been "bred" as early as the Ming Dynasty, has been blocked for more than 160 years, becoming the most pathetic frog at the bottom of the well in the history of scientific and Technological Development in the world. During this period, Western science and technology flourished in the wave of "ideological liberation". For example, Isaac Newton (1643–1727, a famous British physicist) "three laws of object motion" opened the era of classical physics; James Watt (1736–1819, British inventor and entrepreneur) steam engine gave birth to the world's first industrial revolution; James Clerk Maxwell (1831.1879, British physicist and mathematician) formed classical electrodynamics and classical statistical mechanics; Michael Faraday (1791.1867, British physicist and chemist) established the law of electromagnetic induction... and so on [1].

As far as the development of Rail Transit Intelligent Technology is concerned, this book naturally needs to make a brief description of the development of modern rail transit in the world, so as to help readers think before and after.

Y. Khan, *Principles of Intelligent Rail Transit*, Advances in High-speed Rail Technology,
https://doi.org/10.1007/978-981-19-6072-7_1

1.1 A Brief Introduction to the Development History of Modern Rail Transit in the World

As early as A.D. 60, Heron (A.D. 10–70, ancient Roman mathematician and engineer) had created the prototype of steam engine—a hollow steel ball driven by steam. The steam is discharged from the two opposite bends of the steel ball to form a torque to make the ball rotate—commonly known as "steam ball". This is a simple mechanical technology method that human beings first invented to convert steam into power. For this purpose, Hiro also wrote a monograph called the theory and application of compressed air, which remains in the world [2].

Although Thomas Severy (seventeenth century, British engineer) in 1698, Thomas Newcomen (1664–1729, British engineer) in 1712 and James Watt (1769) all made their own contributions to the development of early steam engine technology, in the end, James Watt made more prominent contributions. This is because James Watt gradually discovered the technical defects of this steam engine by using the existing scientific theory at that time. So in 1776, he built the first steam engine with practical value.

The importance of the steam engine was beyond people's imagination at that time, and Watt's improvement of the mechanical structure of the original steam engine greatly improved the energy conversion efficiency of the steam engine and had a high practical value for industrial production at that time.

Soon, the steam engine was widely used in industrial production and became the power unit of almost all machines, which changed people's production mode, greatly promoted technological progress and then opened the prelude to the western industrial revolution. One of the most important is that the centrifugal governor invented by Watt is an important technology for the speed control of early power machinery. Ever since James Watt's governor solves the problem of adjusting and stabilizing the rotating speed of the machinery driven by the steam engine, which is also the basis of the groundbreaking theoretical creation and technological progress in the development history of human automation technology [1].

After Watt solved the technical problems of mechanical speed stability and automatic regulation, the world's first steam locomotive came out. That is, in 1804, Richard Trevisk (British mining technician) manufactured the world's first steam locomotive "New Castle" according to the principle of Watt's steam engine and its governor. Its mass is 6.5 T and can tow 30t goods. The diameter of moving wheel is 4ft (about 1219 mm) and the gauge is 4ft8in (about 1423 mm).

Although the earliest steam locomotive had simple structure, low thermal efficiency, bulky body and other auxiliary technologies of the locomotive were extremely simple; it was a fashionable means of transportation for the British nobility at that time.

In the nineteenth century, although the train did not run as fast as a horse, it was able to carry large quantities of goods for long-distance transportation, which was already very prominent at that time. In 1814, British railway train operation has become a mature means of rail transit.

In 1894, after the United Kingdom, Germany successfully developed and put into operation the first gasoline diesel locomotive, which ushered in a new era of train technology.

Near the twentieth century, western steam locomotives have begun to "drive into" the history museum, and railway transportation has begun a new era of electric locomotives and diesel locomotives.

Since 1900, the United States followed Europe and built railways on a large scale until 1945.

In 1924, Germany, the United States, France and other countries successfully developed various diesel locomotives, which were widely used in the world. In 1941, Switzerland successfully developed a new type of oil turbine locomotive, which uses diesel as fuel. Because of its simple structure, small vibration and good operation performance, it was widely used in western industrial countries at that time.

According to historical data, in 1941, the total length of Railways in the world reached 1.26 million kilometers. This is the most vigorous period in the history of rail transit technology in the world.

However, from 1946 to 1964, with the development and popularization of automobile technology, European and American countries led by the United States began to dismantle redundant Railways. This is due to the "door-to-door" advantage of cars and the economic benefits of social input/output, which leads to the temporary "stagnation" of the development of rail transit technology, and people's contempt for rail transit.

Since the late 1960s, with the development of a new round of rail transit technology (including the rise of new technologies such as traction mode, new energy, rail structure, tunnel and bridge network, body design, running speed and so on), the overall development trend of transportation tools in the world has undergone earth-shaking changes. With the endless emergence of new technologies, rail transit has returned to the "high level" of the historical process.

In 1964, Japan built the world's first high-speed railway with a speed of 200 km/h, which caused a sensation all over the world. Countries all over the world began to pay attention to the new development of rail transit and imitated Japanese technology to build high-speed trains and high-speed railways [3].

Meanwhile, Japan and Germany quickly became the "birthplace" of high-speed railway technology.

Faced with the rapid development of rail transit technology accompanied by electronics, computers, communications, control and even intelligent high technology, we Chinese people have been filled with emotion. Catching up with and surpassing the world's advanced scientific and technological level has become the direction of efforts hidden in the hearts of the Chinese people and the infinite driving force for the Chinese people to work hard. The following is a brief introduction to the development history of rail transit technology in China.

1.2 Brief History of Rail Transit Development in China

The development of rail transit in China can be viewed from two periods of history. The first period is a hard journey before the birth of new China. During this period, it can be said that it was "terrible". After a period of hard "Climbing", it was just beginning to have an "opportunity to rise", and it was always in a sad situation in the face of continuous wars and foreign aggression. The second period is since the founding of new China. A new page has been opened in the profound history of the Chinese nation. China stands in the forest of nations in the world with a new majestic posture and has launched the great cause of traffic development since then.

1.2.1 Sad Situation of Science and Technology in the Qing Dynasty of Old China

In 1864, the first railway appeared in China. It was built by the British at outside Xuanwu Gate in Beijing. It is about 500 m long. It is a narrow-gauge railway, which cannot be operated commercially at all. During its trial implementation, it also made a big joke.

Li Yuerui (1862–1927, former head of the Ministry of works, in charge of railway and mining Affairs) recorded in his notebook novel "spring ice room wild ride": "In July of the fourth year of Tongzhi, Durand, a British man, laid a small railway with a length of miles on the flat ground outside the Yongding gate of the capital, and drove it with a car as fast as flying. The capital was surprised and unheard of. It was a monster. The whole country was crazy and almost changed greatly. The commander of the infantry army ordered the Yamen to dismantle it, and the group of doubts began to stop. This matter could be the beginning of the introduction of the railway into our country before the Songhu railway started".

At first, after the completion of the railway, Empress Dowager Cixi (1835–1908, the actual ruler of the late Qing Dynasty), the actual controller of the Qing Dynasty, was very interested in the new thing of "Railway" and held the "opening ceremony" in a "solemn" manner.

At that time, the train was moving fast and loud, which frightened people in Beijing who had never seen a train and accused it of being a monster. Empress Dowager Cixi did not like such novel things. In addition, the people did not accept trains, so she immediately sent someone to dismantle the small railway. The first railway in China ended in farce [4].

At that time, although the Qing Dynasty had begun to admit its backwardness, it was still unwilling to accept new things in the western world, and few people in the country were interested in Western learning.

The book "Atlas of the sea country" written by Wei Yuan (1794–1857, the Enlightenment thinker, politician and writer of the Qing Dynasty) was regarded as a demon book by the Qing Dynasty at that time. On the contrary, it was regarded as a treasure by the Japanese and became a key reading for the Meiji Restoration of Japan. At that time, the Westernization Movement had just begun, and the huge resistance also made the Westernization school have no time to take into account the development of railways [5].

In 1876, the British built another railway in China—Wusong railway. This time, it was secretly built in cooperation with merchants of the Qing Dynasty without the permission of the Qing government. This is also the first railway ever operated in China. After more than a year of railway operation, it was forced to stop due to various contradictions. The Qing government began negotiations with Britain. The result of the negotiations was that the Qing government redeemed the railway with 285,000 Liang (equivalent to 4599.9 kg) of silver. Then, the railway tracks were destroyed and all thrown into the sea. Although the Qing Dynasty had accepted the railway at this time, the railway still did not operate.

A year later, Ding Richang, a senior member of the Westernization Movement, built a railway in Keelung mining area, Taiwan, mainly for transporting coal mines. This is also the first railway built by China itself.

In 1881, China made the first simple steam locomotive "rocket". When the Tang Xu railway was opened to traffic in that year, Chinese workers tried to make it with old materials such as mine hoisting boiler and channel iron of shaft frame by relying on several design drawings of Claude William kinder (1852–1936, British engineer), who was then an engineer of Tangshan Kaiping Mining Bureau. This is the construction process of the "Rocket" steam train.

In 1888, after the mediation of Li Hongzhang (a famous official of the late Qing Dynasty), Empress Dowager Cixi rejected the objection and approved Li Hongzhang's proposal to build a railway in Ziguangge, Zhongnanhai. The narrow-gauge railway runs from Beihai in the imperial palace to Zhongnanhai, with a total length of 1510 m. After the completion of the railway, Li Hongzhang imported a steam locomotive and six passenger cars from France as Cixi's imperial car.

In 1909, the Beijing–Zhangjiakou Railway was completed. This is the first independently designed and built railway in China. It has a total length of more than 200 km and took more than four years to complete. It is a pioneering work in Railway Engineering Technology in New China and is of great significance.

The above is a historical tragedy of China's rail transit technology from the end of the nineteenth century to the beginning of the twentieth century.

1.2.2 Rail Transit Construction After the Founding of New China

After heavy historical suffering, the Chinese nation's heavy history books have finally been opened the most glorious page by the Communist Party of China under the leadership of the great man Mao Zedong: The People's Republic of China was founded, and the Chinese people have stood up since then!

However, due to the poverty and weakness of the Chinese nation for hundreds of years, we have to move forward with a heavy load.

At the beginning of the founding of new China, transportation was very backward, and the total mileage of Railways in China was only 22,000 km.

According to the statistics at the beginning of the founding of new China in 1949, there were 4069 locomotives in China at that time, which were produced by more than 30 manufacturers from 9 countries. There were more than 198 types of locomotives. Therefore, they were jokingly known as the "Universal Locomotive Museum".

With the development of railway transportation in New China, the demand for locomotives is increasing. However, since China's railway traction power started with steam locomotive at that time, it can only follow the technical route from transformation, imitation to self-designed step by step.

In 1958, the first diesel locomotive designed by imitating the T3 electric drive diesel locomotive of the former Soviet Union was successfully manufactured in Dalian Locomotive Factory. The speed of the locomotive is only 100 km/h.

In 1964, the speed of "Dongfanghong 1" diesel locomotive produced in China has reached 120 km/h. Since then, "Dongfanghong" series diesel locomotives have come out one after another, and the train speed and other technical performance have also been greatly improved.

Meanwhile, China's Southwest Jiaotong University as the main research unit has carried out research on maglev train technology since 1986. In 1989, the National University of Defense Technology developed China's first maglev test vehicle. In 1990, the maglev team of Southwest Jiaotong University successfully studied the maglev model vehicle composed of four small electromagnets and realized the stable suspension of the model vehicle and the drive based on linear motor. In 1994, the research team successfully developed China's first maglev vehicle and realized the stable suspension and operation of the system. Then, in 1995, China's first maglev train test line was completed in Southwest Jiaotong University. However, there is still a certain gap between China and Germany in the research and application of magnetic levitation technology.

China began to produce "Shenzhou" diesel locomotives in 2000. At the same time, the development and production of electric locomotives in China are also carried out simultaneously. Among them, "Shaoshan" series locomotives are representative, and their running speed and overall technical performance are improving rapidly. In 2007, China has absorbed "Siemens" technology and started to produce "Harmony" electric locomotive.

In 2013, based on CRH380BL technology platform, CRH3A EMU (electric multiple units) was jointly designed and produced by CNR (China CNR Corporation Limited) Changke Co., Ltd. and Tangche company. The maximum design speed of CRH3A EMU is 250 km/h. CRH3C electric multiple units followed by CRH3A electric multiple units adopt power distribution, with 8-car formation, 4 motors and 4 trailers for each train, and the maximum operating speed is 350 km/h [6].

In 2006, the Shanghai Maglev Train demonstration operation line using German technology was put into operation, which is also the first commercial maglev line in China. Shanghai maglev train has 4 trains, including 1 domestic train.

In 2017, the highest speed suspended monorail train in China went offline in "CRRC Sifang" and entered the stage of type test and trial operation. Its design speed is 80 km per hour, and the maximum operating speed is 70 km per hour. The suspended monorail train developed by CRRC Sifang applies some high-speed rail technology to improve operation safety and passenger comfort. It is suitable for light traffic in scenic spots, mountains and cities.

In 2017, an "Automatic rail Rapid Transit (ART)" developed by CRRC Zhuzhou Electric Locomotive Research Institute Co., Ltd., made its debut in Zhuzhou. This is an intelligent virtual rail train.

This new means of transportation, which integrates the respective advantages of modern tram and bus, is a "cross-border work". It subverts people's traditional understanding of urban transportation and brings a new choice and experience for people to solve the travel difficulties in large- and medium-sized cities. In fact, this is an image recognition technology, which enables the train to follow the ground traffic sign line independently. Therefore, it is called intelligent rail express system. The "track" does not have a physical track, but it is only two parallel color lines, so it is called "virtual track".

In 2018, the highest speed suspended monorail train in China officially entered the trial operation stage in Qingdao.

The power machine used in this kind of air train is a new type of permanent magnet motor. Its crawling ability is three times that of ordinary trains. Even on rough and steep road sections, it will make passengers feel very comfortable. The train adopts double-layer shock absorber with low noise. The biggest advantage of air train is that it runs in the air, saves ground space and has low cost. In the operation process, even in case of rain, snow and other bad weather, there is no need to worry about derailment and other accidents. At the same time, the train is also equipped with whole train monitoring system, battery rescue, coupling rescue and other devices, which can easily, quickly and safely leave the dangerous scene in case of emergency.

In fact, in the 1980s, with the support of the German federal government, this new rail transit system has been developed. Under the guidance of this advanced concept, Germany has already born a new public transport system such as H-Bahn (air rail). Japan has also begun to build Chiba-Dangui suspension train operation line since 2006. This kind of technology can be an extension and supplement of the original transportation. This new rail transit system has the function of environmental protection (no exhaust emission, no noise pollution) and has no great impact on the original buildings and environment.

Shanghai also plans to build suspended air trains in the city center, which is intended to connect several core business districts in this way. Its design scheme is being demonstrated.

Of course, in the future, there may be "vacuum tube bullet trains" with newer concepts and speeds of more than 20,000 km per hour, and even other emerging technologies. However, these are still in the conceptual stage, and there is still a considerable period of time from reality. Therefore, I will not make redundant statements for the time being [7].

At present, China's rail transit has also become the forefront of the application of advanced technologies, such as information, intelligence, energy conservation, safety, environmental protection and other technologies which have been widely used in rail transit. Driven by the wave of rail transit development, a large number of advanced technologies and new equipment have been promoted, which has effectively promoted the high-quality development of modern rail transit system characterized by safe, efficient, green, intelligent and sustainable development.

China's rail transit construction has gone through an extraordinary process. By the end of July 2020, China's railway business mileage has reached 141,400 km, ranking second in the world; The mileage of high-speed rail is 36000 km, ranking first in the world. The national "Outline of Powerful Nation Railway Advance Plannin" also defines the development blueprint of China's railway in the next 30 years: by 2035, the modern railway network will be completed first. The railway network has the functions of internal and external interconnection, smooth inter-regional multi-channel, efficient connectivity of provincial capitals, rapid access to prefectures and cities, basic coverage of counties and smooth hub connection. At the same time, the network facilities should be upgraded intelligently and have sufficient effective supply capacity. The total length of China's railway network has reached about 200,000 km, including about 70,000 km of high-speed rail. Cities with a population of more than 200,000 have achieved railway coverage, and cities with a population of more than 500,000 have achieved high-speed rail access. By 2050, a more advanced and perfect modern railway network will be built.

As an important part of the overall transportation system, urban rail transit has also developed rapidly in recent decades. As of May 1, 2020, 47 cities have opened urban rail transit in China (41 in mainland China alone except Hong Kong, Macao and Taiwan). By the end of 2019, a total of 3982 urban rail transit stations had been put into operation in mainland China, including 354 transfer stations. There are 28 cities with transfer stations. According to incomplete statistics, a total of 317 depots and parking lots have been put into operation. By the end of 2019, China's mainland has opened 208 urban rail transit operation lines, with a total length of 6730.27 km, including 5187.02 km of subway, 255.40 km of light rail, 98.50 km of monorail, 715.61 km of urban regional express rail, 405.64 km of modern tram, 57.90 km of maglev and 10.20 km of automatic rapid transit system. According to the line laying mode, the underground line is 4366.5 km, the ground line is 920.3 km and the overhead line is 1449.5 km 4 km.

All these are miracles in the world, which can only be achieved by the Chinese people under the leadership of the Communist Party of China.

1.3 Disasters of World Rail Transit

However, while rail transit provides convenience and speed for people to travel and transport goods, it also brings people a lot of painful memories: the disaster caused by traffic accidents [8].

1.3.1 Examples of Train Accidents in the World

In the early nineteenth century, steam engine train was the most common form of long-distance transportation in the world. Until now, the train is still a very safe and reliable mode of transportation. However, due to the immaturity of rail transit safety guarantee technology, accidents have occurred frequently.

In 1833, the first recorded train accident in New Jersey almost made people want to kill former President John Quincy Adams. However, compared with the train accident in New Jersey in 1833, the train accident in the world after that is the most deadly and unforgettable memory.

On December 12, 1917, a French military train derailed in the St. Michel River valley near the Mont Cenis, killing about 700 people.

In 1944, because the Italian steam engine used inferior coal, the carbon monoxide produced by coal combustion filled the heavily overloaded train carriage. As a result, a disastrous accident occurred in Balvano, killing 520 people.

On April 3, 1955, a train in Mexico fell into a valley in Guadalajara, killing 600 people.

On June 6, 1981, just when a bridge had been damaged by flood and cyclone, when a passenger train in India was passing the bridge in Bihar, the passenger train fell into the water due to the sudden collapse of the bridge, killing about 800 people. It is even estimated that more than 1000 passengers may have been drowned by the river.

In January 1985, due to excessive speed, an express passenger train derailed near Awash town, Ethiopia, causing part of the train's trunk to fall to the bottom of the river canyon. 400 passengers died and another 500 were seriously injured.

On June 4, 1989, the train from the former Soviet Union to Уфа (Russian city) exploded in the gubishev railway section, and the shock wave shook the whole town of Аша (Russian town). This is the most serious accident in the history of Russia and the former Soviet Union. The accident was caused by a huge gas leak from a nearby pipeline. When two passenger trains met here, the friction spark ignited the leaked gas and exploded. 625 people were killed, nearly 1000 injured and 37 wagons were destroyed.

On January 19, 2002, the passenger train from Cairo to Luxor exploded in Aiayat, and seven train boxes caught fire. At that time, the bus was seriously overloaded. Except that some people jumped and survived, many people were burned alive. The actual death toll is much higher than the 383 officially reported [9].

On December 26, 2004, Sri Lanka's "Ocean Queen Consort" train was hit by a huge tsunami caused by the Indian Ocean earthquake when it was driving to the coastal railway in Telwatta. Almost all the people on board were drowned in the tsunami. It is estimated that more than 1700 people died in the accident, and only 100 people were rescued in the end. This is the biggest train accident of this century.

On February 28, 2007, due to strong wind, the 5806-passenger train from Urumqi, Xinjiang to Aksu derailed about 42 km from Zhenzhuquan station to hongshanqu station on the southern Xinjiang line, causing 11 carriages to overturn in strong wind.

The accident resulted in the death of four passengers, and the traffic of Nanjing line was interrupted for a time.

On May 23, 2010, the railway line from Shanghai to Guilin was buried due to landslides caused by continuous rainfall. As a result, when the k859 passenger train from Shanghai South Railway Station to Guilin ran between Yujiang and Dongxiang in Jiangxi Province, the locomotive and vehicles 1–9 derailed, thus interrupting the operation of the up and down lines. The accident caused 19 deaths and 71 injuries [10].

On July 23, 2011, in Wenzhou City, Zhejiang Province, the D301 train from Beijing South Station to Fuzhou Station and the D3115 train from Hangzhou station to Fuzhou South Station had a rear end collision accident. The accident has confirmed that a total of six carriages derailed. As a result, 40 people were killed and 172 injured, the train was interrupted for 32 h and 35 min, and the direct economic loss reached 193,716,500 yuan [11].

On December 19, 2017 (18 local time), a passenger train of American Railway Corporation derailed near Interstate 5 in Washington state. The accident resulted in 6 deaths and 77 injuries [12].

Accidents of long-distance rail transit on the ground can be seen frequently, and accidents in urban subway (or light rail) are not uncommon. For example, on December 22, 2009, a sudden power supply trip failure occurred between Shaanxi South Road and People's Square of Shanghai Metro Line 1, resulting in the suspension of trains in this section. At this time, because line 1 operates from Fujin Road Station to Shanghai Railway Station in the north section, after arriving at Shanghai Railway Station and passengers get off, the empty train leaves Shanghai Railway Station and drives hundreds of meters to the tunnel in the direction of Zhongshan North Road. When it is ready to reach the turn back line through the turnout, there is an abnormal situation. That is, when the subsequent train 150 of line 1 from Zhongshan North Road to the railway station runs to the turn back station of Shanghai Railway Station, the signal system should send a speed signal of 20 km/h to train 150 in this track section, but it mistakenly sends a speed signal of 65 km/h. Due to insufficient braking distance, train 150 collided with empty train 170 turning back.

In July 2013, there was a train derailment accident in Spain (see Fig. 1.1). It is said that it was due to the derailment and rollover caused by the driver speeding when the train crossed the curve [2].

In order to achieve the purpose of safe operation, the speed of train operation must be controlled by the track operation management system. Among them, the control command can come from the satellite positioning and communication system, the wireless data packet communication system of the operation management center or be operated by the driver. In order to improve the operation reliability, the train operation speed control should and must accept the constraints of multiple parallel commands, including the fixed-point speed limit warning signal during operation. It seems that there are still major technical functional defects in the Spanish train operation management system.

Fig. 1.1 An example of train derailment accident in Spain

1.3.2 Causes of Train Accidents

Rail transit includes ordinary railway, high-speed railway, urban subway, urban light rail, urban monorail suspension train, virtual rail train and pipeline quasi vacuum supersonic train in the future.

All these types of rail transit have had traffic accidents. To sum up, it is nothing more than collision, rear end collision, derailment, fire, explosion, power failure, lack of sensing signal, control failure, communication interruption, traction failure, carriage auxiliary technology damage and so on.

"Collision, rear end collision, derailment, fire and explosion" are common in ordinary railways. Among them, "collision" will not only occur between trains, but also at the intersection of railway and highway.

The reasons can be divided into two major factors: human (improper management, scheduling and operation) and technology (imperfect). In order to completely overcome any form of accident, we must rely on science and technology to provide the most effective guarantee in order to achieve people's ultimate safety goal.

The causes of the train accident are briefly analyzed as follows.

(1) Improper human scheduling and lack of automatic error correction mechanism

Improper human scheduling is the most common factor causing railway accidents in the early stage, which is easy to cause the rear vehicle to break into the driving section of the front vehicle too early, which is related to the rear end collision of most trains. The direct cause of the "particularly serious railway traffic accident on July 23 at Ningbo-Wenzhou Railway" in 2011 is the confusion of dispatching. With the rapid development of electronic information and communication technology, railway dispatching system is not what it used to be. The possibility of major railway accidents caused by dispatching errors of EMU trains on new high-speed (including fast) railway lines is greatly reduced.

(2) Lack of intelligent technology to correct over speed driving

Because of the high speed and mass of the train, it is necessary for the train to have strong centripetal force when passing through the curved track with large radius. If the speed limit is exceeded, the component force in the horizontal direction of the train gravity will not provide enough centripetal force for the train, and the train will squeeze the outer rail and cause derailment accident. The "April 25 derailment accident of JR Fukuchiyama line in Hyogo Prefecture, Japan" in 2005 and the "July 24 derailment accident of Spanish train" in 2013 were both caused by the train exceeding the speed limit on the curve.

(3) Lack of geological disaster prediction technology and risk avoidance function.

Most of the traditional railway lines are sparsely populated and dangerous mountains and rivers, which are prone to geological disasters. Once the bridge or tunnel collapses due to natural disasters, it is very easy to cause the train to derail, bury or fall off the cliff. The train falling bridge accident of Chengdu-Kunming Railway in 1981 was caused by the collapse of the railway bridge caused by debris flow in Liziyida gully, a tributary of Dadu River. With the maturity of bridge and tunnel engineering technology and satellite monitoring technology, such accidents have been effectively curbed.

(4) The system inspection and control equipment lack redundancy technology.

The large mass of the train has great inertia. Even the speed of tens of kilometers per hour will produce a long "inertial sliding" distance after braking. If the railway system equipment fails and "fails", it is very easy to cause scheduling confusion, and finally, the driver cannot confirm the road conditions ahead, resulting in rear end collision or collision accidents. The "4.11 passenger train rear end collision accident of Beijing-Kowloon Railway" in 2006 is a tragedy caused by the driver's misjudgment caused by the failure of signal light. The cause of Wenzhou bullet train accident is also the failure of signal light equipment.

Illegal construction by railway maintenance personnel is another common reason for trains to collide with pedestrians. For example, "1.23 major off-road traffic accident on Qingdao-Jinan Railway" in 2008 is an example.

In terms of the current technical situation, the vast majority of cases still rely on manual inspection of the track. Once the turnout is found to be faulty or broken, it is necessary to protect it immediately, prohibit all locomotives and vehicles from passing, timely report to the station attendant (shunting district head) and notify the public works and electrical service departments for inspection and repair. Obviously, this is a traditional and outdated maintenance method. If there is a slight negligence, it is bound to hide (latent) the crisis of possible accidents. Nowadays, the protective measures along the railway are relatively perfect, and the use of Viaducts in a large area has overcome the accident of pedestrians crossing the track without authorization from the track engineering structure.

High-speed railway system is more dependent on equipment. Therefore, reducing equipment failure rate and increasing multi-level (multi-redundancy) emergency equipment is one of the important links in high-speed railway construction.

(5) Lack of intelligent identification function for fault diagnosis of train parts.

High-speed train and its high-speed track are equipment integrating a large number of complex and precise parts, and the operation environment is extremely strict. Therefore, it also increases the difficulty of locomotive and vehicle maintenance and railway track maintenance. High-speed trains are like civil airliners. Once parts fail, such as looseness, deformation or falling off, it may lead to train subversion. The "Aishud high-speed railway accident" in Germany in 1998 was caused by the deformation and falling off of the outer shell of the wheel on the rail and blocking the railway turnout, resulting in the derailment of the train.

In addition to the lack of intelligent identification and decision-making function of the running environment of the train itself, accidents sometimes occur because the brake shoe holds the wheel tightly in the running train. When there is a slight fault, it only makes some friction sound, smoke or spark. When the fault is serious, there may be the danger of vehicle holding brake. In case of holding brake, the train will slow down and stop on a slope. In serious cases, it may also cause fire or explosion of dangerous and flammable goods.

At present, the train does not have the intelligent identification and decision-making of the operation state of the power plant and basically depends on the driver to judge it according to the conventional detection instrument (or sensing data) to determine whether it needs to stop, because the inaccuracy (or error) of human judgment will often lead to the abnormal operation of the train, out of control, and even the sudden drop of speed, resulting in the inertial extrusion of the carriage and other accidents endangering the safety of passengers.

In short, a series of safety protection systems of rail transit need to build a multi-level structure, and even if there is a mistake in a certain link, it can be found and corrected in time.

Since the birth of rail transit, its accidents have been common. In order to completely overcome the occurrence of rail transit accidents, it is very important to improve the level of rail transit science and technology. Of course, a more effective "measure" is to use intelligent technology to assist in manual train driving and manual management of rail transit system, gradually replace all manual operations and finally make rail transit technology enter the era of full intelligence.

1.3.3 Basic Elements of Safe Operation of Modern Train

Through the detailed analysis of the tragic accidents in the past, many countries in the world put forward the requirements to ensure the safety of train operation. In order to ensure the safety, real time and comfort of train operation, we need to rely on modern science and technology.

1. **Contents of safety guarantee**

The safe operation of the whole rail transit system needs to rely on subsystems such as communication, dispatching, detection and control to form a complete technical guaranteed chain. Each subsystem has both technical independence and mutual "hook" and intersection, so as to form a safety guarantee system for rail transit operation.

(1) **Train positioning integrity**

Global satellite navigation system has become one of the guarantees for railway operation safety. Although the current accuracy of satellite navigation and positioning has reached the demand that it can provide train operation information more accurately most of the time, in the process of train operation, once the navigation system fails, if it cannot provide alarm information to the train in time, there will be serious potential safety hazards.

The train receiver can determine its own position, running speed and the accuracy of keeping synchronization with the time of satellite navigation system, which depends on the interaction results of various complex factors. Generally speaking, the accuracy of satellite navigation system depends on the pseudo range and pseudo range residual measurement and the quality of satellite ephemeris data. The relevant errors are caused by the control end, space segment and user segment. In order to analyze the influence of various errors on accuracy, a basic assumption is usually made, that is, these error sources can be attributed to the pseudo range of each satellite and can be regarded as the actual equivalent error in the pseudo range.

The clock offset of satellite and receiver and various error sources will affect the geometric distance measurement from satellite to user and then affect the positioning accuracy of user.

The satellite signal will be delayed when passing through the atmosphere, and the signal may be delayed (or advanced) due to reflection (multi-path), selection availability and hardware operation error between the phase center of the user antenna and the code correlation point of the receiver. The total time deviation caused by these factors is

$$\delta t_D = \delta t_{\text{atm}} + \delta t_{\text{noise\&rcs}} + \delta t_{\text{mp}} + \delta t_{\text{hw}} + \delta t_{\text{SA}} \tag{1.1}$$

where δt_{atm} is the delay caused by the atmosphere, $\delta t_{\text{noise\&rcs}}$ is the receiver noise and resolution error, δt_{mp} is the multi-path deviation, δt_{hw} is the receiver hardware deviation, and δt_{SA} is the selection availability deviation. The time equivalent value of pseudo range is the difference between the receiver clock reading when the signal is received and the satellite clock reading when the signal is transmitted.

The occurrence of "integrity" anomaly is caused by satellite or master control station (MCS), which will lead to unpredictable positioning error exceeding the safety allowable range. In general, the number of "integrity" anomalies is only a single digit per year, but once they occur, they may have a huge impact on positioning. Therefore, this is a very important and key technical element for train navigation [13].

The main causes of "integrity" abnormal problems may come from satellite clock, satellite ephemeris, special problems of satellite and master control station, etc.

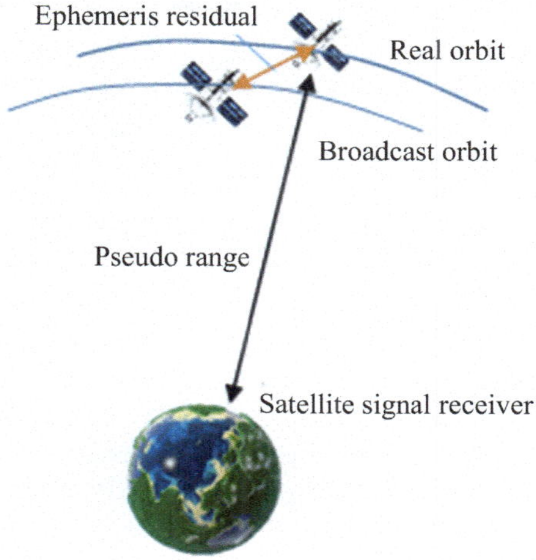

Fig. 1.2 Schematic diagram of satellite ephemeris error

The satellite clock anomaly is caused by the frequency standard problem caused by random dissociation, large jump or the combination of the two conditions. When the beam current or temperature of the frequency standard changes sharply, the master control station will have "Over clock jump".

Satellite ephemeris error refers to the difference between satellite broadcast orbit and real orbit, which is called ephemeris residual. This is because the position information of the satellite is an estimation. Transmitting this estimation information to the user will naturally bring errors. Only with the continuous progress of satellite orbit determination technology will the positioning accuracy be gradually improved.

The principle of satellite ephemeris anomaly is shown in Fig. 1.2. It can also be said that the anomaly of the satellite ephemeris is the result of the influence of various resistances on the satellite in orbit. At the same time, people cannot accurately determine the specific impact of these resistances on the satellite position.

At present, the estimation of satellite ephemeris anomalies is still very difficult. This is because the primary and key issue is the need to depend on the accuracy of the satellite navigation system.

In GPS (Global Positioning System), BLOCK I satellite (the first generation GPS communication satellite in the United States) is relatively prone to ephemeris anomalies. When a solar eclipse occurs, the satellite must rely on its internal power supply because it is blocked from the sun. As satellites age, they will stop running once there is no additional power collected from the sun by solar panels. Therefore, this anomaly usually occurs when the satellite leaves the shadow area of the earth and the view of the solar plate to regain the tracking of the sun after the solar eclipse. At this time, the satellite adjusts its operating attitude by igniting the booster, which will increase the deviation. The internal of BLOCK I satellite lacks strengthening measures to resist space environmental radiation, so it is prone to "bit damage" and "word damage" affecting navigation messages and C-field tuning. This is also because the C-field tuning register aligned with cesium beam will be disturbed by solar radiation. At the same time, in order to overcome the "bit damage", when the BLOCK I satellite changes some "bits" (those bits used to count the alignment direction of the cesium beam), it will lead to a deviation of several kilometers in a few minutes. BLOCK I satellite is also prone to P-code offset, which will lead to the loss of P-code tracking and the deviation of positioning results in the order of several kilometers. After BLOCK I, the US BLOCK II/IIA/IIR satellite has overcome this defect. China's BeiDou system has obvious late development advantages and has surpassed GPS in these technical performances.

The problems of the master control station are mostly caused by hardware, software and human factors. Once there was an accident in the main control station, which caused the error in the covariance calculation of Kalman filter. Due to the existence of these outliers, thousands of meters of deviation may have occurred before the fault is detected. At present, the GPS ground monitoring network has not been able to cover all satellites for 24 h. Therefore, if the "incompleteness" problem occurs, it is likely that it will not be found immediately.

In order to minimize the possibility of "integrity" abnormalities, the master control station technology is still in the process of continuous improvement. The specific measures are install redundant hardware and solid software, and train operators to reduce human errors as much as possible. However, only relying on the master control station alarm can never meet the technical requirements of "timeliness", because sending alarm information requires a certain response time (time lag). Therefore, the method of independent detection of "positioning integrity" and satellite anomalies by users (vehicle and dispatching inspection and control in rail transit system) needs to be improved accordingly. This is the technical advantage of China's BeiDou system after the global network is deployed [14].

After the completion of the global network, China's BeiDou satellite system has provided a reliable technical guarantee for the accurate positioning of the rail transit system.

(2) **Train dispatching management**

In the train dispatching management, the most prominent problem is the lack of protection section in the station area. The so-called protection section means that the train dispatching management center is not allowed to send driving instructions to the section where the signal is missing without the guarantee of safety information.

For example, in relay interlocking system, the outbound signal or route signal goes out due to red light fault. This situation, especially at night, may have the most tragic consequences. In older (technologically obsolete) systems, red lights that go off at the end of the receiving route (or at the end of the transfer route from one garage to the next) are usually ignored. In this case, vehicle dispatching accidents are likely to occur. At this time, the dispatching management center is absolutely not allowed to cancel the departure ban (send driving instructions to the protection section) [15].

(3) **Influence of geological disasters on safe operation of trains**

With the increasing popularity of high-speed railway, many new high-speed railway lines need to pass through mountainous areas and even debris flow prone areas. The common medium strength debris flow will pose a serious threat to the operation safety of high-speed trains on the bridge [16].

This is mainly because the impact load of debris flow will cause large transverse stress response of adjacent bridge beams and rails. The lateral displacement and acceleration of the bridge caused by debris flow impact load are much larger than those caused by train driving excitation. When the main frequency component of bridge response is close to the transverse bending frequency of the whole bridge, it will pose a great threat to the safety of train operation.

(4) **Influence of human factors on safe train operation**

Human factors are one of the key factors affecting the safe operation of subway trains. Strengthening the management of human factors can effectively protect the life safety of passengers, which has high practical significance. How to manage scientifically has become a common concern for the safe operation of trains.

Among them, the train driver is an important factor affecting the safety of train operation. At present, some drivers have problems such as weak safety awareness and poor professional quality, resulting in errors in emergency affairs processing and on-site judgment, unable to effectively control emergencies, and finally lead to risk events.

In addition, at this stage, the posts of Metro dispatching commander mainly include power dispatching, traffic dispatching and environmental control dispatching. In train operation, the role of train dispatching is very prominent. Taking train dispatching as an example, the main content of train dispatching is to customize the schematic diagram of daily train operation, and organize and plan train operation, which is the key to safe and stable train operation. If there is a deviation in the work of train dispatching, it will lead to the inaccurate departure time of the train and eventually lead to an accident.

In addition, under the guidance of relevant departments, train maintenance personnel are responsible for various maintenance and inspection of the train, which can further prolong the operation life of the train and ensure driving safety. If the maintenance personnel have deviation in their work, it will lead to problems in the operation of the train and eventually evolve into a safety accident [17].

2 Basic Technical Methods to Ensure Safe Train Operation

With the development of science and technology, many intelligent technologies have been integrated into the rail transit system, especially the high-speed rail system. It can be said that high-speed rail technology has been accompanied by intelligent technology since the first day of its birth. From the vehicle design and manufacturing of high-speed railway to every process of operation, it is closely connected with power mechanical engineering, engineering materials, communication theory and technology, detection and control, signal processing and intelligent algorithm, artificial intelligence, etc. The intersection of these disciplines has formed the support for rail transit technology and finally realized today's advanced and safe rail transit tools, which has brought unprecedented advanced productivity and high convenience to people's life.

(1) **Basic methods to ensure train positioning**

When the satellite navigation system fails, it is of great significance to provide the integrity of alarm ability to the train and its dispatching system. This involves the autonomous integrity monitoring technology of satellite signal receiver.

For example, detection algorithms deal with the integrity of linear Gaussian systems, but in the real operating environment, the observation noise distribution is often difficult to be accurately described by a single deterministic distribution. At this time, a Gaussian mixture particle filter RAIM algorithm can be used. The Gaussian mixture model is introduced. Through the expectation maximum algorithm, the observation noise of the satellite is approximately expressed in the form of Gaussian sum from the unknown noise. The particle filter method is used to analyze and judge the autonomous integrity of the satellite positioning receiver under non-Gaussian conditions, and then, the faulty satellite is isolated on the basis of identifying the satellite fault time by the test statistics. Thus, it provides an important positioning technical guarantee for the safety of train operation in complex environment.

(2) **Optimization of train dispatching management**

Considering that many railways in the world have a protection section with a length of about several hundred meters behind the inbound signal, route signal, outbound signal and passing signal, it is necessary to check the condition of this section (allowing to enter the station when it is idle, or stopping to enter the station when it is not idle) before issuing the command to allow to follow and start the route.

After adopting the protection section, the calculation of the braking distance length of these systems is based on the kinematic dynamics requirements that the train can reach complete stop during automatic emergency braking.

(3) **Technical methods to overcome geological disasters**

For the load impact caused by geological disasters, while increasing the detection of geological movement state, we can use intelligent algorithm to determine the impact of shock wave on the train according to the different responses brought by different impact forces at different times of geological load impact and then make the train operation decision (change the running speed or emergency stop). For example, in the case of the same bridge parameters, track parameters and track irregularity spectrum (mathematical description reflecting track irregularity), it is necessary to study the joint impact of train speed and debris flow impact intensity on train operation safety.

(4) **Methods to overcome the influence of human factors on train operation safety**

In addition to strengthening the driver's online safety monitoring mechanism and strengthening the early warning of adverse factors, we should also improve the train operation management system based on our unit, that is, we need to establish a top-down safety management network system and improve the safety accident analysis system. At the same time, more importantly, with the high development of artificial intelligence, it is necessary to gradually popularize and promote the application of artificial intelligence technology in train operation, including the popularization of automatic train driving or intelligent auxiliary train driving technology.

With the continuous improvement of Rail Transit Intelligent Technology, it is finally possible to make the once common accidents gradually become history. Finally, the full intelligent technology in rail transit system can be expected soon.

1.4 Enlightenment from the Development History of Rail Transit

The development history of rail transit is a history accompanied by the development of modern science and technology. It is a history of the comprehensive application of scientific theories and technological inventions that are constantly improving and progressing.

1.4.1 Rail Transit is Inseparable from the Support of Scientific Basic Theory and Technological Innovation

When everyone talks about rail transit, they almost invariably mention Watt's steam engine because it plays an unshakable role in power engineering. In fact, the technical improvement of steam engine has gone through a long time. The earliest steam engine not only has low operation efficiency, but also does not solve the problems of speed stability and regulation technology of transmission machinery, so it cannot be directly popularized and applied in the industrial field.

From 1765 to 1790, Watt made a series of improvements and technical inventions to the original steam engine.

It should be pointed out that the centrifugal governor, which occupies an important technical position in the popularization and application of early steam engine, is an important invention achievement of dynamic mechanical motion control.

The mathematical description of speed control of steam-driven rotating machinery is not complex, and it can be described as follows

$$M_z - M_f = \frac{\eta \rho Q P}{\omega} \tag{1.2}$$

where M_z and M_f are the output torque and load moment of force of steam engine, respectively; Q, P and ρ are steam flow, pressure and density, respectively; η is the working efficiency of steam engine; ω is the angular velocity of rotating machinery driven by steam engine.

When the angular velocity ω (rad/s) is expressed by the number of revolutions n (r/min) commonly used in engineering, formula (1.2) can be transformed into

$$n = \frac{30\eta\rho QP}{(M_z - M_f)\pi} \tag{1.3}$$

It can be seen from formula (1.3) that M_z, M_f, P and ρ are parameters (variables) that affect the rotating speed of rotating machinery, and Q is the parameter variable of M_z, M_f, P and ρ. Therefore, the number of revolutions n can be effectively controlled (adjusted) as long as the flow Q is changed. This is the basic theoretical basis of Watt's invention of centrifugal governor.

Watt mastered the formation principle of rotating speed of rotating machinery driven by steam and also realized that the description of formula (1.3) alone is not enough. The key problem is how to detect the rotating speed and feed it back to the steam regulating valve in real time? This is the wonderful and genius of Watt's invention of centrifugal governor. Watt's genius is that he invented the principle model of centrifugal governor before others and found the theoretical principle of the model.

Driven by the steam working host, there is a corresponding relationship between the speed of the speed detection wheel and the speed of the working host. When the rotational speed increases, the rotational speed of the two symmetrical "centrifugal flying pendulums" accelerates and deviates from the rotating shaft under the action of the increase of centrifugal force, so pull down the sleeve at the upper end, causing the right end of the lever to lift up, reducing the valve opening of the steam regulating valve, thereby reducing the steam flow into the steam engine working chamber, and the rotational speed of the working host decreases accordingly. On the contrary, when the speed decreases, the speed of the two symmetrical "flying pendulums" slows down and approaches the rotating shaft under the action of the reduction of centrifugal force, so the sleeve at the upper end is lifted upward, causing the right end of the lever to push down, increasing the opening of the steam regulating valve, and then increasing the steam flow into the working chamber of the steam engine, and the speed of the working host increases accordingly (see Fig. 1.3) [1].

Using the classical control theory, we can make a perfect explanation for the speed detection and regulation of "centrifugal flying pendulum", and the centrifugal regulator can well realize the proportional and integral regulation of rotation speed.

In other words, when the "centrifugal flying pendulum" changes the upper and lower positions of the sleeve with the change of mechanical speed, the opening of the steam control valve can be adjusted according to the proportional relationship of the mechanical transmission structure. At the same time, the deviation signal of "centrifugal flying pendulum" can be used to overcome the static deviation of mechanical speed in the adjustment process and finally realize proportional integral

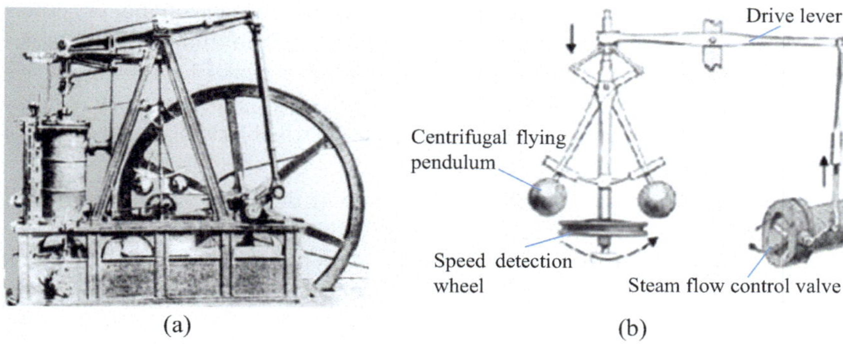

Fig. 1.3 centrifugal governor invented by Watt

adjustment. Its transfer function $G(s)$ is,

$$G(s) = K_0\left(1 + \frac{1}{T_i s}\right) \tag{1.4}$$

Among them, K_0 is proportional transfer coefficient, T_i is integral transfer coefficient, and s is Laplace operator.

It is precisely because of Watt's invention of the "centrifugal flying pendulum" governor that the world's first steam power device with practical value was manufactured. Since then, mankind has ushered in a new era of energy utilization and marked the beginning of the world's first industrial revolution. In memory of the great inventor, later generations set the unit of power as "Watt (W)".

It is precisely because Watt invented the "centrifugal flying pendulum governor" through his genius idea in the process of analyzing and describing the working principle of steam engine speed regulation, and formed a new proportional/integral regulation method, which finally created a good theoretical and experimental basis for the formation of subsequent classical control theory.

Of course, Watt did not clearly put forward the transfer function expression of the governor at that time, but there is no doubt that Watt has been "subconsciously" applying this transfer function to the invention and creation of the regulator. The later classical control theory first got very important hints and inspiration from the principle of Watt's invention of governor [1].

With Watt's steam engine and the appearance of its associated governor, the early train and vehicle manufacturing technology really has practical value. At the same time, it also means that mankind began to take the first step toward industrial automation technology.

Of course, this is also the substantive content of the world's first industrial revolution, that is, the mechanical industrial revolution.

The first industrial revolution brought mankind from the manual labor era dominated by manpower in the past 2 million years to the steam machinery era of large-scale production of modern machines, marking the first peak of industrial development in history. As the steam engine is widely used in industrial production, it has become the power unit of almost all machines, which has changed people's production and lifestyle.

However, human industrial technology does not stay in the innovation of mechanical technology. Then, in 1831, Michael Faraday discovered the phenomenon of electromagnetic induction through a series of major experiments, put forward the law of electromagnetic induction, established the theory of electromagnetic induction, and personally invented (produced) the world's first original generator: Faraday disk generator according to the principle of electromagnetic induction.

Electromagnetic induction is one of the most important discoveries in electromagnetics. It reveals the relationship between electrical and magnetic phenomena. Moreover, later generations invented all kinds of generators and motors according to the battery induction phenomenon (Law).

This also proves the important guiding significance of basic theory to technological invention: A basic scientific theory is the new source of scientific and technological development.

Since then, the second industrial revolution of mankind has been started. The second industrial revolution is based on the background that mankind has entered the "electrical age".

The following new technologies such as electric power, internal combustion engine, new materials, telegraph communication and radio highlighted their important role in the second industrial revolution. Therefore, these modern technologies began to appear on the stage of history and laid the foundation for the production and lifestyle under today's high-tech conditions. At the same time, there will be the scientific theoretical basis and technological innovation thinking for the subsequent emergence of electric locomotive.

The third industrial revolution began in the early stage after the Second World War, reached a climax from the mid-1950s to the early 1970s and entered a new stage after the 1970s. The third industrial revolution made the global exchange of information and resources more rapid, most countries and regions were involved in the process of globalization, the world political and economic pattern was further established, and the development of human civilization reached an unprecedented level. Electronic computer, information and digital communication technology are the landmark achievements of the rapid development of global science and technology.

The third industrial revolution caused the transformation of various elements of productive forces. From then on, it strengthened the trend of non-material industrial structure and intelligent production process and realized the modernization of production and life. With the development and wide application of computer technology, various "man–machine control systems" have been gradually formed to accelerate the realization of production automation, office automation and family life automation. It indicates that human society will enter another higher-level automation era from the era of mechanization and electrification.

The fourth industrial revolution began in the twenty-first century. It is a technological revolution focusing on artificial intelligence, clean energy, robotics, quantum information technology, controllable nuclear fusion, virtual reality and biotechnology. This is a new green industrial revolution. Its essence and feature are to greatly improve the utilization rate of resources. In this fourth green industrial revolution, China stood on the same starting line with developed countries for the first time.

In combing the process of human scientific and technological progress, people have been used to using "industrial revolution" to divide each stage of progress and development. In fact, the so-called industrial revolution is the scientific and technological revolution of mankind. Future generations should understand its "revolutionary impact" on nature, society and mankind from the essence of "industrial revolution" (scientific and technological revolution) [1].

There is no doubt that the changes brought about by scientific progress and the emergence of new technologies will affect the whole world.

In the final analysis, today's rail transit has such a relatively perfect technology, which is the result of the repeated industrial revolution in the world, and it is inseparable from the guidance of many scientific theories and the support of technological innovation.

1.4.2 Institutional Guarantee for China's Rail Transit Development

As mentioned above, in 1769, the importance of Watt's steam engine was beyond people's imagination. The steam engine was soon widely used in industrial production and became the power unit of almost all machines, which changed people's mode of production, greatly promoted technological progress and opened the prelude to the industrial revolution. The application and promotion of steam engine have accelerated the pace of entering the industrial era. Steam locomotive has become an important symbol and key tool of cultural and social progress in this era. It can be said that in history, a series of scientific and technological achievements focusing on the invention, creation and improvement of steam power devices have contributed to the production and life of human society, reaching a phased high point in the history of scientific and technological development.

At the same time, China still stays in the era of cattle and horses pulling carts and donkey grinding. It is difficult to see the technological breakthrough of "water mill", because at this time, the Chinese people generally lack theoretical knowledge such as dynamics and electricity, and the feudal social system seriously restricts the innovative thinking of the whole Chinese nation.

In fact, a book in ancient China (printed in 1280) recorded that a "double acting pull rod" piston bellows had been invented since the Ming Dynasty, which had a

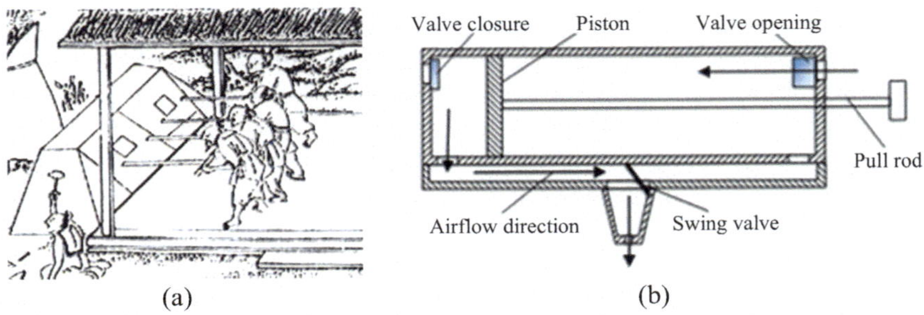

Fig. 1.4 Ancient Chinese pull rod piston bellows

"revolutionary" effect on the smelting technology at that time. In fact, this is an original blowing machine, that is, the embryonic form of the air-blower (see Fig. 1.4 (a)).

The two ends of the bellows are, respectively, provided with an air inlet, the mouth is provided with a valve (open in the forward direction and close in the reverse direction), one side of the bellows is provided with an air duct, the front and rear ends of the air duct are, respectively, provided with an air outlet, and the middle of the air duct is provided with a swing valve. By stretching out the pull rod outside the bellows, the piston is driven to move back and forth to promote the opening and closing of the valve, so the valve will swing with the wind to form a blast channel (see Fig. 1.4 (b)).

Until the 1970s, this kind of bellows still existed in a large number of small manual workshops in urban and rural areas of China. Due to the limitation of manufacturing capacity, electric blowers are not popular in China, resulting in this original bellows still becoming an important "blowing" equipment in manual workshops.

In terms of working principle, the "mechanical work/energy" conversion process of bellows is: push the pull rod to make the piston work, cause the air compression to be converted into the pressure kinetic energy of air and provide air (oxygen) to chemical reaction devices (such as ironmaking furnace), or do work as an aerodynamic force. The "mechanical work/energy" conversion process of steam engine is that the push of steam pressure potential energy on the piston is converted into the kinetic energy of the piston to promote other mechanical movements.

In other words, the pull rod piston bellows in ancient China and Watt's steam engine are two opposite "mechanical work/energy" converters. The former (ancient Chinese bellows) appeared 500 years before Watt's perfect steam engine. Unfortunately, after hundreds of years, there was no "genius" figure in ancient China who could get the inspiration of "reverse thinking" from the working principle of pull rod piston bellows: "The forward/reverse rotation of wind force" can also generate the motion thrust of "forward/reverse direction" to push the pull rod. Once there is this "reverse thinking", it is natural to think that steam has the same effect as wind. In this way, the history of the Ming Dynasty seems to have another "colorful chapter".

After all, the regret of history is a thing of the past. However, people should be inspired from the historical process of the pursuit of "independent spirit and free thought" in the modern western world and the emergence of the first industrial revolution: Innovative thinking is a precious "gift" for future generations [1].

Although modern western technology is only a physical development, modern western civilization has also undergone metaphysical changes. The outstanding performance lies in the discovery, perfection and maturity of many scientific theories based on technology. It can be said that civilization has created the leading position of Western science and technology in the world since the industrial revolution. This leadership was mainly laid in the "Renaissance" beginning in the sixteenth century, that is, humanistic spirit replaced metaphysical theology, scientific rationality replaced superstition and paranoia, and free will broke through the closure of autocracy. It shows that without the reform of social system, there will be no emancipation of people's minds, and civilization, development and innovation can only become empty words.

At the beginning of the twentieth century, the revolutionary pioneers of the Chinese nation were fighting hard to overthrow the decadent rule of the Qing Dynasty. After the outbreak of the Revolution of 1911, the Qing Dynasty was overthrown and the Republic of China was officially established in January 1912.

Just as the new dawn of history rose slowly from the horizon, the sky was covered with dark clouds. The Japanese militarists, who had long coveted the land and wealth of China, took advantage of the fact that the Chinese nation had not yet "recovered" from the "suffering of long illness", and their carefully planned aggressive plot was finally carried out against China. They are most shocking in the human world attempted to bring the Chinese nation to extinction.

Since then, the Chinese nation has spent decades in the smoke of war. Here, where does the Chinese nation have a place in the history of modern scientific and technological development in the world?

Until 1949, Mao Zedong declared to the world on the Tiananmen Gate Tower: "the Chinese people have stood up from now on!" A new page in the history of the Chinese nation has finally opened.

In the 70 years since the founding of new China, the Chinese people have experienced a historical leap from "riding a donkey to Beijing", to "taking a train to Lhasa", to 30,000 km high-speed railway covering 80% of major cities, and even the "Fuxing" high-speed train realizing a "land flight" with a speed of 350 km per hour.

At present (by the end of July 2020), China's total railway mileage has jumped to the second place in the world, second only to the United States.

The number, scale and mileage of subway construction in China have also ranked first in the world, and 47 cities have been opened. Among them, the cities with the most subway mileage are Shanghai and Beijing. In addition, Guangzhou, Shenzhen, Foshan and Dongguan are the cities in Guangdong Province that have subways. The total length of subways in these four cities is 800 km.

In addition to the operation mileage of railway transportation and urban subway (including urban light rail) far ahead of other countries in the world, the automation and intellectualization of rail transportation have begun to gradually surpass the technical level of the original developed countries, and the innovative thinking and practice in some fields are ahead of other countries and reach the world advanced level. This is reflected in the fact that China's rail transit has caught up with the western level in vehicle design and manufacturing technology, especially in the communication, scheduling, detection and control and auxiliary artificial intelligence of rail transit system. This is undoubtedly the embodiment of the superiority of the national social system.

However, future generations of the Chinese nation must never forget the scars of thousands of years of regret and bullying. Today, under the guarantee of a correct political and social system, while developing the economy, we must strive to surpass the "great powers" that once invaded and enslaved the Chinese nation in the field of science and technology, so as to comfort the pioneer souls who sacrificed their lives for the rejuvenation of the Chinese nation and write the most magnificent chapter of history belonging to the Chinese people [1].

On the basis of reviewing the development process of rail transit in the world, this book expounds the basic principle of Rail Transit Intelligent Technology System Theory and the thinking "context" of technology research and development. For the further development and expansion of Rail Transit Intelligent Technology in China, it is necessary to deeply understand and master the complete technology of rail transit system and its future development direction.

References

1. Zhang X (2017) Dialectical law of science and technology innovation. CHINA INTL.CUL.TURE PRESS, Hong Kong
2. Zhang X, Lu D (2014) Invention profiling and creative thinking theory. Shanghai Jiao Tong University Press, Shanghai
3. Sun X (1998) Flying dragon on land—history of train development. Chn Sci Technol Monthly 9:60–62 (In Chinese)
4. Li Y (1995) Spring ice room wild riding. Shanxi Ancient Books Publishing House, Taiyuan
5. Wei Y (2011) Atlas and annals of overseas countries. Yuelu Publishing House, Changsha
6. Li Y (2017) Observing modern China from the perspective of speed—taking ships and trains as examples. Jinan J (Philos Soci Sci) 11:105–114 (In Chinese)
7. Editorial Department of the Journal (2019) Forging ahead for 70 years, brilliant achievement shine the world—New China has made remarkable achievements of communications and transportation industry in the past 70 years. Transp Enterprise Manage 9(12):1–2 (In Chinese)
8. Editorial Department of the Journal (2003) Top ten train accidents in the world. Overseas Chin J Bagui 4:56 (In Chinese)
9. Shikuang (2002) What is it like to die in a fire—a record of the extraordinary train accident in Egypt. Henan Fire Prot 4:12–14 (In Chinese)
10. Li J (2010) Saving lives—a record of the rescue of the "5•23" train derailment accident of Jiangxi fire force. China Fire 11:6–9 (In Chinese)
11. Chen J (2011) An interpretation of the reporting framework on "Ningbo-Wenzhou high speed railway rear end collision accident" – Taking the report of Xinhua net as an example. News Dissemination 11:74–77 (In Chinese)
12. China News Network (2018) Train derailment in Washington State, USA, resulting in 3 deaths and 100 injuries. Urban Mass Transit 1:6 (In Chinese)

13. Wang W, Li X, Liu J (2010) The principle and application of in door GPS. Geomatics Spatial Inf Technol 33(3):116–119 (In Chinese)
14. Shi Y (2019) Research on integrity monitoring method for satellite based train positioning under non-gaussian noise conditions. Dissertation, Beijing Jiaotong University
15. Wang J (2016) Train operation safety: World railway experience. Railway Signal Commun Eng 13(2):103–104 (In Chinese)
16. Wen Z (2019) Dynamic responses of coupled train–track–bridge system subjected to debris flow impacts and running safety nalysis of high-speed train. Dissertation, Southwest Jiaotong University
17. Hou H (2015) Influence of human factors on subway train operation safety and preventive measures. Manage Adm 22(5):236–238 (In Chinese)

Chapter 2
Intelligent Monitoring of Rail Transit System

Rail transit needs intelligent communication and monitoring (or detection and control) to realize the safe and stable operation of trains. This is not only an important condition for the sustainable development of rail transit, but also a backbone technology to ensure the safety and real-time operation of rail transit system.

It can be said that various artificial intelligence algorithms and technologies cannot be realized in the rail transit system without the most basic physical transmission system, because this is the most fundamental physical layer for the realization of the whole rail transit intelligent technology. All monitoring information and train dispatching instructions of rail transit can be delivered through this physical layer. This makes the communication between the original train and the ground more and more demanding. How to apply a series of safer, more reliable and more efficient modern technologies to the rail transit communication and monitoring system has become the core issue of concern in this technical field, that is, the improvement of the function of the signal system. Therefore, the realization of intelligent technology of rail transit system directly involves three academic fields: communication, detection and control.

As an important communication and monitoring medium in the vehicle operation process of traditional rail transit system, signal system is mainly responsible for train route management, real-time monitoring and commanding the overall situation for macro dispatching.

At present, the train control system CBTC (Communication-based train control) based on communication technology has its mature technical advantages. It has the characteristics of two-way, large amount of transmission information and fast transmission speed. Therefore, it has long become the mainstream technology in practical application. At the same time, the signals in CBTC mainly rely on wireless local area network (WLAN) technology, which has the advantages of high transmission rate and strong compatibility.

It can be said that CBTC has a certain degree of intelligent concept, and the actual program already contains the basic intelligent analysis and judgment function. However, from a higher level, the existing CBTC technical scheme does not really

Y. Khan, *Principles of Intelligent Rail Transit*, Advances in High-speed Rail Technology,
https://doi.org/10.1007/978-981-19-6072-7_2

introduce modern intelligent recognition algorithms, so it can only be said that CBTC only has the prototype of intelligent technology.

2.1 Basic Working Principle of CBTC

The basic work contents of CBTC include moving block and moving authorization.

2.1.1 Working Principle of Train Moving Block

In the field of rail transit, "block" is to artificially divide the line into multi-section block sections by using the equipment of signal system. This interval can be defined not only as a physical interval, but also as a logical interval. The purpose of block section division is to facilitate the judgment of train headway and improve the efficiency of the whole system.

At present, block sections are basically divided into two types: fixed block and moving block. The fixed block section is fixed, and corresponding protection signal indication will be set between sections. Only when the train leaves the current geographical section, the geographical section will open the route to other trains. The moving block section will change with the speed, position and other line information of the train ahead.

The moving block section is equipped with wireless communication technology, and the front and rear vehicles send position and speed information to the control center. According to the actual situation of the train, the control center keeps the train running at a minimum interval on the premise of ensuring safety, so as to improve the safety and efficiency of the system. The moving section is composed of train length, emergency braking distance and safety protection distance. Normal CBTC trains can only run in the authorized section, and the section is updated in real time with the position and speed of the train itself and the position of the front train.

Train tracking under moving block is mainly divided into relative position mode and relative speed mode.

(1) In the relative position mode, the rear vehicle only considers the position of the front vehicle instead of the speed of the front vehicle, and the interval between the two vehicles is only determined by the speed of the rear vehicle. The tracking distance S between the controlled train and the front train is:

$$S = L + VT + \frac{V^2}{2a} + S_0 \tag{2.1}$$

where L is the length of the controlled train, V is the speed of the controlled train before braking, T is the driving time at a constant speed, a is the braking acceleration of the controlled train, and S_0 is the safe distance to the rear of the front train.

(2) In the relative speed mode, the rear vehicle should consider not only the position of the front vehicle, but also the speed of the front vehicle. The tracking interval between them will change according to the speed of the two vehicles. The tracking distance S between the controlled train and the front train is:

$$S = L + V_2T_2 + \frac{V_2^2}{2a_2} - V_1T_1 - \frac{V_1^2}{2a_1} + S_0 \tag{2.2}$$

where V_2 and V_1 are the speed before braking of the controlled train and the front vehicle, respectively, T_2 and T_1 are the driving time of the controlled train and the front vehicle at a constant speed, respectively, and a_2 and a_1 are the braking acceleration of the controlled train and the front vehicle, respectively.

2.1.2 *Working Principle of Train Movement Authorization*

In urban rail transit, the train-ground wireless communication system controls the train by assigning mobile authorization. Movement authorization is a section of line with special meaning. The starting point of the line starts from the rear of the train, and the endpoint of the line is the obstacle in front. Obstacles include the endpoint of the route, turnout, train in front, etc.

The generation steps of mobile authorization are:

(1) The train in the controlled line area operates according to the normal timetable, and the on-board controller calculates the current position and speed of the train according to the beacon and speed measurement at the trackside.
(2) The on-board controller sends the position and running direction of the train to the area controller through the trackside AP (access point of trackside wireless unit) and judges whether it is necessary to apply for extending the current movement authorization.
(3) The area controller calculates the movement authorization of each train according to the current position and speed of the train and other line states.
(4) After the calculation, the area controller sends the movement authorization to the on-board controller of each train and sends the position information of the train to the automatic train supervision system, so that the automatic train supervision system can monitor all vehicles in the line.

In the whole process, the regional controller interacts with subsystems such as automatic supervision and automatic protection in real time and periodically send the train information to the area controller through the on-board equipment of the data communication system. At the same time, the regional controller will also calculate the movement authorization of each train according to the track operation status and respond to the on-board equipment.

2.2 Basic Analysis and Evaluation About CBTC

Before elaborating the rail transit intelligent technology system, it is necessary to make a brief analysis and evaluation of the existing CBTC.

2.2.1 Overview of the Development Process of WLAN

In the development of WLAN, many technologies and protocols once emerged, such as IrDA, Blue Tooth, HyperLAN2 and so on.

However, up to now, IEEE 802.11 series standard protocols have been widely promoted and commercialized in the field of WLAN, and WLAN is also defined as wireless LAN based on IEEE 802.11 standard protocol.

In the early 1990s, in order to meet the growing demand for WLAN, IEEE established a special 802.11 working group to study and customize the standard protocol of WLAN and launched the first generation WLAN protocol IEEE 802.11-1997 in June 1997. The protocol defines that the physical layer works in the 2.4G band of ISM, and the data transmission rate is designed as 2 Mbps. The protocol is not widely used because its design in rate and transmission distance cannot meet people's needs.

Subsequently, IEEE introduced IEEE 802.11a and IEEE 802.11b in 1999.

IEEE 802.11a works in the industrial scientific medical band (ISM band) of 5 GHz and selects orthogonal frequency division multiplexing (OFDM) technology, which can effectively reduce the impact of multi-path attenuation and improve the utilization of spectrum, so that the physical layer rate of IEEE 802.11a can reach 54 Mbps.

IEEE 802.11b still works in the ISM band of 2.4 GHz, but the technical improvement is made on the basis of IEEE 802.11 to make the communication rate of IEEE 802.11b reach 11 Mbps.

OFMD is a multi-carrier modulation technology, which mainly divides the specified channel into several subchannels and uses one subcarrier for modulation on each subchannel, and each subcarrier is transmitted in parallel, which can effectively improve the spectrum utilization of the channel.

Although the access rate provided by IEEE 802.11b is lower than that of IEEE 802.11a, the development progress of 5 GHz chip was too slow at that time. When the chip was launched, IEEE 802.11b has been widely used. Because IEEE 802.11a is not compatible with IEEE 802.11b, and the price of 5 GHz chip is high and limited by local regulations, IEEE 802.11a is not widely used.

In early 2000, the IEEE 802.11 g working group began to develop a protocol standard that can not only provide 54 Mbps rate, but also downward compatible with

IEEE 802.11b. The first IEEE 802.11 g draft was proposed in November 2001, which officially became a standard in 2003. IEEE 802.11 g is compatible with IEEE 802.11b and continues to use the 2.4 GHz band. In order to achieve the rate of 54 Mbps, IEEE 802.11 g borrows the achievements of IEEE 802.11a and adopts orthogonal frequency division multiplexing (OFDM) technology in the 2.4 GHz band. The introduction of IEEE 802.11 g met people's demand for bandwidth at that time and played a great role in promoting the development of WLAN.

For the rapid development of WLAN, the rate of 54 Mbps cannot meet the needs of users. In 2002, a new IEEE working group, IEEE 802.11n (TGN, task group n), was established to study a faster WLAN technology with the goal of reaching 100 Mbps.

In the seven-year development process, the rate of IEEE 802.11n has improved from 100 Mbps originally designed to 600 Mbps.

IEEE 802.11n adopts dual band working mode, supports 2.4 and 5 GHz, and is compatible with IEEE 802.11a/b/g.

Subsequently, IEEE 802.11ac was developed and officially launched in 2013. IEEE 802.11ac operates in 5 GHz band and is backward compatible with IEEE 802.11n and IEEE 802.11a. IEEE 802.11ac follows many technologies of IEEE 802.11n and makes technical improvements to make the rate reach 1.3 Gbps.

Compared with the previous standard protocols, IEEE 802.11n mainly has the following advantages: more subcarriers, higher coding rate, shorter GI (guard interval), wider channel, more spatial streams and message aggregation function of MAC layer.

2.2.2 Technical Limitations of the Original CBTC

Since the original WLAN standard was not designed for high-speed scenes, the original CBTC has some limitations, which are embodied in the following aspects [1].

1. **Poor Mobility**

For rail transit system, especially high-speed railway, when the train speed is too high, the original communication standards used in rail transit system will often lead to a sharp decline in communication performance, which cannot meet the requirements of rail transit communication system for high reliability and low time delay.

2. **Multiple Interference Sources**

WLAN works in the 2.4 GHz open frequency band, which is also the working frequency band of many civil equipment, which is bound to interfere with vehicle ground wireless communication transmission.

3. **Small Antenna Coverage**

Since the coverage of wireless access point (AP) is only about 200 m, when the train is running at high speed, frequent handoff is required, which is bound to affect the overall communication quality.

4. **Lack of Priority Guarantee Mechanism**

In addition to safety-related CBTC train control information, the information transmitted in urban rail wireless communication system shall also have passenger information system (PIS) and video monitoring system (VMS). If all these systems work in one frequency band, the reliable transmission of safety-related train control information cannot be guaranteed without establishing priority. In order to ensure the reliability of train safety information transmission, the current solution is to use independent networking, which leads to the non-sharing of resources and spectrum competition. At the same time, the construction investment and later maintenance cost of multiple networks will inevitably be increased.

In order to solve the above technical limitations, a new communication technology, the so-called long-term evolution (LTE) technology, has been introduced into the existing operation system. LTE has obvious advantages over WLAN:

(1) Proprietary frequency band, interference within controllable range;
(2) Multi-level service quality (SQ) control can give priority to ensuring the transmission of CBTC;
(3) It can support 350 km/h mobility;
(4) The coverage distance is long, the coverage area radius can reach 1 km, and the cross-area switching smoothness is strong.

However, although the traditional CBTC technology still has technical defects, that is, the communication in the system is based on the mutual exchange between vehicles and ground. All trains transmit their own status information to the ground control center for unified dispatching through the server. In this way, there must be a problem in the control link mode: Once the ground control center is attacked, the whole system will be on the verge of paralysis, which will eventually lead to accidents. The "July 23 Ningbo Wenzhou line" major railway traffic accident in 2011 was caused by serious design defects in the equipment of the train control center, that is, the human responsibility accident caused by poor emergency response after equipment failure caused by lightning stroke.

It can be predicted that the future CBTC will be a new intelligent communication and monitoring system. This book is to explain the scientific theories and methods of proliferating multi-layer intelligent algorithms and their functions on the basis of CBTC along the direction of this innovative thinking.

2.3 System Wide Intelligent Communication and Monitoring

The overall structure of the whole system communication and intelligent control of rail transit is shown in Fig. 2.1. Among them, "Trackside equipment" and "Adjacent trackside equipment" are used to monitor the condition of subgrade and track, and also undertake the detection of train speed and acceleration in operation, so as to

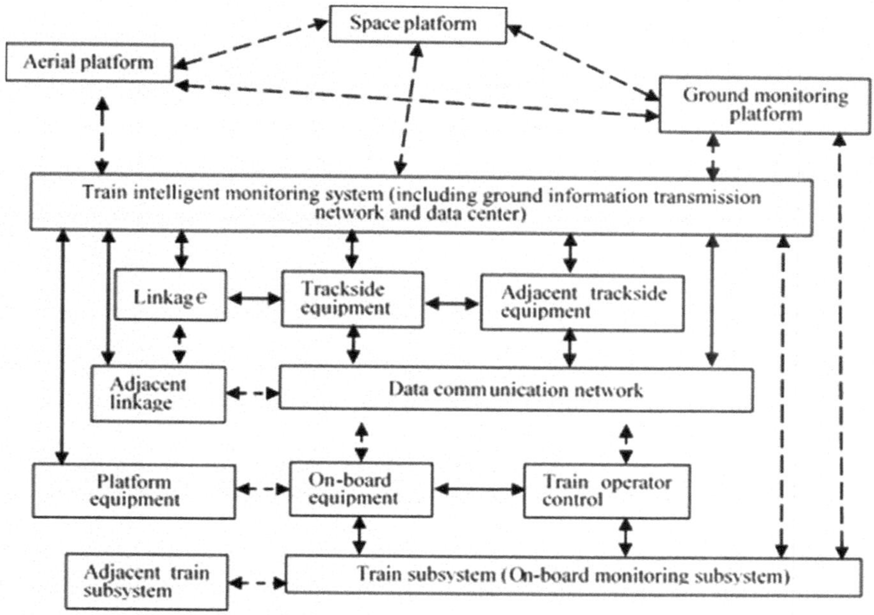

Fig. 2.1 Structure block diagram of rail transit intelligent monitoring system

provide real-time information of subgrade, track condition and basic state parameters of train operation to "Train intelligent monitoring system" and "Train subsystem" in advance. "Train subsystem" and "Adjacent train subsystem", that is, the on-board control subsystem of the train itself, can receive status information from relevant detection platforms and data centers. "Aerial platform" and "Space platform" are monitoring systems installed on air and space vehicles, respectively. In the figure, the solid line is a wired link. The dotted line is a wireless link [2].

2.3.1 Data Communication System

Data communication system (DCS) focuses on the two-way data communication network in urban rail transit system. Among them, wired network provides data transmission channel between ground control center, station, trackside and other equipment. Wireless network provides continuous two-way wireless data transmission between train and ground control equipment. The general data communication system network is a special communication network, which adopts independent networking mode, and the network structure adopts double redundancy design to provide two independent network transmission channels for urban rail transit system data. The function of the system is mainly divided into three parts.

1. **Communication Transmission Channel**

DCS equipment can realize the data communication capability between point-to-point and point-to-multi-point, and can classify the data based on point-to-point, that is, classify the bandwidth according to the data requirements of each service.

2. **Network Management**

DCS equipment adopts SNMP (Simple Network Management Protocol) to realize management, mainly including network performance, configuration, fault and other management functions.

3. **Network Security**

The communication network system has a complete set of security policies and meets specific standards in terms of wireless communication security management function and functional requirements of transmission channel.

Dedicated Short-range Communication (DSRC) included in the whole system communication shown in Fig. 2.1 is a group of protocols and standards dedicated to short-range or medium-range wireless communication between vehicles. It mainly adopts 5.9 GHz frequency band and has low transmission delay characteristics. The essence of DSRC is a set of protocols. It is published in the form of protocol stack. The involved standards mainly include IEEE802.11p, IEEE1609 and other protocol series.

IEEE 802.11p is a communication protocol extended by IEEE 802.11 standard. The communication protocol is mainly used in the wireless communication of vehicle electronics. In terms of setting, it is extended from IEEE 802.11 to meet the relevant applications of Intelligent Transportation Systems (ITS). Its application level includes data exchange between high-speed vehicles and between vehicles and ITS standard roadside infrastructure in the 5.9 GHz (5.85–5.925 GHz) band.

IEEE 802.11p extends the traditional wireless short-range network technology to realize more advanced handoff scheme, mobile operation, enhanced security, identification and peer-to-peer authentication. Most importantly, communication on the specified on-board frequency will serve as the basis of DSRC or on-board communication. On-board communication can be carried out between vehicles or between vehicles and roadside infrastructure network.

IEEE 1609 standard is a high-level standard based on IEEE 802.11p communication protocol.

In addition, DSRC adds a special processing mechanism for secure messages.

As shown in Fig. 2.2, DSRC protocol stack mainly includes four layers: physical layer, data link layer (including MAC layer, MAC sublayer extension and LLC sublayer), network transmission layer and application layer.

The data link layer is mainly divided into two sublayers: media access control (MAC) and its extension and logical link control (LLC).

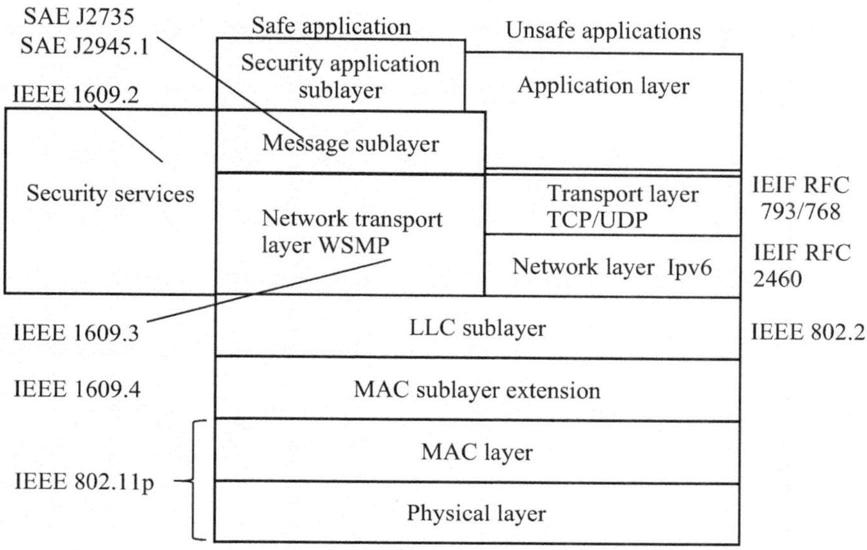

Fig. 2.2 DSRC protocol stack structure

Among them, MAC defines how data packets are transmitted on the media and formulates the competition rules that should be followed when accessing the channel. Its purpose is to make the access nodes compete for resources more efficiently and fairly, so as to achieve the effect of fair channel allocation and improved resource utilization. In the IEEE802.11 protocol cluster, the channel competition rules can be divided into two categories: One is the session rules that require nodes to take certain measures before exchanging information on behalf of the third layer and the other is frame-to-frame rules suitable for individual transmission. For wireless channels, the media access mode of IEEE802.11 is carrier interception multiple access with collision avoidance. The main principle of this mechanism is to monitor the state of the channel. If it is detected that the channel state is busy, it will back off until the channel becomes idle. In addition, the mechanism also adds ACK (acknowledgment) characters to ensure the successful reception of data.

The logical link control (LLC) sublayer uses IEEE802.2 protocol and adds subnetwork access protocol to supplement it. It mainly encapsulates the requests and replies of IP packets and address resolution protocol. In addition, ieee1609.3 protocol also supports no-acknowledgment and no-connection service with no-numbered message frame.

Fig. 2.3 shows the format of LLC sublayer protocol data unit.

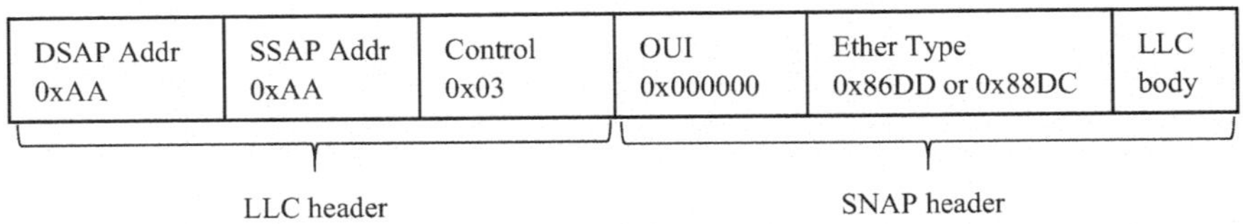

Fig. 2.3 Format of LLC sublayer protocol data unit

2.3.2 Status Information Wireless Sensor Network

In order to ensure the safety of rail transit operation, system communication and control must obtain the operation state information of rail transit system by means of multi-source perception and carry out corresponding state analysis and fault diagnosis, so as to provide technical and decision support for the operation guarantee of rail transit system.

At present, the status of rail transit system mainly depends on the combination of manual track inspection, comprehensive monitoring, train inspection and on-board detection equipment to carry out "regular physical examination" on the operation status of rail transit system, process and analyze the detection data offline, and then estimate the operation status to determine the "safety life cycle". Therefore, the existing detection methods are obviously inefficient and do not have real-time performance.

Because the detection cycle is generally formulated according to the state statistical law and the operation experience of field experts, it is obviously lack of "predictability" for the occurrence of sudden faults. At the same time, the detection period and train operation period cannot be "overlapped—carried out in parallel", which is bound to reduce the operation efficiency of rail transit system. In order to improve the economy, real time, reliability and safety of rail transit system operation status monitoring, it is necessary to form a wired status information monitoring and transmission system, which adopts monitoring and transmission technologies such as video, fiber Bragg grating and stress–strain to realize online monitoring of rail transit system infrastructure and real-time transmission of monitoring information, In order to improve the real-time acquisition of rail transit infrastructure status information. However, the online monitoring system based on wired communication still has many difficulties in practical application, such as high construction cost, difficult construction, low adaptability and poor reliability.

Obviously, wireless sensor networks (WSN) have the ability to accurately perceive the monitoring objects and stably transmit information. With the development of wireless sensor technology, wireless communication and data processing technology, the advantages of wireless sensor technology are becoming more and more prominent [3].

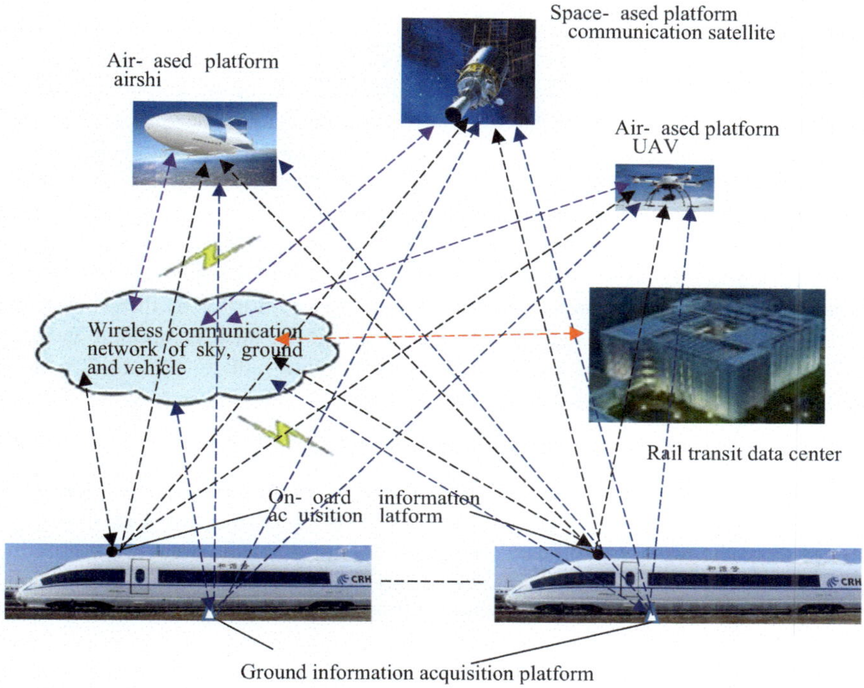

Fig. 2.4 Rail transit wireless sensor system architecture

1. **Wireless Sensor System Architecture**

The wireless monitoring system for rail transit system status is composed of space-based platform, air-based platform, on-board monitoring platform, ground monitoring platform, ground information transmission network and rail transit data center (see Fig. 2.4).

(1) **Space-based platform**

The space-based platform realizes the full coverage monitoring and information transmission of the rail transit system through China's BeiDou satellite navigation system, so as to ensure the real-time accessibility of information in a large range. The space-based platform is mainly responsible for the transmission of key information and instructions such as train positioning.

(2) **Air-based platform**

The air-based platform is composed of airships residing in the stratosphere and low altitude UAVs, which form a "high-low coordination" monitoring and communication subsystem. Airship can make up for the technical defect that the wide area sparse road network is difficult to wait for regional safety monitoring all day and provide an all-round and information-based system operation and safety guarantee capability. Its monitoring coverage can reach 7×10^5 km^2.

UAV has the technical characteristics of rapid deployment, real-time monitoring, simple operation and remote operation, so it meets the technical requirements of local enhanced monitoring. For areas with few trackside monitoring facilities, regular patrol inspection along the line can be implemented. In case of sudden failure, UAV can arrive over the site for the first time to obtain fault information, which is particularly effective for temporary local monitoring.

(3) **On-board information acquisition platform**

The on-board information acquisition platform mainly monitors the service status of key components of the train (bogie, axle and electrical equipment, etc.) and the operation status of the train (speed, acceleration, temperature, etc.) in real time through the sensors installed in key parts of the train, and upload the monitoring information to the on-board data center for processing and analysis, or send it to the data center for data processing through the ground information transmission platform with the help of the technical mode of vehicle ground communication.

The basic physical composition of locomotive on-board information acquisition platform is shown in Fig. 2.5.

(4) **Ground information acquisition platform**

The ground information acquisition platform is the monitoring platform for the service status of track infrastructure and basic parameters of train operation. A large number of sensors are arranged along and around the track to monitor and upload

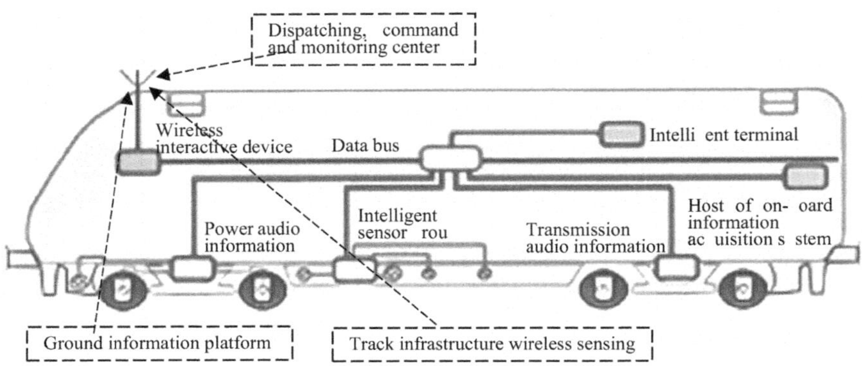

Fig. 2.5 Basic physical structure of locomotive on-board information acquisition platform

the service status of online and offline infrastructure, traction power supply system and train motion parameters in operation in real time.

(5) **Sky, ground and vehicle wireless communication network**

Wireless communication network can realize high redundancy interaction of monitoring data of sky (air), ground and vehicle, and transmit it to rail transit data center in real time.

(6) **Rail transit data center**

Rail transit data center is a platform for information storage, processing, operation state analysis and control command output.

The monitoring wireless/wired network of rail transit system based on the integration of air, space, vehicle and ground can monitor the operation status of rail transit in a wide range, all-weather and systematically, and provide technical guarantee for the efficient and safe operation of rail transit system [4].

2. **Distribution Characteristics of Wireless Sensor Nodes**

The network system of WSN usually includes sensor nodes, sink nodes, gateway devices, management terminals and so on. The task of the sensor node is to collect the environmental data required in the monitoring area and send the environmental data. The sink node collects the data of all sensor nodes and uploads it to the gateway device. The gateway device transmits the data to the management terminal through the Internet or other networks.

Because WSN is limited by energy, network, computing and storage resources, compared with the existing mature networks such as Internet and cellular communication network, WSN has the following characteristics.

(1) **There are many sensor nodes and wide coverage**

In order to accurately monitor the environmental information, a large number of sensor nodes are often arranged in the monitoring area with high density, so WSN has high requirements for multi-point communication performance.

(2) **Limited resources**

The sensor node is powered by battery. Even the acquisition unit with energy will be limited by the battery capacity and the environment, so the power consumption of the sensor node must be limited. In order to deploy sensor nodes on a large scale, the cost of nodes must be controlled low, and the storage space and memory space must be limited. Therefore, the design of communication protocol should not be too complex.

(3) **Self-organizing function**

Because the sensor nodes of WSN are arranged randomly, it requires that nodes can detect each other and automatically build a reasonable wireless communication network.

(4) **Network topology is easy to change**

Due to the influence of node energy depletion, node damage, weak signal and other factors, WSN needs to have the function of timely adjusting network topology and dynamically updating routing.

(5) **Closely related to the application**

Different from other networks, WSN has many applications. In order to give full play to the best performance of WSN, WSN should have the function of corresponding design and optimization according to the technical requirements of applications.

(6) **Uneven flow distribution**

The purpose of WSN is to collect the environment information in the area and upload it to the management terminal, so most of the data will be sent to the sink node. This will make the nodes around the sink node heavy load and consume more energy, which will shorten the network life.

Thanks to the rapid development of microelectromechanical system (MEMS), system on chip (SOC), wireless communication and Low-Power Embedded system, WSN has become a multi-disciplinary research hotspot [5].

3. **Topology of Dedicated Wireless Sensor Network for Rail Transit**

Since the monitoring area of the existing rail transit is relatively independent from the existing wireless communication system, the wireless sensor node responsible for information sensing and transmission is far away from the rail transit dedicated base station. In addition, the energy storage of wireless sensor nodes is very limited, and its communication energy consumption is mainly determined by the amount of data of monitoring information and its transmission distance. In addition, the communication distance of wireless sensor nodes is also limited. It is obviously not economical and feasible for sensor nodes in the monitoring area to directly communicate with base stations. Therefore, it is necessary to divide the whole monitoring area into multiple submonitoring sections. A sink node is deployed at the edge of each submonitoring section to receive the monitoring data of the submonitoring section, and the information in all submonitoring sections is forwarded by the sink node to the base station. On the one hand, this information transmission method can ensure that the information perceived by each sensor node can be sent to the base station smoothly. On the other hand, it also greatly reduces the energy consumption of sensor nodes and then improves the life cycle of wireless sensor networks.

The special wireless sensor network for rail transit condition monitoring is a two-layer system structure, which is divided into subnet layer and backbone network layer (see Fig. 2.6).

The subnetwork layer is composed of several submonitoring areas, the wireless transmission network of each submonitoring area is composed of several wireless sensor nodes and their wireless sink nodes, and the sink node is responsible for collecting the information perceived by all sensor nodes in the subarea. The backbone layer consists of several wireless sink nodes and their base stations. The sink node is

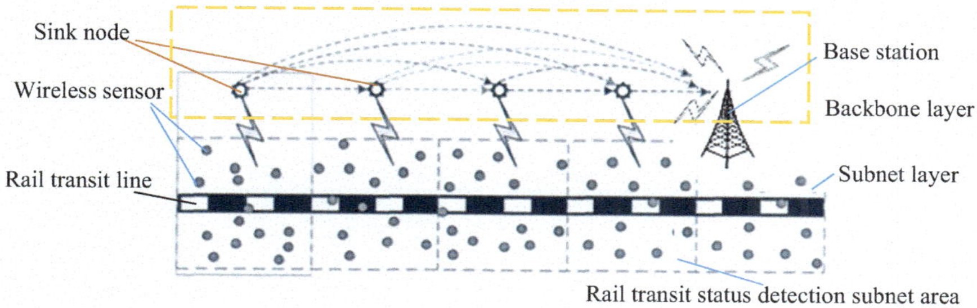

Fig. 2.6 Double layer structure of rail transit wireless detection system

responsible for forwarding the information collected from the subnet layer to the base station. The service status information of rail transit infrastructure and its operating environment is first perceived by wireless sensor nodes in the subnet. Then, the sensor node in each submonitoring area sends the sensing information to the corresponding sink node. Finally, the sink node sends all the sensing information to the base station, and then, the base station sends it to the remote server by wired or wireless for the diagnosis and prediction of the state of the rail transit system.

(1) Subnet structure and routing protocol of Wireless Sensor Networks

The subnetwork communication structure in each submonitoring area of the special wireless sensor network for rail transit condition monitoring is composed of several sensor nodes and their sink nodes. On the one hand, sensor nodes are responsible for sensing the service status information of infrastructure and operating environment. On the other hand, it is responsible for sending the sensing information to the sink node.

Due to the limited energy storage of sensor nodes, the energy efficiency of sensor nodes is mainly considered when selecting and optimizing routing protocols (see Fig. 2.7).

The communication energy consumption of wireless sensor nodes depends on the amount of data transmitted and its transmission distance. As can be seen from Fig. 2.6, the distance between sensor nodes and sink nodes is different. When each sensor node transmits data of the same scale to the sink node, it will lead to the imbalance of energy consumption of each sensor node, and the nodes far away will fail prematurely due to energy depletion. In order to improve the energy efficiency of each sensor node in the subnet, it is necessary to adopt clustering communication routing protocol in the subnet.

Firstly, all sensor nodes in the submonitoring area are divided into different clusters. Then, select one node in the cluster as the cluster head, and the other nodes are members of the cluster group. Among them, the cluster members are responsible for sensing the information of the key components of the rail transit system and sending the sensing information to the cluster head node. Then, after receiving the information, the cluster head node performs simple processing and analysis, and forwards the information to the sink node.

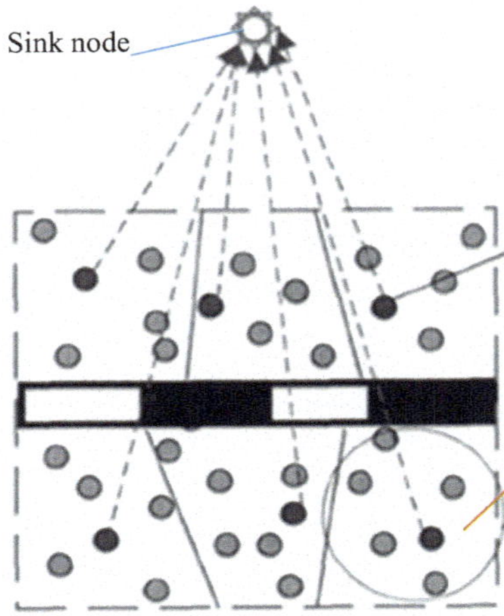

Fig. 2.7 Rail transit wireless monitoring subnet structure

Using clustering communication routing protocol, the selection of cluster head and clustering results will have an impact on the total energy consumption of all nodes and the balance of energy consumption among nodes. Therefore, the clustering communication routing protocol can be optimized to improve the energy utilization efficiency of nodes in the subnet and the life cycle of the subnet.

In addition, the communication energy consumption of wireless sensor nodes is also related to the amount of data transmitted. The cluster head collects information from cluster member nodes and then processes the data before sending it, which can reduce the amount of data transmission and corresponding communication energy consumption. Of course, this will increase the corresponding data processing energy consumption. Therefore, the total energy consumption of cluster head nodes can be effectively reduced through adjustment and trade-off, so as to improve the overall energy utilization efficiency of the subnet.

(2) **Backbone structure and routing protocol of Wireless Sensor Networks**

The backbone network communication structure of rail transit condition monitoring wireless sensor network is composed of several sink nodes and their base stations. Each sink node and base station are arranged linearly along the track. The sink node is responsible for transmitting the information collected from the submonitoring area to the base station.

In wireless sensor networks, the energy of each sink node is also limited. Therefore, when selecting and optimizing the backbone network information transmission routing protocol, it is also necessary to give priority to the energy use efficiency of nodes (see Fig. 2.6).

As can be seen from Fig. 2.6, the distance between each node and the base station is very large. It is difficult for the sink node on the side far from the base station to communicate directly with the base station. Even if it can communicate directly, it will cause imbalance in energy consumption among nodes. At this time, in the process of backbone network information transmission, the routing protocol based on multi-hop communication can well solve this kind of problem.

The routing of multi-hop communication links and the amount of data communicated on each link will affect the balance of energy consumption between nodes and then affect the life cycle of the backbone network. At the same time, the number and communication sequence of multi-hop links will affect the delay of data transmission and the real-time performance of information transmission in the backbone network.

In addition, there are many kinds of monitoring objects in the wireless sensor network for rail transit condition monitoring. The data type, bandwidth requirements for data transmission and communication requirements (priority, importance, etc.) of each monitoring object are different from each other, which is bound to increase the demand for the total bandwidth. However, in the detection and communication system of rail transit, the bandwidth resources of dedicated wireless sensor network are very limited, which is difficult to meet the real-time and reliable transmission of all monitoring business information. Therefore, it is necessary to analyze the bandwidth and communication requirements of each monitoring object, so as to allocate the limited bandwidth resources to the most urgent monitoring objects, so as to improve the utilization efficiency of bandwidth resources and ensure the reliable transmission of key information [6].

4. **Characteristics and classification of wireless sensor network protocols**

(1) **Characteristics of wireless sensor network protocol**

The function of routing protocol is to find the optimized path of source node and destination node, and forward data packets correctly along the optimized path. As the core technology of WSN, routing protocol is closely related to the performance of the whole network. In general, the sensor nodes in WSN have equal status in the network. Each sensor node coordinates with each other through distributed algorithms and forms a wireless network through self-organization. Limited by the energy of sensor nodes in WSN, the communication energy consumption W_{WSN} of sensor nodes in wireless communication is directly proportional to the power of effective transmission distance L_{WSN}, i.e.

$$W_{\mathrm{WSN}} = kL_{\mathrm{WSN}}^{n} \tag{2.3}$$

where k is the scale factor, $n \in [2, 4]$.

In order to reduce the energy consumption of data transmission, the effective transmission radius of nodes must be controlled. At the same time, in order to realize the wide coverage of WSN, we must use multi-hop to transmit data. Therefore, the routing protocol of WSN should be designed and added with corresponding support according to the characteristics of WSN, and the general protocol cannot be used. WSN routing protocols often have the following characteristics.

1) **Data center properties**

The design concept of WSN should be data centric, and the data of all nodes will be transmitted to the sink node. Users mainly focus on the data collected by the sensing nodes in the monitoring area. At the same time, the protocol design needs to take into account the management and processing of data.

2) **High efficiency characteristics of energy**

The so-called high efficiency characteristics of energy, that is, how to reduce the energy consumption and increase the network life cycle when discovering routing and transmitting data under limited energy, is the focus of routing protocol design.

3) **Scalability**

The routing protocol must be scalable. When new nodes are added or nodes fail, the network can make timely adjustments to adapt to the dynamic changes of network topology.

4) **Multi-hop communication**

Multi-hop communication can greatly increase the coverage of WSN and reduce the energy consumption of long-distance communication. Multi-hop communication also helps to realize data fusion. Merging and compression according to certain standards in the transmission process can further reduce the load and energy consumption of the network.

5) **Relevance of application**

When dealing with different application scenarios, routing protocols should be optimized and improved according to the environment, so as to make WSN give full play to its advantages and improve the efficiency of the network.

(2) **Basic classification of wireless sensor network protocols**

Due to the application characteristics of WSN routing protocol, it is of great significance to design a unique routing protocol for WSN. With the deepening of WSN research, in order to meet different application environments, various routing protocols have been proposed by researchers. These protocols have their own characteristics. In order to summarize and analyze these routing protocols, researchers classify them according to some specific standards.

1) **Classification according to the communication mode of node participation**

According to the way nodes participate in communication, WSN routing protocols can be divided into the following categories.

① **Plane routing protocol**

The protocol has equal status for all sensor nodes in WSN, the routing function is the same, and there is no special node. Sensor nodes cooperate with each other to complete data acquisition, processing and forwarding in a preset working mode. The advantages of this routing protocol are simple structure, less computing resources, good robustness and easy maintenance. It is suitable for small-scale WSN. The disadvantage is that when the scale of WSN is large, there is no special node to optimize the network, the number of transmission hops is too many, and the establishment and maintenance of routing table will occupy a lot of energy and network resources. The classic plane routing protocols include SPIN, DD, GBR and so on.

② **Hierarchical routing protocol**

The sensor nodes of such protocols are divided into different clusters, and then, the qualified sensor nodes are selected as the cluster head nodes. The cluster head node is responsible for collecting the monitoring data of the sensor nodes in the cluster, reducing the amount of transmission information through data aggregation and fusion, and finally, the cluster head node transmits the processed data to the sink node. Compared with plane routing protocol, hierarchical routing protocol can meet the scalability of routing protocol, effectively reduce the power consumption of sensor nodes, and prolong the network life cycle. In such protocols, the energy consumption of cluster head nodes is large. Protocols often use specific clustering algorithms to regularly select cluster heads to balance the energy consumption. Classic hierarchical routing protocols include LEACH, HEED, TEEN, etc.

2) **Classification of path acquisition methods according to source nodes**

According to the way the source node obtains the path, WSN routing protocols can be divided into the following categories.

① **Active routing protocol**

This kind of protocol is also called a priori routing protocol or table-driven routing protocol. The protocol should establish the path of the target node before data transmission. All nodes in the network should periodically broadcast routing information and maintain a routing table to all known destination nodes. When the network topology changes, the routing table will be updated accordingly. If the source node wants to transmit data, it can immediately obtain the route to the node from the route table, which takes less time. However, active routing protocol has a certain application in small-scale WSN. It consumes too much energy, network, computing and storage resources to establish and maintain routing table in large-scale WSN. Typical active routing protocols include Batman-adv, OLSRv2, Babel and so on.

② **On-demand routing protocol**

This is a routing protocol that finds routes only when data needs to be transmitted. Only when each node needs to use, it can establish and maintain its own route. The topology and routing table are established on-demand, avoiding periodic route broadcasting to maintain a large routing table and topology. The energy, network, computing and storage resources required by the on-demand routing protocol are small, but if there is no established link or link failure in the routing table before data transmission, there will be a large time delay in calculating the routing. Typical on-demand routing protocols include AODV, DSR, TORA, etc.

5. **Key Technologies of Rail Transit Wireless Sensing**

With the support of the BeiDou satellite navigation system and the advanced technology of the wireless communication base station in China, the information sensing of rail traffic becomes the most critical link to ensure the stability and reliability of the whole system. These sensors include the induction and acquisition of mechanical stress, hydraulic transmission pressure, temperature, thermal image, voiceprint, image and other information. Moreover, a series of preprocessing must be carried out for all sensing information before it can be delivered to the back-end processor for analysis and evaluation. In this process, it needs to be solved effectively through wireless sensor network.

The key technologies of wireless sensor networks mainly include network topology control, network protocol, synchronization technology, positioning technology, data fusion and so on.

(1) **Network topology control**

Wireless sensor networks have the characteristics of large scale and easy failure of nodes. How to form a good topology in a large number of sensor nodes and quickly form a new topology when nodes die has become the focus of network topology control research. Good topology requires reducing unnecessary wireless communication in the network to prolong the network life. Network topology control can be divided into hierarchical topology control and node power control.

(2) **Network protocol**

The ability of sensor nodes (processing, storage, communication and energy) is limited, which puts forward higher requirements for reducing the complexity of the protocol. The network layer protocol, namely the routing protocol, determines the transmission path of monitoring data to the sink node; the data link layer mainly accepts the media access control (MAC), which is used to coordinate the node's access mode and time to the wireless channel.

Among them, the life cycle of the dedicated wireless network subnet for rail transit condition monitoring is determined by the number of communication rounds of the earliest failed node. Therefore, in the process of routing protocol optimization, we should take into account the minimization of the total energy consumption of all nodes and the balance of energy consumption among nodes. In the cluster communication routing protocol, the establishment and solution of optimization models such as cluster head optimization and cluster generation are obviously the key technologies to be solved urgently. At the same time, the backbone network of rail transit condition monitoring wireless sensor network is a typical linear sensor network. The routing protocol of multi-hop communication can improve the balance of energy consumption between sink nodes, but it will increase the corresponding communication delay. In the routing protocol optimization based on multi-hop communication, the overall efficiency of the backbone network can be improved by taking into account the needs of different nodes for the life cycle and communication delay of the backbone network.

(3) **Synchronization technology**

In the application of wireless sensor networks to measure the speed of objects, it is necessary to know the time difference detected by each sensor node, which requires time synchronization technology. Many MAC protocols also put forward high requirements for time synchronization. NTP protocol widely used on the Internet is only suitable for wired networks with stable structure; GPS is expensive and cannot be used in indoor and underwater environments. How to realize time synchronization only by using the simple communication between sensor nodes is the research focus of synchronization technology.

(4) **Positioning technology**

Wireless sensor network is a bridge between the physical world and people. It needs to quantify the physical information of the physical world and then provide it to users. In the condition monitoring of rail transit, the monitoring data without location information is meaningless.

(5) **Data fusion**

Sensor network is a data-centric wireless self-organizing network. Users do not care about the observation data of a specific sensor node, but about the index of a physical information in a certain area. And transmitting redundant monitoring data will unnecessarily increase network traffic and waste valuable network energy. Data fusion means that sensor nodes synthesize the data of multiple nodes to remove redundant information in the process of data acquisition. Data fusion improves the efficiency of network energy utilization, but increases the average network delay and reduces the robustness of the system.

In the rail transit wireless sensor network, before the cluster head node communicates with the sink node, using the data fusion algorithm to reduce the amount of communication data can effectively reduce the communication energy consumption of the cluster head node, but it will increase the corresponding processing energy consumption at the same time. These are two contradictory aspects. The change of data fusion rate will affect the relationship between data processing energy consumption and communication energy consumption of cluster head nodes and then affect the energy efficiency of cluster head nodes and even subnetworks. At the same time, the cluster group size corresponding to different cluster head nodes and their distance to the sink node are also different and change with the rotation of cluster head and the update of clustering. In the process of subnetwork data fusion optimization of rail transit wireless sensor network, it is particularly important to adaptively optimize the cluster head data fusion rate according to the optimization results of clustering routing protocol.

(6) **Bandwidth resource allocation**

In the wireless sensor network for rail transit condition monitoring, the bandwidth resources are limited and dynamic, and the total bandwidth demand of the monitoring object is very large, and the bandwidth and communication requirements of different services are also different. How to dynamically allocate limited bandwidth resources to improve the overall utilization efficiency of bandwidth is obviously an important key technology.

(7) Other key technologies

Other key technologies of wireless sensor networks include network security, data management, wireless communication technology, embedded operating system, application layer technology and so on.

The above key technologies can only be solved by relying on the support of many high-end technologies such as hardware structure and software algorithm. The following focuses on the basic algorithm of hierarchical routing in wireless sensor networks.

6 The Basic Algorithm of Hierarchical Routing in Wireless Sensor Networks

Hierarchical routing algorithm is divided into three stages: cluster head selection, cluster formation and stable data transmission. At the same time, the energy consumption is allocated to each sensor node through the periodic rotation of the cluster head.

In the cluster head selection stage, a certain number of nodes are randomly selected as cluster heads in the detection subnet area. In the cluster formation stage, the cluster is formed according to the distance principle. Ordinary nodes select the cluster head closest to themselves to join its cluster. In the stable data transmission stage, the nodes in the cluster send the data to the cluster head in their own TDMA time slot, and the cluster head sends the data to the sink node after data fusion.

(1) Cluster head selection

Cluster head selection adopts distributed algorithm, and node n decides whether to be selected as cluster head in this round according to Formula (2.4).

$$T(n) = \begin{cases} \frac{P}{1-P\cdot[r \bmod (1,P)]}; & n \in G \\ 0; & \text{otherwise} \end{cases} \tag{2.4}$$

where P is the desired percentage of cluster heads, usually 5%; r represents the number of current rounds; mod$(1, P)$ is the remainder function, specifically, the remainder of $1/P$; G is the set of nodes that have not served as cluster heads in $1/P$ rounds before r. If a node has been selected as a cluster head in the previous $1/P$ round, its $T(n)$ value is 0, that is, it no longer acts as a cluster head. This is to balance the energy consumption of the network, because the energy consumed by the cluster head is much larger than that of the nodes in the cluster. The node n generates a random number ranging from 0 to 1. If $T(n)$ is greater than the random number, node n is selected as the cluster head. Otherwise, it acts as a node in the cluster. After selecting the cluster head, node n broadcasts the message to the whole network at a constant power through CSMA/CA mechanism.

(2) **Cluster formation**

After receiving the broadcast message from the cluster head, according to the RSSI (received signal strength indicator) of the received broadcast information, calculate the distance to the cluster head, and then select the cluster head closest to itself to apply for joining the cluster. According to the RSSI value, adjust the amplification power of the transmitting circuit to reduce the energy consumption of data communication. After receiving the information of all nodes in the cluster, the cluster head allocates TDMA time slots to each node.

(3) **Data transmission**

In a non-self-TDMA slot, the node is in a sleep state to save energy. When its own TDMA slot arrives, the node wakes up, opens the sensor, collects data, and sends the data to the cluster head. After collecting the information of all nodes in the cluster, the cluster head performs data fusion, and then transmits the data to the sink node by single hop.

7. **Adaptive Adjustment Routing Algorithm for Cluster Head Radius in Wireless Sensor Networks**

The node distribution model in the detection area of the adaptive adjustment routing algorithm for the cluster head radius of wireless sensor networks is shown in Fig. 2.8.

It is assumed that the sensor nodes are randomly deployed in a square area of $M \times M$, the location information is known, and the sink node is located outside the

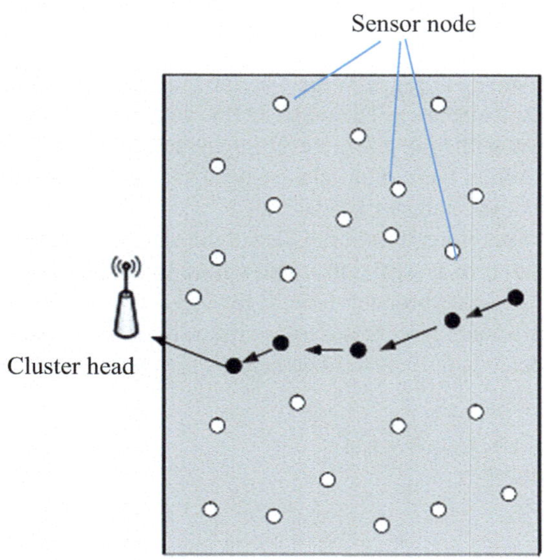

Fig. 2.8 Distribution of nodes in the detection area

whole monitoring area. At the same time, the node location will not change after deployment.

The software and hardware of sensor nodes should meet three conditions: The transmission power of wireless module is adjustable. The communication link is symmetrical. Each node has a unique node number.

The model area is equipped with various acoustic, optical and electrical sensor nodes, and the data receiving terminal is placed in a safe area far away from the monitoring area. In order to perform tasks for a long time, sensor networks must have a long network life. Therefore, efficient routing algorithms must be adopted to slow down the speed of network energy consumption, so as to prolong the network life.

(1) **Algorithmic thought**

Any routing algorithm will produce the fluctuation and uneven distribution of the number of cluster heads in a certain range, which is mainly due to the lack of mutual influence between cluster heads. There are two reasons.

1) **Two nodes close to each other become cluster heads at the same time**

When cluster heads are too concentrated in a certain area, it will lead to a large amount of overlap of coverage. Obviously, corresponding methods must be taken to deal with this obviously unreasonable cluster head distribution. For example, the "exclusive" mechanism is adopted. That is, when a node becomes a cluster head, in order to ensure the uniform distribution of cluster heads, no new cluster heads should be generated within the radius R_M with the cluster head as the center.

2) **There is no cluster head in a larger area**

As shown in Fig. 2.9, there are no cluster heads in most areas on the right side of the monitoring area. The nodes in this area must communicate with the cluster heads far away with high power, resulting in the waste of network energy. The general solution is: When there is no cluster head in the "nearby range" of a node, the node itself becomes a cluster head.

Therefore, the overall idea of the algorithm is: When the node becomes the cluster head, broadcast the message within the radius R. A node that does not receive any cluster head broadcast message and is not a cluster head becomes a cluster head. The key to such problems is to determine the size of parameter R. The adaptive adjustment mechanism of cluster head radius is to solve this problem.

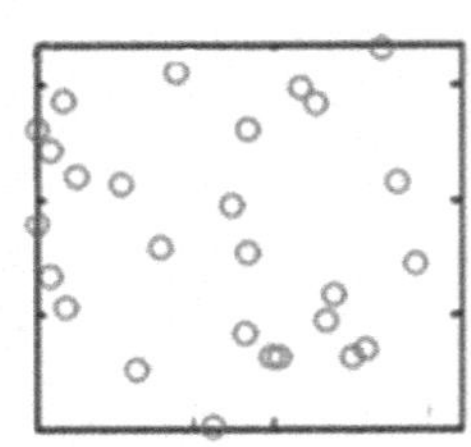

Fig. 2.9 Legend of node distribution

(2) Method for determining the optimal number of cluster heads

Hierarchical routing protocol divides the network sensor nodes into multiple clusters, and each cluster selects a node as the cluster head. The number of clusters directly affects the network performance. When the number of clusters is too large, it will increase the traffic between the network and the sink node, resulting in the rapid consumption of the whole network energy. When the number of clusters is too small, the network coverage of a single cluster head is large, and the amount of data sending and receiving tasks is relatively large, resulting in the "death" of the cluster head too quickly. Therefore, it is necessary to determine the minimum number of cluster heads, that is, the optimal number of cluster heads.

Assuming that the total number of nodes is N and the number of cluster heads is K, each cluster head contains N/K nodes on average. The energy consumption of cluster head includes three parts: receiving sensor data from nodes in the cluster, data fusion and data communication with sink nodes. Since the cluster head is generally far away from the sink node, according to the first-order energy model, the energy E_{CH} consumed by the cluster head in each round is

$$E_{\text{CH}} = l\left[\frac{N}{K}(E_{\text{elec}} + E_{\text{DA}}) + \varepsilon_{\text{mp}} d_{\text{toBS}}^4\right] \tag{2.5}$$

where l is the length of data information, and the unit of measurement is bit; d_{toBS} is the distance from the cluster head to the sink node, and the unit of measurement is m; E_{elec} is the energy required by the RF circuit to send or receive 1 byte of data, $E_{\text{elec}} = 50$ nJ/bit; E_{DA} is the energy consumed by the cluster head to send information to the sink node in each round; ε_{mp} is the multi-path attenuation coefficient, $\varepsilon_{\text{mp}} = 0.0013\ \text{pJ} \cdot \text{m}^4/\text{bit}$.

Since the non-cluster head node only needs to exchange data with the cluster head, and the communication distance is generally short, the wireless channel energy consumption model is free diffusion. The energy consumed by non-cluster head nodes in each round can be expressed as E_{noCH}

$$E_{\text{noCH}} = l\left[E_{\text{elec}} + \varepsilon_{\text{fs}} d_{\text{toCH}}^2\right] \tag{2.6}$$

where d_{toCH} is the distance from the node in the cluster to the cluster head, and the unit of measurement is m; ε_{fs} is the diffusion coefficient of free space, $\varepsilon_{\text{fs}} = 10\ \text{pJ} \cdot \text{m}^2/\text{bit}$.

Assuming that the range area of a single cluster is $\frac{M^2}{K}$, since the node distribution density function is $\rho(x, y)$, the square root of the distance from the non-cluster head node to the cluster head node is

$$\begin{aligned} E\left[d_{\text{toCH}}^2\right] &= \iint (x^2 + y^2)\rho(x, y)\mathrm{d}x\mathrm{d}y \\ &= \iint r^2 \rho(r, \theta) r \mathrm{d}r \mathrm{d}\theta \end{aligned} \tag{2.7}$$

Suppose that the region is a circle with radius $R_M = \frac{M}{\sqrt{\pi K}}$, and $\rho(r, \theta)$ is a constant about r and θ. Then, Formula (2.7) can be simplified to

$$E\left[d_{\text{toCH}}^2\right] = \rho \int_{\theta=0}^{\pi} \int_{r=0}^{\frac{M}{\sqrt{\pi K}}} r^3 \mathrm{d}r\mathrm{d}\theta = \frac{\rho M^4}{2\pi K^2} \tag{2.8}$$

Assuming that the nodes are evenly distributed in the whole cluster, then $\rho = \frac{K}{M^2}$, and

$$E\left[d_{\text{toCH}}^2\right] = \frac{M^2}{2\pi K} \tag{2.9}$$

Therefore, the energy consumed by a single non-cluster head node is

$$E_{\text{noCH}} = l\left[E_{\text{elec}} + \varepsilon_{\text{fs}}\frac{M^2}{2\pi K}\right] \tag{2.10}$$

The energy consumption of a single cluster is

$$E_{\text{cluster}} = E_{\text{CH}} + \left(\frac{N}{K} - 1\right)E_{\text{noCH}} \approx E_{\text{CH}} + E_{\text{noCH}} \tag{2.11}$$

In each round, the total energy consumed by the network is

$$\begin{aligned} E_{\text{total}} &= K E_{\text{cluster}} \\ &= l\left(E_{\text{elec}} N + E_{\text{DA}} N + K\varepsilon_{\text{mp}} d_{\text{toBS}}^4 + \varepsilon_{\text{fs}}\frac{M^2 N}{2\pi K}\right) \end{aligned} \tag{2.12}$$

Take the derivative of Eq. (2.12) and make it equal to zero, and then, the optimal number of cluster heads is

$$K_{\text{opt}} = \frac{M}{d_{\text{toBS}}^2}\sqrt{\frac{N\varepsilon_{\text{fs}}}{2\pi\varepsilon_{\text{mp}}}} \tag{2.13}$$

Here, assuming that the total number of nodes is $N = 100$, the side length of the monitoring square area is $M = 100$, and the distance from the cluster head to the sink node is $d_{\text{toBS}} \in [75, 185]$, the optimal number of cluster heads is $K_{\text{opt}} \in [3, 6]$.

(3) **Adaptive adjustment mechanism of cluster head radius**

In the process of selecting cluster heads, when the distributed random probability rotation mode is adopted, the cluster heads are easy to be concentrated in a certain area of the network, resulting in the overlap of coverage areas and the imbalance of network coverage. At this time, the cluster head radius adaptive adjustment mechanism can be used to solve this kind of problem.

Assuming that in the ideal case, the optimal number of cluster heads just covers the whole network, there is the following formula

$$\pi R^2 \times [N \cdot P] = M^2 \Rightarrow R = \frac{M}{\sqrt{\pi \times [N \cdot P]}} \tag{2.14}$$

where R is the coverage radius of the cluster head; P is the optimal cluster head percentage (generally $3\% \sim 6\%$); $[N \cdot P]$ is the rounding of $N \cdot P$ operation.

In order to reduce the overlap of cluster head coverage, the distance between two cluster heads should be $2R$, that is, there are no other cluster heads in the circle with cluster head as the center and radius R. In this case, when the cluster head broadcasts a message, in order not to cover the whole network, the cluster head will adaptively adjust the broadcast radius to R.

The process of cluster head selection based on radius adaptive adjustment mechanism is as follows.

1) After the node becomes a temporary cluster head, after a random delay for a period of time, broadcast a Be-head message with r as the radius, which contains the remaining energy information of the cluster head.
2) If the temporary cluster head does not receive the broadcast message from other cluster heads, it becomes the final cluster head.
3) After receiving the broadcast message from other cluster heads, compare the residual energy in the message with itself. When its residual energy is low, it will give up becoming a cluster head and broadcast the Quit-Head message. When its residual energy is high, it becomes the final cluster head.
4) If the normal node does not receive any cluster head broadcast information, it becomes the final cluster head message and broadcasts the Be-Head message.

If the distance between two cluster heads is less than $2R$, both cluster heads must receive each other's broadcast messages. According to the adaptive adjustment mechanism of cluster head radius, when one cluster head receives the broadcast message of another cluster head, the node with large residual energy will become the cluster head, and the other node will cancel to become the cluster head.

(4) Cluster head routing algorithm flow

The specific implementation process of cluster head routing can be divided into four stages: cluster head generation, cluster formation, time slot allocation and stable data transmission.

1) Generation stage of cluster head

Nodes should calculate the probability of becoming cluster heads according to their respective distribution characteristics, i.e.

$$T(n) = \begin{cases} \frac{P}{1-P\cdot[r \bmod (1,P)]} \cdot \frac{E_{\text{residual}}}{E_{\text{total}}}; n \in G \\ 0; \quad \text{otherwise} \end{cases} \tag{2.15}$$

where E_{residual} is the residual energy of the node and E_{total} is the initial total energy of the node.

By taking the residual energy of nodes as molecules, the probability of nodes with more residual energy becoming cluster heads can be increased, so as to balance the energy consumption of the network and prolong the service life of the network. If the probability $T(n)$ is greater than the preset threshold T_0, it becomes a temporary cluster head. Temporary cluster heads compete to become the final cluster head according to the adaptive adjustment mechanism of cluster head radius.

2) **Cluster formation stage**

After the cluster head is selected, the common node selects the nearest cluster and sends a Join-Cluster message to apply to join the cluster. In the Join-Cluster message, ordinary nodes need to add their own location information and residual energy information.

3) **Slot allocation**

Considering the large redundancy of wireless sensor network nodes, the cluster head does not necessarily allocate TDMA (time division multiple access) time slots to all members immediately after collecting the information of all member nodes, but runs the "four-way coverage" algorithm. The specific operations are as follows.

Firstly, taking the cluster head as the origin, the coverage of the cluster head is divided into four areas (see Fig. 2.10).

The solid points in the figure are cluster heads, and others are ordinary nodes. According to the collected information of the members in the cluster, the cluster head selects the node with the highest energy in the corresponding area as the working node of this round, and all the other nodes enter the sleep state. The cluster head only allocates TDMA time slots to the working nodes. The advantage of "four-way coverage" is that it can simplify the calculation of cluster head, only need to compare the abscissa and ordinate, and can meet the requirements of coverage. Another advantage of the "four-way coverage" algorithm is that it enhances the response ability of the network to environmental abnormalities. Assuming that a

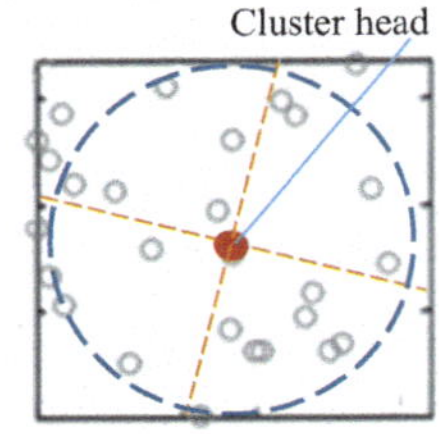

Fig. 2.10 Node distribution of a cluster

cluster has 20 nodes, the cluster head allocates a TDMA time slot to each node, and the node can only send data when its own time slot arrives. In the worst case, node 20 detects an exception, but the information must wait 19 time slots before it can be sent to the cluster head. Using the "four-way coverage" algorithm, the maximum number of time slots waiting for nodes is 3, which improves the real-time performance of the network.

4) **Stable data transmission phase**

The working node wakes up when its TDMA slot arrives and sends sensing data to the cluster head. The cluster head performs data fusion after collecting the data of all working nodes and then sends it directly to the sink node in a single-hop mode.

8. **An Unbalanced Routing Algorithm for Optimizing the Number of Hops in Wireless Sensor Networks**

(1) **Algorithm background**

When the transmission distance is long, the transmission energy consumption of single-hop communication between cluster head and sink node is large. Therefore, multi-hop transmission mode must be adopted to shorten the distance of single-hop transmission. However, in the multi-hop routing algorithm, there will be a "hot spot" problem: The cluster head closer to the sink node will consume more energy per round because it has to bear the task of data forwarding, resulting in too fast death. The "hot spot" problem of multi-hop routing can be solved by optimizing the number of hops in wireless sensor networks and unbalanced routing algorithms. The distribution model applicable to such algorithms is shown in Fig. 2.11.

In this distribution model, many sensor nodes are randomly deployed in a square area. After the cluster deployment is completed, the following assumptions are made

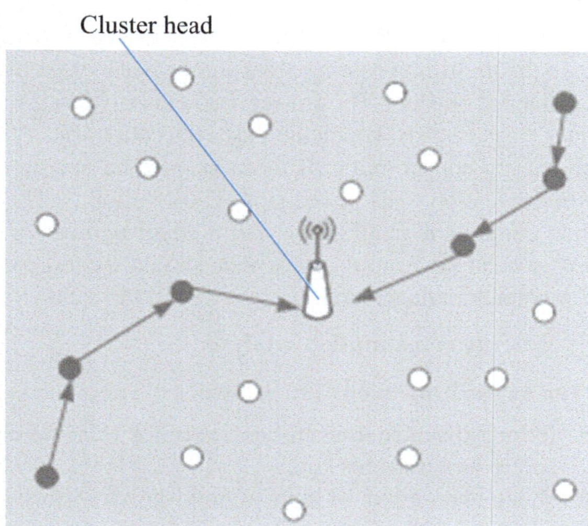

Fig. 2.11 Hop optimization model of wireless sensor

for the software and hardware facilities of the sensor nodes when the location of the sensor nodes does not change:

1) The transmission power of wireless module is adjustable.
2) The communication link is symmetrical.
3) Each node has a unique node number.

The distribution model is suitable for application fields such as environmental monitoring in remote areas. Because the monitoring area is far away, the convergence center cannot directly send the sensing data to the receiving terminal through short-range transmission. Therefore, the existing Internet network must be used. Since the sink node can no longer communicate with the sink center in close distance, the sink center is placed in the center of the whole network, so as to shorten the distance between the sensor cluster head and the sink center, and then reduce the energy consumption of data aggregation.

By constructing "cluster", the whole network is divided into layers. Because the wireless channel model is generally multi-path attenuation model, and the multi-path attenuation model consumes more energy than the free space diffusion model. The cluster head communicates with the sink node through multi-hop mode, which can reduce the network energy consumption.

However, when the cluster head communicates with the sink node through multi-hop mode, the number of cluster heads and the selection of multi-hop path are not reasonable, because the generation process of cluster head has a certain degree of random probability. In other words, multi-hop routing can reduce the network energy consumption compared with single-hop routing, but at the same time, there are new problems, that is, the closer to the cluster head of the sink node, the greater the amount of data forwarded, so the energy consumption is faster than the cluster head far away from the sink node, resulting in the death of the cluster head too quickly. The emergence of this problem will lead to the emergence of routing vacuum, because the nodes close to the sink node have died in a large area.

In the multi-hop routing algorithm, we also need to solve the problem of forming unbalanced clusters: The cluster head close to the sink node needs to relay the data of other cluster heads, which increases the data traffic. Accordingly, in order to balance the energy consumption of the network, the energy consumption for intra-cluster communication must be reduced. The solution is: The cluster head close to the sink node contains a small range and a small number of members in the cluster. The cluster head far from the sink node should try to expand the range and the number of members in the cluster.

(2) **Energy consumption analysis**

It can be mathematically proved that:

1) In long-distance transmission, two-hop transmission consumes less energy than single-hop transmission.
2) When the number of hops in multi-hop transmission is greater than the optimal number of hops, the energy consumption increases with the increase of the number of hops.

The above practical conclusions can be briefly expressed in Fig. 2.12.

As can be seen from Fig. 2.12, when the number of hops gradually increases from 1 to 2, the network energy consumption gradually decreases, and at 2 or 3, the network energy consumption reaches the minimum value. When the number of hops is greater than 3, the energy consumption will gradually increase with the increase of the number of hops. At the same time, it is found that there is an optimal number of hops to minimize the network energy consumption in multi-hop transmission. To solve the above problems, the design goal of the unbalanced routing algorithm for optimizing the number of hops in wireless sensor networks is to find the optimal number of hops in multi-hop transmission and reduce the network energy consumption.

(3) **Algorithm steps**

1) **Optimal hop selection**

As can be seen from Fig. 2.12, multi-hop transmission can reduce the network energy consumption when the number of hops meets certain restrictions, and there is an optimal number of hops to minimize the network energy consumption.

It can be proved that the minimum number of hops for energy consumption is

$$n_{\text{opt}} = \left\lfloor d_{\text{toCH}} \sqrt{\frac{\varepsilon_{\text{mp}}}{2E_{\text{elec}}}} \right\rfloor \tag{2.16}$$

where $\lfloor \cdot \rfloor$ means taking an integer down.

It is calculated that when $n_{\text{opt}} = 350\,\text{m}$, $n_{\text{opt}} = 3$, which is consistent with the results in Fig. 2.12, indicating that the approximate treatment of total energy consumption w does not affect the selection result of optimal hops.

2) **Cluster size distribution**

To solve the problem of energy consumption in multi-hop transmission, a better solution is to form unbalanced clusters. The cluster head closer to the sink node must receive and forward the data of the farther cluster head, which consumes more energy. In order to achieve energy balance, the coverage of cluster heads close to the sink node must be smaller than that far away from the sink node, so as to reduce the energy consumption of receiving and fusing member node data (see Fig. 2.13).

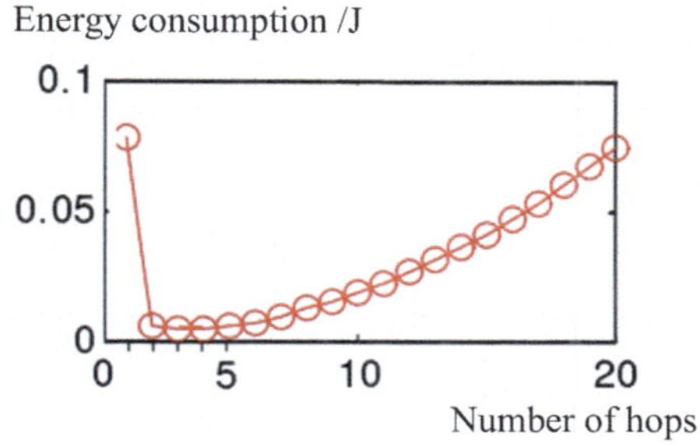

Fig. 2.12 Relationship between energy consumption and hops

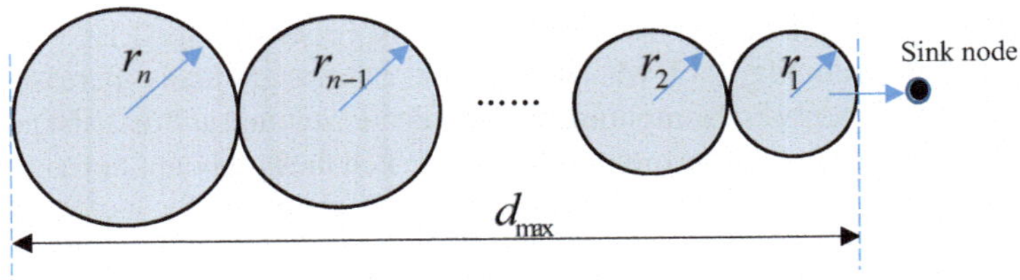

Fig. 2.13 Cluster size distribution

In the figure, if the maximum distance from the node to the sink node in the region is $d_{\max}$, the optimal number of hops is n_{opt}, and the total number of nodes is N, then the number of nodes in each cluster is

$$N_i = \frac{r_i^2 N}{\sum_{i=1}^{n_{\text{opt}}} r_i^2}; \quad i = 1, 2, 3, \ldots n_{\text{opt}} \tag{2.17}$$

The energy consumed by each cluster in a round is

$$E_i = \begin{cases} l\left[(N_i + 2n - 1)E_{\text{elec}} + n\varepsilon_{\text{fs}}(r_i - r_{i-1})^2\right]; & r_i + r_{i-1} < d_0 \\ l\left[(N_i + 2n - 1)E_{\text{elec}} + n\varepsilon_{\text{mp}}(r_i - r_{i-1})^4\right]; & r_i + r_{i-1} \geq d_0 \end{cases} \tag{2.18}$$

where B is the diameter of the range covered by the cluster head close to the sink node, that is $d_0 = 2r_1$.

Because the cluster far from the sink node has a large scale and a large number of nodes, when the initial energy of the nodes is the same, the total energy of the cluster is large. Assuming that the energy of the nodes in the cluster is exhausted at the same time, there is

$$\frac{E_1}{E_i} = \frac{N_1}{N_i} = \frac{r_1^2}{r_i^2}; \quad i = 1, 2, 3, \ldots n_{\text{opt}} \tag{2.19}$$

The distance relationship is as follows

$$\sum_{i=1}^{N} r_i = \frac{d_0}{2}; \quad i = 1, 2, 3, \ldots N \tag{2.20}$$

The radius r of the cluster can be obtained by simultaneous Eqs. (2.17), (2.18), (2.19) and (2.20) [7].

2.3.3 Wireless Sensing and "High-Speed Rail Vehicle Networking"

The connotation of vehicle networking mainly refers to that the on-board equipment on the vehicle can effectively use all vehicle dynamic information in the information network platform through wireless communication technology and provide different functional services during vehicle operation.

The Internet of vehicles shows the following characteristics: The Internet of vehicles can provide guarantee for the distance between vehicles and reduce the probability of vehicle collision accidents. Vehicle networking can help vehicle owners navigate in real time and improve the efficiency of traffic operation through communication with other vehicles and network systems.

The concept of the Internet of vehicles originates from the Internet of Things, that is, the vehicle Internet of Things. It takes the moving vehicle as the object of information perception and realizes the network connection between vehicle and *X* (i.e. vehicle and vehicle, people, road and service platform) with the help of a new generation of information and communication technology. The Internet of vehicles improves the overall intelligent driving level of vehicles, provides users with safe, comfortable, intelligent and efficient transportation services, improves traffic operation efficiency and improves the intelligent level of social transportation services.

"High-speed railway vehicle networking" is a further extension of vehicle networking.

1. **Architecture of "High-Speed Railway Vehicle Networking"**

The architecture of "High-speed railway vehicle networking" consists of three layers: acquisition layer, network layer and application layer.

(1) Acquisition layer

The acquisition layer is responsible for data acquisition, which is completed by various on-board sensors, including vehicle real-time operation parameters, road environment parameters and prediction parameters, such as vehicle speed, direction, position, mileage, engine speed and vehicle temperature. All collected data will be uploaded to the background server for unified processing and analysis to obtain the business data required by users and provide reliable data support for the Internet of vehicles.

(2) Network layer

The main function of the network layer is to provide transparent information transmission services, that is, to realize the summary, analysis, processing and transmission of input and output data. It is generally composed of network server and web services. In addition, satellite positioning signals and on-board sensor signals are uploaded to the background service center, and the server performs statistical management on the data to provide corresponding services for each vehicle. At the same time, it can jointly analyze the data, form various relationships between vehicles, become a local vehicle networking service business and provide efficient, accurate and timely data services for the user group.

(3) Application layer

The application layer is the highest level of the Internet of vehicles, which can provide various vehicle services for networked users. The real-time location data of the vehicle is mainly obtained by the global positioning system and then returned to the Internet of vehicles control center server. After being processed by the network layer, it enters the user's vehicle terminal equipment. After analyzing and processing the positioning data through the terminal equipment, it can provide users with various navigation, communication, monitoring, positioning and other application services.

2. **Physical Composition of "High-Speed Railway Vehicle Networking"**

"High-speed railway vehicle networking" mainly consists of three parts: evolved packet core network, evolved ground wireless access network and on-board equipment. The first two parts and the interface between the systems follow the LTE (long-term evolution) architecture protocol, while the third part takes the on-board equipment as a mobile transceiver device on the basis of the LTE architecture and connects it with the base station (eNodeB) through the air interface. In the terrestrial wireless access network, the connection between enodebs is transmitted through X2 interface and IP protocol at the network layer. The forwarding of data packets and signaling can realize the seamless connection of handoff.

LTE network architecture is E-UTRAN (UMTS terrestrial radio access network, UMTS is Universal Mobile Telecommunications System) removing RNC (radio network controller) network nodes. Its purpose is to simplify the network architecture and reduce the delay, so that the RNC functions are distributed to evolved node B (eNode B) and serving gateway (S-GW).

LTE access network is called evolved UTRAN (E-UTRAN). Compared with the traditional UTRAN architecture, E-UTRAN adopts a more flat network structure.

The E-UTRAN structure includes several eNodeBs. The bottom layer of eNodeBs adopts IP transmission and is logically connected with each other through X2 interface, that is, grid network structure. This design is mainly used to support the mobility of UE in the whole network and ensure the seamless handover of users. Each eNodeB is connected to the mobility management entity (MME) of the evolved packet core (EPC) network through the S1 interface, that is, it is connected to the MME through the S1-MME interface. Through S1-U and S-GW connection, S1-MME and S1-U can be regarded as the control plane and user plane of S1 interface, respectively.

On the EPC side, S-GW is the anchor point in 3GPP mobile network. The MME function is separated from the gateway function and is mainly responsible for handling control signaling such as mobility. This design is conducive to network deployment, the evolution of a single technology and comprehensive and flexible capacity expansion. At the same time, LTE/SAE architecture can also integrate SGSN and MME functions into the same node, so as to realize a general packet core network that supports GSM, WCDMA/HSPA and LTE technologies at the same time.

In “High-speed railway vehicle networking”, the redesign of LTE network structure can be realized after upgrading the existing LTE network equipment and security mechanism.

One eNodeB is connected to multiple mobility management entities or service gateways, which is the evolution of eNodeB at the level of mobile communication system. It adds some functions on the physical layer (PHY), media access control sublayer (MAC), wireless link control sublayer (RLC), packet data convergence protocol sublayer (PDCP) and wireless resource control layer (PRC) of wireless network controller, and all functions of access networks such as mobility management and wireless resource management [8].

The physical model of “High-speed railway vehicle networking” is shown in Fig. 2.14. Through the establishment of “high-speed railway vehicle networking”, it can not only realize the dispatching and real-time information interaction of high-speed railway trains by the monitoring center, but also realize the information interaction between high-speed railway trains.

Two complementary transmission modes are adopted for the communication of “High-speed railway vehicle networking”: One is direct communication, which is based on D2D (Device-to-Device) and directly transmits medium and short distance through Side-link link. It operates in ITS (Intelligent Transportation System) frequency band and mainly works at 5.9 GHz frequency, which is independent of cellular network and the other is network communication, that is, the traditional communication with the help of cellular network adopts Uu Interface and runs in the traditional mobile broadband authorized frequency band to meet the requirements of large amount of data between the network and the terminal.

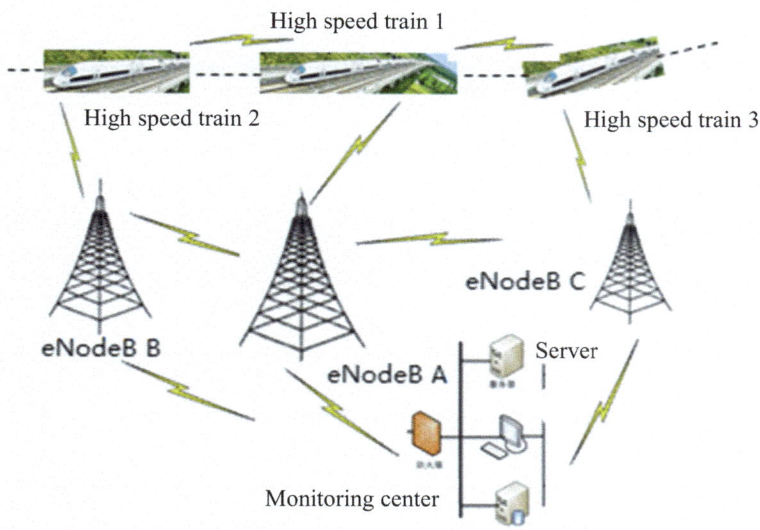

Fig. 2.14 Physical model of “Internet of vehicles” system

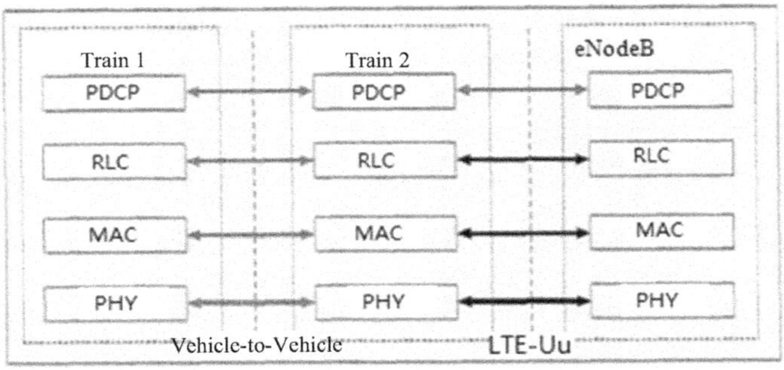

Fig. 2.15 Blank port protocol stack of Vehicle-to-Vehicle and eNodeB

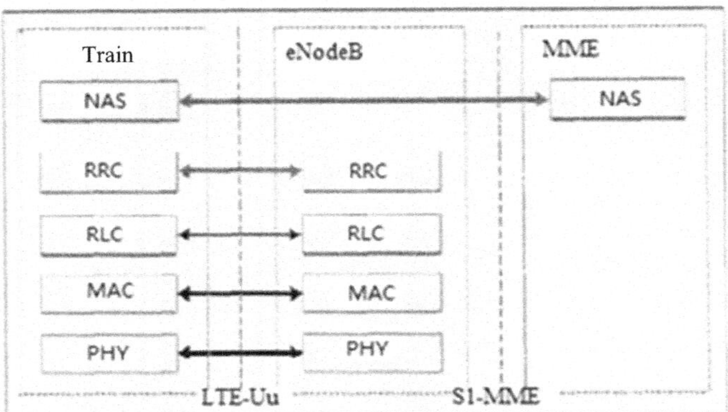

Fig. 2.16 Vehicle-to-Vehicle eNodeB control plane blank port protocol stack

The blank port protocol stack between "High-speed railway vehicle networking" and eNodeB is shown in Fig. 2.15. Among them, the blank port part inherits the LTE blank port protocol, and the user plane protocol includes four levels, all of which are terminated at the eNodeB entity on the network side.

The control plane protocol stack including LTE wireless access is shown in Fig. 2.16.

2.4 Brief Description of Working Principle of Train Subsystem

The so-called train subsystem is the mobile train automatic control system based on CBTC. It has the functions of automatic train protection, automatic driving and automatic operation. It mainly includes signal interlocking, automatic train protection, automatic train supervision, automatic train operation and data communication system, etc.

2.4.1 Signal Interlocking

Signal interlocking means that the annunciator, train detection device, turnout and other equipment information within the control range of the train intelligent control system form a relationship of mutual connection and mutual restriction, so as to ensure the safety of the route train and improve the transportation efficiency.

The system adopts high reliability hardware and redundant structure. The hardware architecture is hierarchical, which can be divided into man–machine dialogue layer, security operation layer and execution presentation layer.

The man–machine dialogue layer is composed of operation display equipment and system diagnosis and maintenance equipment, and adopts redundant structure, which can be combined with automatic train monitoring equipment. The security operation layer is composed of chain computers. The execution layer is implemented by an intelligent unit with CPU or an electronic circuit without CPU.

Signal interlocking mainly has the following functions.

(1) Blocking/unlocking section divides the section of train occupancy detection into logical section and physical section, and has the authority to block/unlock section. When a section is blocked, the interlocking system will not arrange the entry of the section until the section is unlocked.
(2) Switching annunciator display: When the signal interlocking subsystem receives the train state change information provided by the automatic train protection subsystem, it controls the annunciator at the incoming end to display the switching indication.
(3) Operate the turnout, the basic state of turnout is divided into positioning, reverse position and four opening. The interlocking subsystem can control the turnout to complete the switching of several states and has the functions of separate locking and unlocking.
(4) Functions of access road, including route handling (manual handling/automatic handling of automatic train monitoring), route locking and route unlocking.
(5) Protection section function: The interlocking subsystem provides section protection of different paths. When the protection section is consistent with the direction of subsequent routes, they can be locked repeatedly.

2.4.2 Automatic Train Protection Subsystem

The automatic train protection subsystem mainly ensures the operation safety of the train. It is constructed according to the principle of "fault-oriented safety". It is the basis of the automatic train control system. Both automatic train monitoring and automatic train operation rely on the automatic train protection subsystem. Automatic train protection subsystem can realize train interval control, overspeed prevention and other functions. Its specific functions are as follows.

(1) Train speed and position measurement: The system can determine the train speed, position (including the positions at both ends of the train) and operation direction in CBTC area and verify its speed information through redundant speed measurement system.
(2) For the calculation of train safety braking curve, the ground equipment of the automatic train protection subsystem calculates the movement authorization curve and periodically sends it to the on-board equipment of the automatic train protection subsystem. The on-board equipment will build a continuous speed-distance safety braking model to prevent overspeed and keep the safe driving distance between trains within a reasonable range.
(3) For train integrity supervision, the automatic train protection subsystem will obtain the integrity information of continuous supervision of train equipment from relevant vehicle interfaces (once the vehicle integrity is lost, the train will implement emergency braking).
(4) Door status and platform door monitoring. The automatic train protection subsystem will monitor the status of door and platform door in real time and make different responses to temporary emergencies under different driving modes.
(5) The control level is established. When the train enters the CBTC area, the automatic train protection subsystem will obtain the boundary information of the area and check the track equipment to verify whether the conditions for entering the area are met.
(6) Driving mode management: The automatic train protection subsystem supports the conversion of driving mode, that is, the conversion of manual driving mode or controlled (restricted) manual driving.
(7) Communication status supervision and fault handling. The automatic train protection subsystem has the function of supervising the communication status between on-board equipment and ground equipment, the communication status of signal interlocking and the communication status of adjacent ground equipment (once the ground equipment fails, the on-board equipment of the automatic train protection subsystem will immediately generate an alarm, emergency brake the train, and switch the driving mode to the restricted manual driving mode).

2.4.3 Automatic Train Supervision Subsystem

Automatic train supervision subsystem is an important subsystem in automatic train control, which mainly provides real-time monitoring of train operation status in the line. The automatic train supervision subsystem forms a unified network of corresponding equipment distributed in control center, station, parking lot/depot through independent redundant network and displays the status and position information of all trains in the line on the line diagram, so as to realize the automatic adjustment of train grade, departure time and departure time. Ensure that the actual operation of the train conforms to the previously formulated operation schedule. Its main functions are as follows.

(1) System information display: The automatic train supervision subsystem can display the line layout of the whole line, train position information, train number information and the status of main trackside equipment (such as turnout, annunciator and axle recording) of the signal system in real time.
(2) Control function. The control level of automatic train supervision subsystem is divided into control center control, station control and station interlocking control.
(3) Train tracking control: Train tracking within the scope of signal system is divided into communication train tracking and non-communication train tracking. The system can realize automatic switching according to the change of communication state.
(4) Train diagram management and automatic train supervision subsystem generate train diagram according to certain format and can provide the functions of creating, adding, deleting and changing relevant train diagram.
(5) For train operation adjustment, the automatic train supervision subsystem can modify the train service number, train number, destination number and other information, and can automatically adjust the stop time and interval operation time of the train to ensure the punctual operation of the train. At the same time, the automatic train supervision subsystem also provides online train conflict management, terminal departure sequence management and train adjustment mode.

2.4.4 Automatic Train Drive Subsystem

Automatic train operation subsystem is a control system that realizes the functions of automatic train control, platform automatic operation, automatic train operation adjustment and so on under the protection of automatic train protection subsystem. Its main functions are as follows.

(1) Automatic train operation, automatic control of train operation equipment: train startup, acceleration, braking, coasting and other processes, and make the train meet the requirements of comfort and energy saving.
(2) Platform parking control, the subsystem can ensure the accurate parking of the train in the station and adopt one-time continuous braking mode to brake to the target parking point. Finally, when the train stops, maintain the braking command to prevent rolling.
(3) Door/platform door monitoring: After the train stops at the platform, the system will prohibit the train from starting until it is determined that the door/platform door is closed and locked.

(4) For operation adjustment, the subsystem can support the adjustment methods of vehicle shutdown, vehicle detention, vehicle dwell time, etc.
(5) Operation assistance: The subsystem can provide information data about passengers to train broadcasting equipment and provide operation speed and various prompt information to drivers.
(6) Fault diagnosis and alarm. The subsystem has self-diagnosis function. When a fault is detected, it will immediately exit the automatic driving mode and provide alarm signals to the driver and other relevant subsystems.

2.4.5 Data Communication Subsystem

The data communication subsystem constructs a two-way data communication network for the urban rail transit system. Among them, wired network provides data transmission channel between ground control center, station, trackside and other equipment. Wireless network provides continuous two-way wireless data transmission between train and ground control equipment. The network of general data communication system is a special communication network, which adopts independent networking mode, and the network structure adopts double redundancy design to provide two independent network transmission channels for CBTC data. The functions of the system are mainly divided into the following three parts.

(1) Communication transmission channel, the data communication subsystem can realize the data communication capability between point-to-point and point-to-multi-point, classify the data of applications based on the port and classify the bandwidth according to the data requirements of each service.
(2) Network management, the system uses SNMP (Simple Network Management Protocol) to realize management, that is, it has the management functions of network performance implementation, configuration, fault diagnosis and so on.
(3) For network security, a complete set of security policies are formulated, and certain functional standards are met in the requirements of wireless communication security management and transmission channel.

2.5 Communication Coordination Principle Between Vehicles

Vehicle-to-vehicle communication is a new concept of "vehicle networking". It can not only expand the coverage of rail transit information, but also extend the information interaction between vehicles, so as to further improve the safety and reliability of rail transit system.

2.5.1 Communication Technology Structure Between Vehicles

The communication system between vehicles (a special case of "vehicle networking") consists of two parts: ground system and on-board system. Ground system includes automatic train monitoring, traffic management unit TMU (Traffic Management Unit), object controller OC (Object Controller), etc. The core of on-board system is intelligent vehicle equipment IVE (Intelligent Vehicle Equipment). The main functions of each system are as follows.

(1) The automatic train supervision system realizes the operation supervision of the train. The specific implementation is as follows: train information display in the whole line → automatic generation of train timetable → train route arrangement → train operation state detection → auxiliary staff dispatch and control the trains within the control range.
(2) The train management unit is responsible for the management of line data, recording the information of trains in the area, and providing the query function of line train information. The online trains need to conduct version inspection with the train management unit and issue the received temporary speed limit instruction.
(3) The target controller is responsible for collecting the status information of trackside equipment, transmitting the status information of trackside equipment to the control center and the train, and controlling the trackside equipment according to the received command.
(4) The intelligent on-board equipment realizes the direct communication between vehicles and has the functions of train overspeed protection and automatic train driving. At the same time, it has the ability of independent operation and can cooperate with other trains to calculate its own mobile authorization.

2.5.2 Steps of Communication Principle Between Vehicles

In the communication system between vehicles, the main factors affecting the current running state of the train are the position and speed of the front vehicle. The front vehicle is equivalent to a moving obstacle for the rear vehicle. In order to track the front vehicle, the front vehicle status information must be obtained periodically to ensure that the mobile authorization can be updated.

In the traditional CBTC system, trains in a specific area will communicate with the area controller periodically. The train only needs to send its own position and speed information to the regional controller. After the regional controller judges the movement authorization, it sends the front vehicle status information to the rear vehicle. The train does not need to know the specific train around it, but only needs to plan its own driving "rules" according to the movement authorization.

However, once the train loses the information interaction with the regional controller for some reason, the communication system between trains loses the centralized signal processing command of the regional controller and is converted to direct communication between trains. At this time, the train must obtain the information of the front train.

Coordination principle steps: ① Positioning has been completed through beacons, electronic maps and other means, and real-time communication with the automatic train monitoring system of the control center to obtain the operation task; ② The train establishes communication with the train management unit, obtains the latest line information and completes the train registration; ③ Communicate with the target controller to obtain the occupation of the current section and turnout information; ④ In order to operate normally, the train also needs to obtain the ID (identity number) and IP (Internet protocol address) information of the front train, because the identification of the front train cannot be completed by virtue of its own conditions, which requires the assistance of other equipment systems; and ⑤ Analyze which devices can be used as auxiliary devices.

2.5.3 Functional Differentiation of Communication Between Vehicles and Automatic Train Supervision System

The automatic train supervision system is mainly responsible for the train monitoring of the whole line, sequencing the position of all trains in the line, issuing the command of temporary speed limit to the train and communicating periodically with the train management unit, so as to update the train information in the whole line. The plan for this operation is only sent when the train is just launched, not periodic communication. Therefore, from the mechanism of the communication system between vehicles, the automatic train supervision system does not have the conditions to assist the train to identify the front vehicle.

When the train goes online, the train management unit will register the train and provide the train with a database version of the information in the line. It can directly issue the speed limit command issued by the automatic supervision of the train, communicate with the train periodically and obtain the information of all trains in the whole line in real time, so it can assist the train to complete the identification of the train ahead.

In the communication system between trains, the target controller is only responsible for collecting the status information of trackside equipment and indicating and controlling turnout and other trackside equipment to the train according to the automatic train supervision information. Therefore, it does not have the conditions to assist the train to identify the front train.

If only the train itself is considered and other ground equipment is not used for assistance, another solution shall be adopted. That is, trains periodically communicate with all trains in the line, but this will have many disadvantages: ① When there are many trains in the line, establishing communication with each train will lead to channel "congestion" and real-time performance cannot be guaranteed; and ② There will be a lot of useless information, resulting in low information utilization, which makes the logic of train independent judgment complex. Obviously, it is not necessary, because the train operation is mainly affected by the train ahead, and the purpose of collecting all train information is only to obtain the information of the train ahead. In other words, on the premise of not adding new equipment for train sequencing of the whole line, the train management unit can be selected to assist the train to complete the identification of ground equipment, and the periodic communication between the train management unit and the train can be established. Meanwhile, the train management unit stores the information of all trains. It divides the line into virtual train identification areas. The train can obtain the information of all trains in this area, so the train only needs to be identified in each virtual identification area. After the front vehicle identification is completed, adjacent trains can start periodic communication and establish tracking relationship. Therefore, the rear vehicle can update its driving conditions according to the real-time status of the front vehicle and realize independent control relative to the train.

2.6 Intelligent Identification and Decision of Road Conditions

The independent and intelligent identification and decision-making of the train on the road ahead is not only an energy increasing auxiliary technology for CBTC and "communication between vehicles", but also an innovative idea for the gradual intellectualization of rail transit technology. In view of the impossibility of "full coverage" of "trackside equipment", the possible loss and defect of received information, and the possible sudden accident of relevant monitoring system, the train itself has the function of independent intelligent identification and decision-making, which is an important auxiliary technology (means) for the safe operation of rail transit. Especially, the urban "virtual rail transit" that has appeared at present must have this technology.

2.6.1 Train Independent Intelligent Visual Perception System

Intelligent visual perception is an important technology to realize train independent intelligent identification and decision-making. The so-called visual perception, that is, the whole dynamic process of intelligent machine's understanding and understanding of external scenery, is divided into two parts: image sensing (acquisition) and view perception, so it is called "Intelligent visual perception" (IVP). In the past, "computer vision" and "machine vision" were formed by using the term "vision" of

Fig. 2.17 Train visual perception configuration

physiological organs generated in the period when human life science was not developed, which obviously lacked the connotation of deep-seated physical and physiological science. Although there is only one word difference between "visual perception" and "vision", the former has fully demonstrated the important and irreplaceable role of human brain organs in information collection [9].

1. **System hardware configuration**

In order to achieve the independent and intelligent identification and decision-making of the train on the road ahead, the key technology is to configure the intelligent visual sense (also known as "computer vision") for the train, so that the train has the function of independent identification of the road ahead.

In Fig. 2.17, three pinhole cameras need to be configured in front of the train. Among them, 1 and 2 are the right and left cameras set near and below the motor car headlamp, respectively, and 3 is the camera set in the middle of the front windshield frame of the power train.

2. **Basic working process of the system**

Among them, image processing includes image enhancement and filtering, image edge detection, image binarization, image segmentation and so on.

How to quickly identify the "obstacles" in front of the moving train has become a key technology to be solved. The key technology in visual perception recognition is a fast algorithm for extracting object image information features.

3. **Implementation Conditions for Intelligent Recognition of Road Conditions Ahead**

The signal processor in the on-board visual perception system centrally processes the images collected from cameras 1 and 2 (see Fig. 2.17). When the camera is initially set, the coordinate system has been established on the vehicle (see Fig. 2.18), and the internal and external parameters of the two cameras have been calibrated [10].

The dotted line in Fig. 2.18 shows the respective field of view of each camera.

As long as the vehicle route, the on-board visual perception system is in working state immediately: The camera collects the image of the road scene in front in real time, and the left and right views are sent to the signal processor through the

Fig. 2.18 Coordinate setting method for identifying road conditions

image channel in parallel. The algorithm software in the signal processor processes, recognizes and understands the left and right views, and makes accurate judgment according to the operation results.

4. **Object feature extraction and Epipolar Constraint Principle**

The key point of intelligent identification of road obstacles or pits is to calculate the 3D coordinates of spatial feature points. The calculation of 3D coordinates of spatial points based on binocular vision perception system, which is composed of two pinhole cameras, relies mainly on epipolar geometry constraint theory (referred to as outer pole constraints) [11].

5. **Basic Algorithm Steps**

The fundamental algorithm for automatic recognition of road obstacles and pits is implemented by the intelligent algorithm program in the image processing program module. Its detailed steps are as follows.

(1) **Collection of Road Scene Image**

Two pinhole cameras, the binocular image sensor system, real-time acquisition of vehicle front road scene image, and each collected image transmission along with the transport channel corresponding to the signal processor.

(2) **Eliminate Image Distortion**

The left and right views are removed to eliminate the distortion, respectively.

After the coordinates (i_l, j_l) and (i_r, j_r) of all pixels in the computer image coordinate system of the left and right cameras are calculated one by one, the ideal arrangement of pixels on the picture can be obtained, that is, the real scene image can be restored, or the left and right view ideal images reflecting the real scene can be obtained.

So-called normalized virtual imaging plane, which excludes the physical concept of optical imaging, builds the imaging plane in a virtual position at the focal length of the lens located in front of the lens. Compared with the pinhole camera physical imaging plane, it possesses several advantages as follows:

Fig. 2.19 Road front image collected by vehicle in real time

1) The actual distance units instead of pixels, because of the different visual sense of image sensor, the focal length of the lens and the image resolution and other parameters caused by the calibration and calculation difficulty in the more visual sense system.
2) As the optical axis passes through the origin of the coordinate system, there is no deviation of the optical axis and the origin of the physical imaging plane.
3) As a result of normalization, it facilitates the operation of matrix system and assumes the intermediary role of transformation.
4) The coordinates of the imaging points on the normalized virtual imaging plane in the camera coordinate system are the homogeneous coordinates of the two-dimensional coordinates.
5) If the imaging plane is built on the imaginary plane before the lens is not present, it can solve the problem that the mathematical model is not intuitionistic. As a result, the accuracy of the discussion is ensured while the discussion of the problem is simplified.

Figure 2.19 shows the left and right views of the road conditions in front of the vehicle collected by the on-board 1 and 2 cameras and after removing the distortion.

(3) Finding and Matching Common Corners of Obstructions or Pits

To confirm whether there is an obstacle or a pit in front of the vehicle, it is necessary to look for the obstacle or pit and match the common corners points from the left and right view [12].

For this reason, the view window is first intercepted, as shown in Fig. 2.20. The size of the window is expanded along the top, bottom, left, right four directions of the camera axis of the pinhole camera and a certain number of pixels are formed. Among them, the horizontal (column pixel) is suitable for monitoring the width N of the lane; longitudinal (line pixels) is used to identify the operation cycle to the maximum allowed number of pixels for M for the upper limit, so formed is the same as the size and pixel coordinates of their corresponding to each other about two view windows with $M \times N$.

In the intercepted window, the corner detection algorithm based on image gray is adopted, and several corner points are represented by calculating curvature and gradient to represent the most characteristic edge points. Then, in order to find the corner points, under the guidance of epipolar constraint principle, we find the matching corners in the left and right views.

Taking the vehicle shown in Fig. 2.20 as an example, the front vehicle feature point b in the figure is located on the outer polar lines L_l and L_r in the left and right views, respectively. Point b is one of the common feature points of the left and right views automatically found through the above epipolar constraints.

(4) **Determination of 3D Coordinate Values of Object Points**

Determine the coordinate system involved in the 3D coordinates of the object points and their relations, namely the pinhole camera geometry model.

The projection matrix is multiplied by the inner parameter matrix by 4 degrees of freedom and the outer parameter matrix with 6 degrees of freedom, which results in 10 degrees of freedom [13].

(5) **Spacing calculation principle**

The principle of epipolar constraint is used to quickly determine any common feature point on the matched target. According to the corresponding point of the common feature point on the virtual imaging plane, the distance between the common feature point and its own vehicle is calculated directly by using the parallax principle.

1) **Parallax principle**

Binocular vision is a machine structure that simulates binocular vision of animals. Double vision geometry is the theoretical foundations of binocular vision sensing.

In fact, when binocular vision sensing is used, there is a complex functional relation between the measurement error of the measured point and the angle of the two cameras' axis. When the angle between the two cameras' optic axis is fixed, the distance between the coordinates of the measured points in the world coordinate

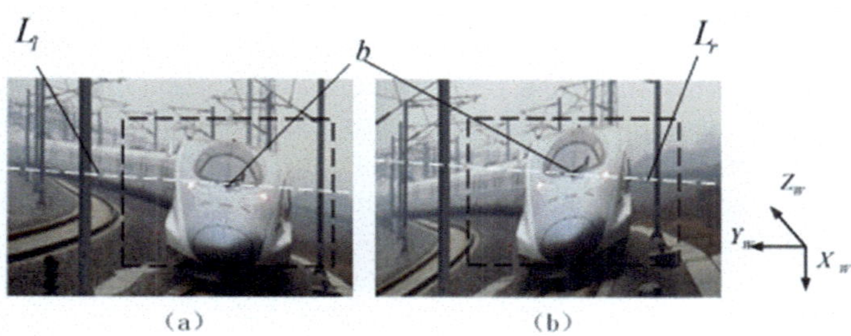

Fig. 2.20 Example of epipolar constraint in binocular view

system and the camera is larger, and the detection error of the detection point distance is greater.

At present, when the binocular pinhole camera is set up in the vehicle vision system, the parallax principle has been considered for distance measurement, so during the installation of two pinhole camera it is required that the two pinhole camera optical axis can be parallel to each other and perpendicular to the optical center of the connection line. Because of the geometric characteristics of the two pinhole cameras in a binocular vision system, binocular parallax principle can be used directly in the detection of vehicle distance.

2) **Vehicle Distance Calculation**

Using the epipolar constraint principle, a common feature point of the matched target is quickly determined. According to the corresponding feature points in the virtual imaging plane, the distance between the common feature points and the vehicle is calculated directly by using the parallax principle [14].

2.6.2 Fast Identification Method for the Train Ahead

Template matching algorithm is one of the effective methods for fast recognition of objects in front [15].

1. **Feature Point State Recognition Theory**

(1) **Algorithm principle**

Collect the scene image in front of the track in real time through the on-board cameras 1 and 2 (see Fig. 2.17), and always compare the current image with the image collected at the previous time. The so-called comparison refers to the spatial alignment of two images containing moving targets recorded by the image sensor from the same scene background, so as to determine the relative translation process of the targets in the two images.

(2) **Threshold Extraction**

In order to realize the accurate decision, the decision threshold is still needed [16].

In view of the sequence of real-time image acquisition, even the same scenery will because of fog, wind, sunlight and other natural environment in a different color contrast, i.e. after two frames may be in the scene the same part to different gray value, gray threshold for each image to calculate the frame to determine the gray level difference is acceptable at what level. Threshold here, you can use the OTSU algorithm.

The OTSU algorithm divides the histogram of the image into two parts: the target and the background with the best threshold, so that the variance between the two classes is taken as the maximum, that is, the maximum separation.

(3) **Similarity measure**

Because various error factors will be introduced in the image acquisition process, it is almost impossible to find a region that can be completely consistent in the two images. Therefore, when comparing the two images, it is necessary to measure by establishing similarity [17].

2. **Algorithm Practice**

The purpose of template matching is to find the best matching relationship between the "field image" to be matched and the preset "template image".

(1) **Spatial domain description of template matching**

Assume that the field image is $s(x, y)$ and the image size is $W \times H$; The template image is $t(x, y)$, and the image size is $w \times h$. x and y are the coordinates of the column and row of pixels, respectively. Under normal circumstances, the pixel size of the field image is often much larger than the template image, that is, $W \geq w$, $H \geq h$.

The basic idea of spatial domain template matching is similar to the spatial domain filtering method used in image enhancement, but the filter operator value of image enhancement is fixed and the size is generally small. The operator used in template matching is the template image itself, which is generally large in size. The spatial domain filtering method basically adopts convolution or correlation method, and template matching can carry out complex operations on the basis of field image and template image, and the operator form is more flexible.

In the spatial domain, the common way to match the template is to use the template as a filter operator to slide through the whole image to find the image area that meets the matching benchmark. This process is sometimes called back project, and the results obtained can be expressed as

$$R(s, t, x, y) = \sum_{m=0}^{w-1} \sum_{n=0}^{h-1} f\{s(x + m, y + n), t(m, n)\} \tag{2.21}$$

where $x \in [0, W - w]$, $y \in [0, H - h]$, the size of the resulting image is $(W - w + 1) \times (H - h + 1)$. m, n is the number of columns and rows moved by the template.

It can be seen that $R(s, t, x, y)$ is related to template $t(x, y)$, field image $s(x, y)$ and current position (x, y).

In the matching process, if the field and template images are fixed, the result can be abbreviated as $R(x, y)$. Finally, the matching position (x, y) is obtained by analyzing the value of $R(x, y)$. According to the different forms of $f\{\cdot\}$ operator, $R(x, y)$ also has different forms, so different template matching algorithms can be extended.

(2) **Identification steps for the train ahead**

1) **Establish vehicle background template database**

Collecting and sorting templates is the basic work of template matching.

Taking advantage of the rapidity of template matching and the characteristics of rotation, reduction/amplification and translational moment invariants, a track template database is established in the on-board system in advance (see Fig. 2.21). The track template shown in the figure is to use an "exclusion" matching algorithm to determine that there is no "stay object with non-track attribute" on the track. Therefore, the scale and calculation of the template database can be greatly reduced, and then, the operation speed of identification and determination (decision) can be improved.

At the same time, the vehicle background template database should be established. This is to finally confirm whether the identified object is a non-orbital retention object, so as to enhance the reliability of identification.

2) **Real-time acquisition of scene image in front of vehicle**

The on-board binocular vision can be configured with corresponding optical amplifier according to the design speed of the train in advance, so that the binocular vision can meet the technical requirements of effective sight distance that the train must have. Generally speaking, the running speed of Metro trains (including urban light rail, etc.) belongs to relatively high-speed vehicles. The recognition of vehicles and road conditions in front by using visual sensing technology must have the ability of "beyond visual range". Therefore, it is inseparable from the optical amplification mechanism, so that the due clarity (image resolution) can be maintained when the visual range is extended [18].

The scene image in front of the vehicle is collected in real time by vehicle binocular vision, which is called the main field of view image.

The main field of view image is intercepted according to the size specification that is most conducive to analysis and operation. As shown in Fig. 2.22, the dotted box in the figure is the main field of view observation window of binocular vision.

3) **Fast recognition algorithm for vehicles ahead**

Make the size of the template $w \times h$, w and h are the number of columns and rows of the template, respectively. The size of the main field of view observation window is $W \times H$, and W and H are the number of columns and rows of the window.

Fig. 2.21 Track template database

Fig. 2.22 Main field of view observation window

Each time a template in the template database is selected to match the observation window. Through the template matching operation, whether there is a vehicle ahead can be quickly determined.

Since the observation window of the main field of view has been greatly reduced compared with the whole view, the target search does not have to start from the pixel (0, 0) in the upper left corner of the view one by one, so the speed is increased by an order of magnitude.

In each matching process, the combination of coarse-matching and fine-matching is used for search: In the coarse-matching stage, a large step size jump is used to calculate several similarities, so that the conventional fine-matching can be carried out near the best matching position of the coarse matching, so the matching speed can be improved.

At the same time, the random walking method is also introduced: While jumping with variable step length at the distance from the starting point of the search, the search position may also be carried out in turns up, down, left and right, and the walking is staggered and maneuvered from the starting point. For example, the most famous typical complex walk is a large number of "diamond search" or "Diamond Search" algorithms used for MPEG motion vector analysis.

In the process of matching the target with the template, first use the "track template" for matching. When it is confirmed that there is no foreign matter with "non-track attribute" in front of the track, the normal running of the train will not be affected. Once the "non-track attribute" foreign matter appears in the field of view, start the deceleration or braking control immediately, and implement "positive identification", that is, read the vehicle background template database to confirm whether the "retention" in front belongs to the retention or "late speed" train. If it is neither

Fig. 2.23 Identification process of long-distance train

stranded nor "late" train, it must be determined as other obstacles and further confirm what objects it is through artificial or space-based platform detection system.

The matching similarity described here can adopt any of the above similarity. The actual measurement shows that the recognition time of the vehicle ahead (1000 m) using the fast template matching algorithm is less than 30 ms.

The objects in the area shown in the white dotted box shown in Fig. 2.23 are the front targets captured by the train intelligent vision system.

(3) **Basic principle of vehicle distance measurement**

Once the vehicle ahead is identified, the distance between the vehicle ahead and its own vehicle can be obtained according to the aforementioned parallax principle and its distance measurement. Mastering the distance between yourself and the vehicle in front is a very important parameter [19].

(4) **Distance control**

The distance control strategy is determined according to the position of the identified train.

When the on-board visual perception system is equipped with speed intelligent linkage control device, the safe distance s can be calculated according to the following formula

$$S = \frac{V^2}{2|a|} \tag{2.22}$$

where V is the current vehicle speed and a is the negative acceleration. For example:

When $V = 250\left(\text{km} \cdot \text{h}^{-1}\right)$, $a = -4\left(\text{m} \cdot \text{s}^{-2}\right)$; Take $S = \left\lceil \frac{V^2}{2|a|} \right\rceil = 62(\text{km})$ as the safe distance;

When $V = 150(\mathrm{km} \cdot \mathrm{h}^{-1})$, $a = -0.9(\mathrm{m} \cdot \mathrm{s}^{-2})$; Take $S = \left\lceil \frac{V^2}{2|a|} \right\rceil \approx 965(\mathrm{km})$ as the safe distance.

Here, considering the strong inertia of high-speed train, the negative acceleration must be "selected" properly, otherwise the rear trailer will "squeeze" due to inertia; $\lceil \cdot \rceil$ means taking an integer up.

In view of the lag of mechanical linkage, the control of safety distance S shall be

$$S = \frac{V^2}{2|a|} + V\tau \tag{2.23}$$

where τ is the lag time of human physiological response, generally taken as $\tau = 3s$; as

When $V = 250(\mathrm{km} \cdot \mathrm{h}^{-1})$,$a = -0.5(\mathrm{m} \cdot \mathrm{s}^{-2})$; Take $S = \left\lceil \frac{V^2}{2|a|} \right\rceil + V\tau \approx 5031(\mathrm{km})$ as the safe distance;

When $V = 150(\mathrm{km} \cdot \mathrm{h}^{-1})$,$a = -0.9(\mathrm{m} \cdot \mathrm{s}^{-2})$; Take $S = \left\lceil \frac{V^2}{2|a|} \right\rceil + V\tau \approx 1090(\mathrm{km})$ is the safe distance, etc. [20]

(5) **Multi-eye recognition of train**

The camera 3 configured for the train, and even more image sensors with different perspectives may be configured to capture image information from a wider perspective, so as to increase the criterion of independent decision-making during the train running. For example, camera 3 can not only directly detect the scene directly in front of the train, but also form a parallax structure on the longitudinal axis coordinate with camera 1 or 2, so as to directly calculate the height value of the measured object.

References

1. Peng Y (2018) Research on detection and defense of urban rail CBTC system man-in-the-middle attack. Dissertation, Beijing Jiaotong University
2. Chen P (2019) Design and implementation of C-V2X technology in urban rail transit collaborative driving. Dissertation, Beijing University of Posts and Telecommunications
3. Ma X (2018) Resource allocation optimization of wireless sensor networks for railway system condition monitoring. Dissertation, Beijing Jiaotong University
4. Zhao W (2016) Research on innovation mechanism of high speed railway EMU in China. Dissertation, Beijing Jiaotong University
5. Zhang Q (2021) Design and implementation of low-power routing protocol for wireless sensor network supporting network convergence. Dissertation, North University of China
6. Wu X, Zhang C, Zhang R et al (2019) Clustering routing protocol based on improved PSO algorithm in WSN. J Commun 40(12):114–123 (In Chinese)
7. Bai D (2013) Research and improvement of clustering routing algorithms for wireless sensor networks. Dissertation, Chongqing University
8. Xu H (2019) Application of wireless communication technology in rail transit communication and signal service. China New Telecommun 8:1–2 (In Chinese)
9. Zhang X, Ying J (2011) Principles of automobiles intelligent technology. Shanghai Jiao Tong University Press, Shanghai

10. Zhang X, Ying J (2010) Principles of intelligent visual perception. China Water Conservancy and Hydropower Press, Beijing
11. Zhang X, Ying J (2009) Intelligent detection based on visual perception. Science Press, Beijing
12. Zhang X, Mahammad Mansoor K (2018) Principles of intelligent automobiles. Springer Nature & Shanghai Jiao Tong University Press, Berlin & Shanghai
13. Ji C, Zhang X, Ying J et al (2006) PRIC key technologies used in storage yard based on machine vision. WSEAS Trans Inf Sci Appl 3(7):1341–1346
14. Zhang X, Ying J, Hu Z et al (2010) Device and method for intelligent inspection and control of safe vehicle distance. Chinese Invention Patent CN 201010283951.6
15. Ying J, Zhang X, Men P et al (2006) A method for automatic recognition and fast tracking of target image. Chinese Invention Patent CN 200410068030.2
16. Ying J, Zhang X, Wu J et al (2004) A method to improve the speed of automatic recognition and tracking of target image. Chinese Invention Patent CN 200410089272.x
17. Ying J, Zhang X, Sun Z et al (2004) A method to reduce the error rate of automatic target recognition and fast tracking. Chinese Invention Patent CN200410084298.5
18. Zhang X, Mahammad Mansoor K (2012) Intelligent visual perception tutorial. China Water Conservancy and Hydropower Press, Beijing
19. Zhang X, Jiao D, Wang H et al (2009) Vehicle intelligent device for automatic recognition of road pits and obstacles. Chinese Invention Patent CN200910049885.3
20. Zhang X, Jiao D, Ying J et al (2009) Vehicle intelligent method for automatic recognition of road pits and obstacles. Chinese Invention Patent CN200910049884.9

Chapter 3
Intelligent Sensing and Identification of Train Power System

Each component of the train power system and its operation status cannot be set with traditional contact sensors (devices) to achieve information collection and transmission. In particular, the moving parts cannot use the contact sensing technology at all. Even if the micro-sensor can be "placed" on the moving parts, the wireless transmission method must be used to transmit the sensing signal. It can be imagined that the biggest difficulty in achieving such a "micro"-level in the current scientific and technological level lies in the power supply technology of micro-wireless sensors. Therefore, the real-time state detection of train power plant (together with the existing mechanical transmission device) needs to introduce non-contact sensing technologies such as light, sound and magnetism.

But how? It is not easy. The following only describes the condition monitoring technology of high-speed railway train power system (applicable to bullet trains, light rail, Metro trains, etc.).

3.1 Intelligent Sensing Structure and Control of Power Unit of Electric Locomotive

Electric locomotive refers to the train that obtains electric energy from the power supply network (catenary) or power supply rail and drives the vehicle through the motor. The electric energy required for the operation of electric locomotive is provided by the power supply system of electrified railway.

Electric locomotive has many advantages, and its comprehensive performance is much better than steam locomotive and diesel locomotive. It is not only widely used in the operation of trunk railway, but also serves almost all urban rail transit. Nowadays, most countries in the world have launched the construction of electrified railways, popularized the use of electric locomotives and vigorously developed more advanced electric locomotives.

Y. Khan, *Principles of Intelligent Rail Transit*, Advances in High-speed Rail Technology,
https://doi.org/10.1007/978-981-19-6072-7_3

High-speed railway power belongs to a kind of electric traction device, and its required electric energy is supplied by the catenary or the third rail of the electrified railway power supply system. Therefore, it is a locomotive without its own power energy.

The electric multiple units rely on the electric energy transmitted from the outside. The electric energy comes from the ordinary power grid and then is transmitted to the catenary through the voltage transformation of the substation. The pantograph of the locomotive contacts the catenary, so as to lead the power to the locomotive and then supply the motor to drive the multiple units through voltage transformation and rectification.

3.1.1 Composition Basis of Intelligent Sensing of High-Speed Railway Power Plant

The power unit of high-speed railway is composed of traction motor, traction transformer, traction converter, bogie, braking system, pantograph, traction control system and train network control system (see Fig. 3.1) [1].

For high-speed railway power plant, except that the electrical parameters of traction motor, transformer and converter can directly obtain the parameter values such as voltage and current through the detection points of input cables, all mechanical parameters (including torque of traction motor) and thermodynamic parameters are not easily (or even impossible) obtained by contact sensing technology, so it is necessary to introduce non-contact sensor technology.

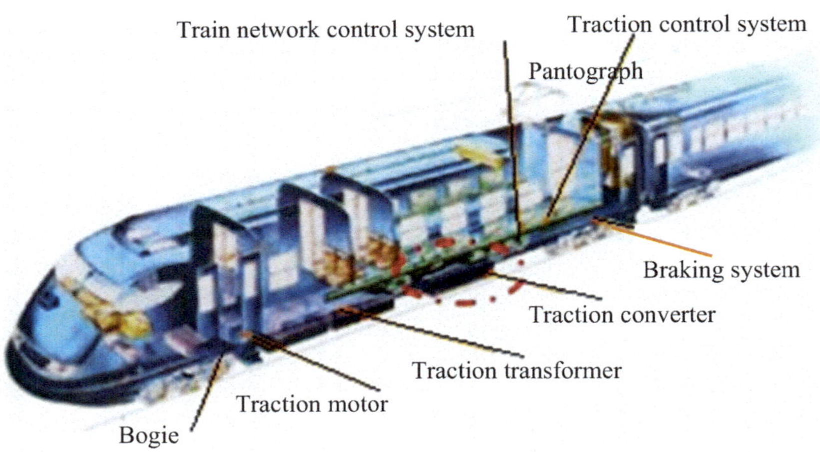

Fig. 3.1 Composition of high-speed railway power plant

3.1.2 Intelligent Control of Train Traction

In general, the train traction control system is mainly composed of vehicle dynamic parameter acquisition system (sensing system), signal processor and actuator. As shown in Fig. 3.2, the signal processor mainly includes sensing signal input interface, conditioning circuit, mathematical model and control program, reference model and its parameter database, calculator, safety circuit, output interface, etc.

Train power and steering control is a real-time multi-task control system. The development trend of the control system is to simplify the hardware structure and enrich the software functions, so as to improve the reliability and intelligence of the system and facilitate the expansion of the system functions. Figure 3.3 shows a representative system main control process.

3.2 Intelligent Identification of Abnormal Conditions of Power System

Voiceprint recognition is a new non-contact sensing technology. The so-called voiceprint is to analyze the acoustic spectrum of audio signal by electroacoustic method. In view of the relative stability and variability of the acoustic characteristics of the audio signal, the spatial/frequency-domain characteristics of the audio signal can be obtained through the analysis of the sound signal spectrum, and then the relevant physical phenomena and characteristics existing (or occurring) in the source of the audio signal can be determined. Then, the real-time characteristics (or abnormal conditions) of the sound source are determined by artificial intelligence algorithm.

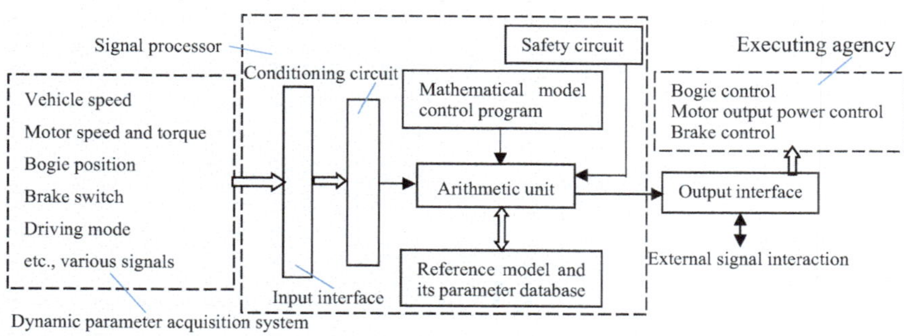

Fig. 3.2 Schematic block diagram of train control system

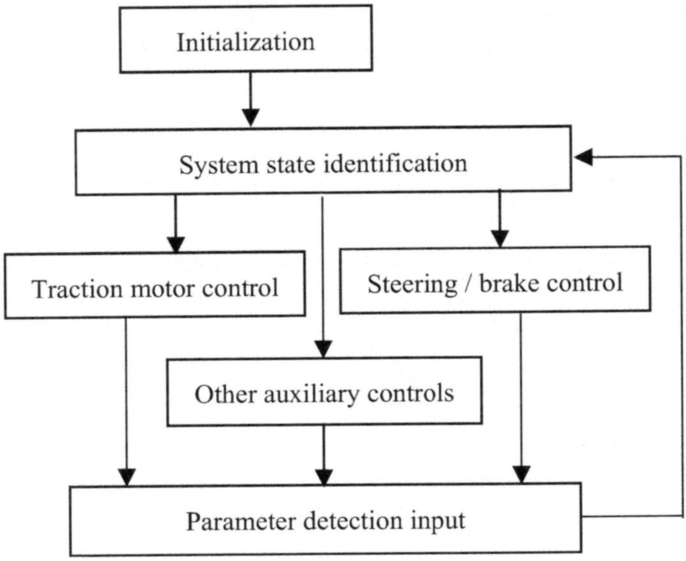

Fig. 3.3 System main control block diagram

3.2.1 Composition of Identification System

Intelligent recognition system for abnormal conditions of power and transmission system includes sound sensor, signal preprocessing, signal processor and voice player. The principle structure block diagram of the device system is shown in Fig. 3.4.

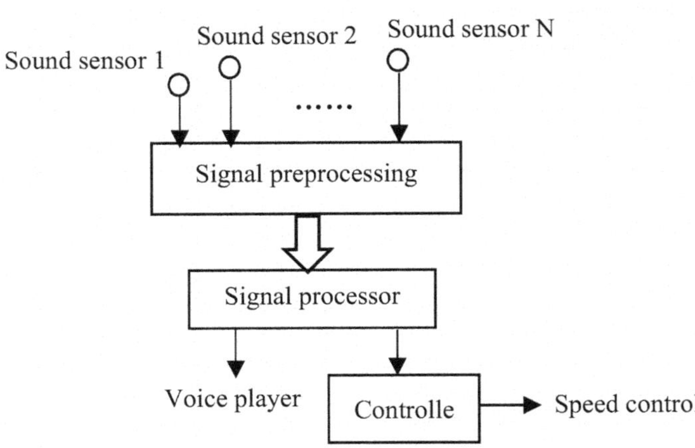

Fig. 3.4 Principle and structure of sound recognition of power and transmission system

As shown in Fig. 3.5, the sound sensor 1 in the identification device system is set close to the traction motor to sense the operation sound signal of the traction motor; the sound sensor 2 is set at a position close to the transmission mechanism until the sound sensor N is set at another position close to the transmission mechanism, which is used to sense the sound signals at multiple places of the transmission mechanism [2].

The signal processor in the identification device includes input interface, A/D converter, operation module, memory module and output interface. The structural principle is shown in Fig. 3.6. The operation module is responsible for processing, analyzing and judging the received sound frequency signal; the memory module registers the sample feature database.

The controller in the device includes a D/A converter and a voltage amplifier. After the decision command output by the arithmetic unit is transmitted to the controller, it is converted into an analog signal by the D/A converter. The analog voltage signal is amplified by the voltage amplifier to promote the corresponding servo mechanism to achieve the purpose of control.

The basic working process of the intelligent identification device for abnormal conditions of the power plant system is the sound signals of the traction motor and the

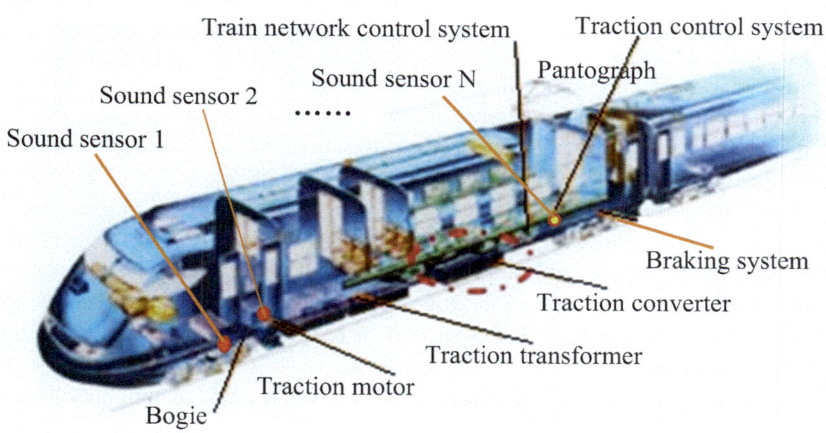

Fig. 3.5 Sound sensor setting diagram

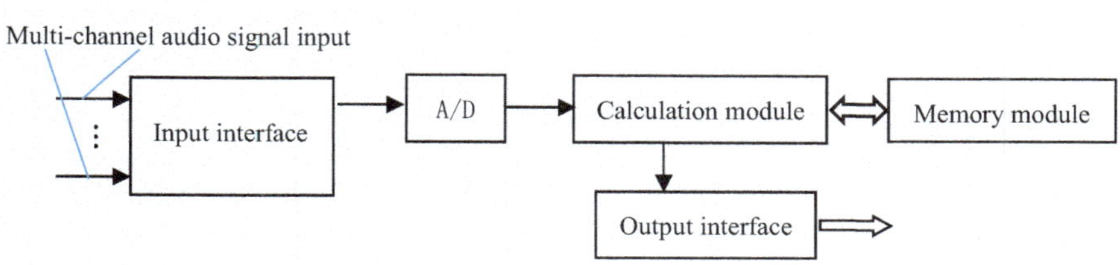

Fig. 3.6 Schematic diagram of signal processor structure

transmission mechanism are detected in real time by the sound sensor 1 and the sound sensor 2, respectively. The sound signals detected by the two sound sensors are sent to the signal processor in parallel after preprocessing and amplification respectively. After the signal processor deeply processes and analyzes the received audio signals, once it finds that these signals contain abnormal audio information, the processor will immediately send control instructions to the controller to control the driving of the vehicle, and the processor will send warning information (screen display and voice warning) to the train driver. Among them, the processing and analysis of audio signal are the core algorithm software of the device [3].

There are two traditional methods to identify and judge the abnormal phenomena of train power mechanism.

(1) System dynamic model combined with experiment

This method is based on the establishment of a mathematical model to accurately describe the actual operating conditions of engine and transmission mechanism, and then a large number of experiments are carried out. After the standard mode of power system operation state based on the power spectrum characteristics of current signal is given, the mode recognition method based on grey correlation calculation is used to identify the working conditions of engine and its transmission mechanism.

(2) Feature information fusion recognition method

The so-called feature information fusion recognition method is to identify and judge the signal representing the phenomenon through signal feature extraction and information fusion from the perspective of monitoring signal (parameters).

It must be pointed out that the former needs to rely on the accurate mathematical model of the operating state of the power system and establish the standard mode of the operating state of the power system based on the power spectrum characteristics of the current signal. However, it is extremely difficult to establish the accurate mathematical model of the actual operating condition of the power system, so its scope of application is limited to a certain extent. The latter relies too much on the sensing technology of parameter detection, and the redundancy of information fusion is obviously greatly reduced.

The following describes the basic theory and technical method of audio intelligent recognition algorithm suitable for abnormal conditions of power system.

3.2.2 Intelligent Algorithm Theory of Audio Recognition

The theory and technology of audio recognition is to analyze and screen the abnormal audio signals contained in the collected sound signals of the tested object, so as to determine whether the monitored object has operating faults and determine the type of faults.

A variety of basic theories have been formed for sound signal feature analysis. As far as rail transit is concerned, its core content is to extract and classify the characteristic parameters of vehicle driving sound signal. How to analyze and classify sound signals can be divided into time-domain characteristic parameters and frequency-domain characteristic parameters.

1. **Time-domain feature extraction method**

(1) **Zero-crossing rate**

The zero-crossing rate refers to the number of times a signal passes through the zero level in unit time. The short-time zero-crossing rate indicates the number of times the time-domain waveform passes through the zero level in a frame of sound signal. Zero-crossing rate is one of the most basic time-domain feature extraction methods of sound signals [4].

The calculation method of zero-crossing rate is

$$\begin{aligned} Z_n &= \sum_{m=-\infty}^{\infty} |\mathrm{sign}[x(m)] - \mathrm{sign}[x(m-1)]|w(n-m) \\ &= |\mathrm{sign}[x(n)] - \mathrm{sign}[x(n-1)]|w(n) \end{aligned} \tag{3.1}$$

In the formula

$$\mathrm{sign}[x(n)] = \begin{cases} 1, x(n) \geq 0 \\ -1, x(n) < 0 \end{cases} \tag{3.2}$$

$$w(n) = \begin{cases} \frac{1}{2N}, 0 < n \leq N \\ 0, \quad \text{other} \end{cases} \tag{3.3}$$

where $x(n)$ is the sampling value of the signal, $w(n)$ is the rectangular window, and N is the window length, generally 256.

In sound signal recognition, the zero-crossing rate is used to judge the starting and ending point of the target sound signal, which is mainly used when the environmental noise has little impact on the sound signal.

(2) **Short-time frame energy**

The energy of sound signal changes with time. The short-time frame energy of sound signal is a characteristic parameter reflecting the change of energy amplitude. It is defined that the short-time average energy of a single frame sound signal marked by t is $\mathrm{En}(t)$ [5].

The calculation formula is as follows:

$$\mathrm{En}(t) = \frac{1}{L}\sum_{t=0}^{L-1} |s_t(\tau)|^2 \tag{3.4}$$

where t is also called the sequence number of the frame, $s_t(\tau)$ is the sampling value of the sound signal, L is the frame length, and $L = 256$ is taken.

(3) Short-time average amplitude

The theoretical principle of short-time frame energy is to directly increase the amplitude difference between two adjacent sampling values with amplitude difference by calculating the square of all sampling values. However, with the increase of amplitude difference, the reasonable selection of time window width has become a problem that needs to be solved effectively. Because only the selected time window that is relatively wide can effectively smooth the change of square amplitude in the sampling interval, this will make $\mathrm{En}(t)$ unable to fully reflect the characteristics of real-time change of sound signal energy.

Therefore, another parameter that can represent the time-varying characteristics of short-term energy of sound signal needs to be introduced, which is short-term average amplitude. The calculation formula is

$$M_n = \sum_{m=-\infty}^{\infty} |x(m)|w(n-m) = |x(n)|w(n) \tag{3.5}$$

Select a frame of sound signal and calculate the sum of the absolute value of the sampled value of the frame signal to obtain the short-time average amplitude of the frame sound signal. If the average amplitude is calculated for each frame signal in the complete sound signal, the short-term average amplitude characteristics of the complete sound signal can be obtained. Compared with the short-time frame energy, the short-time average amplitude has similar characteristics, but because there is no square operation step, the result is closer to the actual change law of sound signal energy.

(4) Energy zero ratio

The detection method of energy zero ratio is based on the following: In the speech signal, the energy in the period of speech is raised upward, while the zero-crossing rate is different. The amplitude in the period of speech is relatively depressed downward, because there are a large number of high-frequency components in Gaussian random noise, high frequency means a high short-term zero-crossing rate, and the speech frequency is mainly concentrated below 3 kHz. Therefore, when the background noise is large, the amplitude of speech energy is large, while the amplitude of zero-crossing rate is small. Only when the amplitude of energy in the noise period is small, the amplitude of zero-crossing rate of speech will be large. Therefore, comparing the energy value with the zero-crossing rate value, it can highlight the period of speech. The amplitude of the noise period must be reduced in order to further open the amplitude gap between the period of speech and the noise period and then better identify the endpoint of the period of speech.

When the time series of the sound signal is $x(n)$, after windowing and framing processing, the sound signal of the i-th frame is $x_i(m)$ and the frame length is N, and then, the energy of each frame is

$$\mathrm{AMP}_i = \sum_{m=1}^{N} x_i^2(m) \tag{3.6}$$

An improved energy calculation relationship is introduced here

$$\mathrm{LE}_i = \log_{10}(1 + \mathrm{AMP}_i/a) \tag{3.7}$$

where a is a constant.

When a takes a larger value and AMP_i amplitude changes greatly, LE_i will be effectively reduced. Therefore, proper selection of a can effectively distinguish between ambient noise and target sound signal.

In order to ensure the stability of zero-crossing rate calculation and reduce some small zero drift that may exist in the sound signal, when the windowed and framing sound signal $x_i(m)$ is input, the central clipping processing is required to obtain

$$\tilde{x}_i(m) = \begin{cases} x_i(m), & |x_i(m)| > \delta \\ 0, & |x_i(m)| < \delta \end{cases} \tag{3.8}$$

where δ is a small value. After the center clipping, calculate the zero-crossing rate Z_i of each frame to obtain

$$Z_i = \sum_{m=1}^{N} \left|\mathrm{sign}[\tilde{x}_i(m)] - \mathrm{sign}[\tilde{x}_i(m-1)]\right| \tag{3.9}$$

where

$$\mathrm{sign}[\tilde{x}_i(m)] = \begin{cases} +1, & \tilde{x}_i(m) \geq 0 \\ -1, & \tilde{x}_i(m) < 0 \end{cases} \tag{3.10}$$

According to the improved energy calculation value and zero-crossing rate, the energy zero ratio can be given

$$E_i = \mathrm{LE}_i/(Z_i + b) \tag{3.11}$$

where $b = 1$ can prevent the energy zero ratio E_i at Z_i=0 from overflowing.

2. **Frequency-domain feature extraction method**

(1) **Wavelet sub-band energy**

Wavelet transform is a time–frequency analysis thinning method based on short-time Fourier transform. Because the window shape can be changed, the wavelet transform can carry out the localization analysis of time (space) frequency. Through the expansion, contraction and translation operation, the signal can be multi-scale refined analysis, so as to realize the purpose of time subdivision at high frequency and frequency subdivision at low frequency. The wavelet transform can carry out detailed analysis of different positions of the signal according to the specific requirements of time–frequency signal analysis. The problem that Fourier transform cannot effectively analyze non-stationary signals is solved.

1) **Continuous wavelet basis function**

Let $\psi(t)$ be a square integrable function, i.e., $\psi(t) \in L^2(R)$, if its Fourier transform $\psi(\omega)$ satisfies the following conditions

$$C_\psi = \int_R \frac{|\psi(\omega)|^2}{|\omega|} d\omega < \infty \tag{3.12}$$

Then, $\psi(t)$ is called the mother wavelet function, and Eq. (3.12) is the admissible condition of the wavelet function.

After expanding, shrinking and translating $\psi(t)$, the following wavelet basis function $\psi_{\alpha,\tau}(t)$ can be obtained

$$\psi_{\alpha,\tau}(t) = \frac{1}{\sqrt{\alpha}}\psi\left(\frac{t-\tau}{\alpha}\right) \quad \alpha, \tau \in R; \alpha > 0 \tag{3.13}$$

where α is the expansion and contraction parameter and τ is the translation parameter. Since both α and τ take continuously varying values, we call $\psi_{\alpha,\tau}(t)$ continuous wavelet basis function.

2) **Continuous wavelet-transform**

Let $f(t) \in L^2(R)$, and the result of continuous wavelet transform is $\mathrm{WT}_f(\alpha, \tau)$, then $\mathrm{WT}_f(\alpha, \tau)$ is shown in the following formula

$$\mathrm{WT}_f(\alpha, \tau) = \langle f(t), \psi_{\alpha,\tau}(t) \rangle = \frac{1}{\sqrt{\alpha}} \int_R f(t)\psi^*\left(\frac{t-\tau}{\alpha}\right)\mathrm{d}t \tag{3.14}$$

When $\psi(\omega)$ satisfies the admissible condition, its inverse transformation formula is

$$f(t) = \frac{1}{C_\psi} \int_0^{+\infty} \frac{\mathrm{d}\alpha}{\alpha^2} \int_{-\infty}^{+\infty} \mathrm{WT}_f(\alpha, \tau)\psi_{\alpha,\tau}(t)\mathrm{d}t \tag{3.15}$$

3) **Discrete wavelet-transform**

In order to reduce the redundancy of wavelet-transform coefficients, α and τ in $\psi_{\alpha,\tau}(t) = \frac{1}{\sqrt{a}}\psi\left(\frac{t-\tau}{a}\right)$ are discretized. The expansion and contraction parameter α is discretized by power series, that is, let $a = a_0^m$, $a_0 > 0$, $m = Z$, and the obtained discrete wavelet function $\psi_{\alpha_0^j, k\tau_0}(t)$ is

$$\psi_{\alpha_0^j,k\tau_0}(t) = a_0^{\frac{-j}{2}}\psi\left[a_0^{-j}(t-\tau)\right], \quad j = 0, 1, 2, \ldots \tag{3.16}$$

In order to prevent information loss, τ is discretized uniformly, that is, $\tau = ka_0^j\tau_0$, so the wavelet function can be modified to

$$\begin{aligned}\psi_{\alpha_0^j,k\tau_0}(t) &= a_0^{\frac{-j}{2}}\psi\left[a_0^{-j}\left(t - k\alpha_0^j\tau_0\right)\right] \\ &= a_0^{\frac{-j}{2}}\psi\left[a_0^{-j}t - k\tau_0\right], \quad j = 0, 1, 2, \ldots\end{aligned} \tag{3.17}$$

Then, the discrete wavelet transform can be expressed as

$$\mathrm{WT}_f\left(\alpha_0^j, k\tau_0\right) = \int f(t)\psi^*_{\alpha_0^j,k\tau_0}(t)\mathrm{d}t, \quad j = 0, 1, 2, \ldots, k \in Z \tag{3.18}$$

When analyzing the sound signal, the sound signal is usually discretized, so the discrete wavelet transform can be used to process the sound signal.

4) **Dyadic wavelet-transform**

At present, $a_0 = 2$, $\tau = 1$ is often used in discrete wavelet-transform, and the resulting discrete wavelet is called dyadic wavelet, that is

$$\psi_{j,k}(t) = 2^{\frac{-j}{2}}\psi\left(2^j t - k\right), \quad j, k \in Z \tag{3.19}$$

If $\psi_{j,k}(t) \in L^2(R)$, the relationship between two constants is $0 < A < B < \infty$, and

$$A \le \sum_{j\in Z}\left|\psi\left(2^{-j}\omega\right)\right|^2 \le B \tag{3.20}$$

Then, $\psi_{j,k}(t)$ is called a dyadic wavelet. If $A = B$, it has the most stable condition. The formula of dyadic wavelet transform is

$$W_{2^j}f(t) = \langle f(t), \psi_{2^j}(k)\rangle = \frac{1}{2^j}\int_R f(t)\psi^*\left(2^{-j}t - k\right)\mathrm{d}t \tag{3.21}$$

Different from discrete wavelet transform, dyadic wavelet transform only discretizes the expansion and contraction parameters, but does not change the continuity of translation parameters. Therefore, dyadic wavelet transform will not destroy the translation invariance of sound signal in time domain.

5) **Wavelet sub-band energy**

The wavelet sub-band energy is used as the characteristic parameter, so the wavelet coefficient can obtain the frequency location information of time and sound waveform, which cannot be obtained by Fourier transform [6].

At this time, a set of filters is required, that is

$$\psi_{T,\alpha}(t) = \alpha^{-\frac{1}{2}}\psi\left(t - \frac{T}{\alpha}\right), \quad \psi \in L^2 \tag{3.22}$$

where L^2 is a finite energy space.

The discrete wavelet-transform coefficients of signal s can be expressed as

$$D(j,k) = 2^{-\frac{j}{2}} \sum_i s(i)\psi^*(2^{-j}i - k) \tag{3.23}$$

where $s(i)$ is the sampling value of the signal, $\psi^*(i)$ is the wavelet filter bank, and $D(j,k)$ is the discrete wavelet-transform coefficient.

After wavelet transform, the sound signal is divided into different sub-bands, and the wavelet coefficients in each sub-band represent the energy distribution of the corresponding sub-band.

Wavelet transform can be used as a filter bank including low-pass filter and high-pass filter to filter the speech signal [7].

(2) **Traditional energy entropy ratio detection method**

The energy entropy ratio detection method, that is, the spectral entropy of the sound signal reaches the maximum in the equal probability distribution, indicating that the average uncertainty of the source is the maximum in the equal probability distribution (maximum discrete entropy theorem) [8].

For normal driving noise and normal driving sound, its normalized spectral probability density function distribution is relatively average, so its spectral entropy is large. For the sound signal of abnormal train running, its normalized spectral probability density function is unevenly distributed, so its spectral entropy is lower than that of normal noise and normal driving sound, while the sound signal energy of abnormal event during abnormal driving is greater than that of normal driving. The difference between normal driving and abnormal driving is increased by the ratio of energy and spectral entropy, so as to judge whether there is an exception.

If the sound signal is represented as $x(n)$, and the i-th frame sound signal obtained after windowing and framing processing is $x_i(m)$, it is represented as $X_i(k)$ after Fourier transform, subscript i is represented as the i-th frame sound signal, and k is represented as the k-th spectral line. At this time, the short-time energy is

$$E_i = \sum_{k=0}^{N/2} X_i(k) \cdot X_i^*(k) \tag{3.24}$$

where k is the FFT (fast Fourier transform) length, and only the positive frequency part is taken here.

Suppose, for a certain spectral line, the energy spectrum of k is

$$Y_i(k) = X_i(k) \cdot X_i^*(k) \tag{3.25}$$

Then, the normalized spectral probability density of each frequency component of the corresponding signal is defined as

$$p_i(k) = \frac{Y_i(k)}{\sum_{l=0}^{N/2} Y_i(l)} = \frac{Y_i(k)}{E_i} \tag{3.26}$$

The short-time spectral entropy of the audio frame is defined as

$$H_i = -\sum_{k=0}^{N/2} p_i(k) \log p_i(k) \tag{3.27}$$

If the function of entropy is satisfied with the following relation

$$H(p_1, p_2, \ldots p_q) = H(1/q, 1/q, \ldots 1/q) = \log q \tag{3.28}$$

then the corresponding energy entropy ratio is

$$\mathrm{EEF}_i = \sqrt{1 + |E_i / H_i|} \tag{3.29}$$

(3) **Sub-band energy entropy ratio detection method**

The principle of sub-band energy entropy ratio detection method is consistent with that of energy entropy ratio detection method. The difference is that each frame sound signal is redivided into several sub-bands, and the spectral entropy of each sub-band is calculated, respectively, so as to reduce the interference of noise on the amplitude of each spectral line [9].

Let each sub-band contains 4 spectral lines and a total of N sub-bands, so that the sub-band energy of the m-th sub-band in frame i is

$$E(m, i) = \sum_{k=4(m-1)}^{4(m-1)+3} Y_i(k), \ 1 \le m \le N \tag{3.30}$$

Accordingly, the probability $p(m, i)$ of sub-band energy and sub-band spectral entropy $H(i)$ are

$$p(m, i) = \frac{E(m, i)}{\sum_{m=1}^{N} E(m, i)}, \quad 1 \le m \le N \tag{3.31}$$

$$H(i) = -\sum_{m=1}^{N} p(m, i) \log p(m, i) \tag{3.32}$$

The above traditional sound signal feature extraction methods in time domain and frequency domain can be used to collect and establish the audio information features of the train under normal operation to form the corresponding database.

Generally speaking, the traditional time-domain and frequency-domain characteristic analysis methods for sound signals are not suitable for the analysis of abnormal sound characteristics emitted by railway trains. However, people can extract features of train audio signals in time domain or frequency domain in advance to establish the state feature database of normal train operation, so they can be used as learning samples of dynamic audio signals, and then compare and analyze fault information [10].

After years of research, many intelligent identification technologies have been formed for abnormal working conditions of train power plant. In fact, the most prominent and effective algorithm is support vector machines (SVM).

3. **Support Vector Machines**

Support vector machines (SVM) is a learning machine. It is developed on the basis of statistical learning theory aiming at solving machine learning problems with limited samples. In other words, SVM is based on the VC dimension theory of statistical learning theory and the principle of structural risk minimization. According to the limited sample information, SVM looks for the best compromise between the complexity of the model and learning ability and hopes to obtain the best generalization ability. SVM can effectively avoid many problems in classical learning methods, such as over learning, under learning, "dimension disaster" and falling into local minima. SVM is developed from the optimal classification surface in the case of linear separability. It adopts the strategy of keeping the empirical risk value fixed and minimizing the confidence range.

For the intelligent identification of abnormal phenomena of power and transmission system, the identification and judgment of system operating condition information can be realized through the processing and analysis of sound signals generated by the system.

Among them, support vector machine is one of the effective methods to realize this process [11].

We need to focus on the optimization criteria in multi-layer perceptual and meridional basis function networks.

(1) **ε-insensitive loss function**

In multi-layer perceptron and meridional basis function networks, the square loss function (least square method) has been used as the optimization standard of these networks. The main reason for using this standard is to consider mathematically, that is, for the convenience of calculation. However, the least square estimator is very sensitive to the abnormal "occurrence" performance (e.g., it will get a large and almost impossible observation for a normal model), and it will perform poorly when the inherent distribution of additive noise has a long tail distribution. In order to overcome these limitations, a robust estimator is needed, which is insensitive to small changes in the model.

With robustness as the design objective, any numerical measurement of robustness must take into account the maximum performance degradation due to a deviation of the normal noise model. According to this view, an optimal robust estimation is the biggest performance degradation of the minimization process, so it is a minimax process. When the probability density function of additive noise is symmetrical about the origin, the minimax process of solving the nonlinear regression problem uses the absolute error as the minimization quantity; that is, the loss function has the following form:

$$L(d, y) = |d - y| \tag{3.33}$$

where d is the desired response and y is the output value of the estimator.

In order to construct a support vector machine to approximate the expected response d, the extension of the loss function can be described as follows:

$$L_\varepsilon(d, y) = \begin{cases} |d - y| - \varepsilon; & \text{If } |d - y| \geq \varepsilon \\ 0; & \text{else} \end{cases} \tag{3.34}$$

where ε is the specified parameter, and the loss function $L_\varepsilon(d, y)$ is called ε-insensitive loss function. If the absolute deviation between the estimated output y and the expected output d is less than ε, it is equal to zero; otherwise, it is equal to the absolute value of the deviation minus ε. The loss function of Eq. (3.33) is a special case of ε-insensitive loss function at $\varepsilon = 0$. Figure 3.7 illustrates the dependence of $L_\varepsilon(d, y)$ and error $d - y$.

(2) **Support vector machine for nonlinear regression**

Considering the nonlinear regression model, the dependence of scalar d on vector can be described as

$$d = f(\mathbf{x}) + v \tag{3.35}$$

where $f(\mathbf{x})$ is a nonlinear scalar valued function, and its first implementation is recorded as d. The additive noise term v is statistically independent of the input vector $\mathbf{x}$, and the statistical properties of function $f(\mathbf{x})$ and noise v are unknown. All available information is a set of training data $\{(x_i, d_i)\}_{i=1}^N$, where x_i is the primary

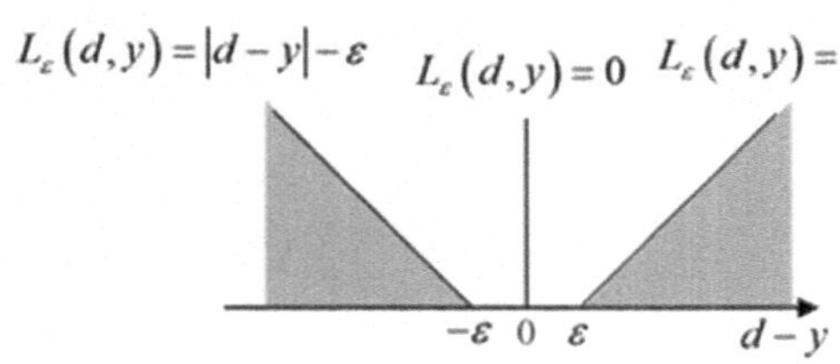

Fig. 3.7 ε-geometric meaning of insensitive loss function

sampling value of input vector **X** and d_i is the corresponding value of model output d. The problem required is to provide an estimate of d dependence on **x**.

Further assume that the estimate of d is **y**, which is obtained by the extension of a set of nonlinear basis functions $\{\varphi_j(\mathbf{x})\}_{j=0}^{m}$, that is

$$y = \sum_{j=0}^{m} w_j \varphi_j(\mathbf{x}) = \mathbf{w}^{\mathrm{T}} \mathbf{0}(\mathbf{x}) \tag{3.36}$$

where $\mathbf{0}(\mathbf{x}) = \left[\, \varphi_0(\mathbf{x})\ \varphi_1(\mathbf{x}) \ldots \varphi_m(\mathbf{x}) \,\right]^T, \mathbf{w} = \left[\, w_0\ w_1 \ldots w_m \,\right]^{\mathrm{T}}$

Assuming $\varphi_0(\mathbf{x}) = 1$, the weight w_0 represents offset b, and the problem of demand solution is to minimize empirical risk, that is

$$R_{\text{emp}} = \frac{1}{N} \sum_{i=1}^{N} L_\varepsilon(d_i, y_i) \tag{3.37}$$

It satisfies the following inequality

$$\|\mathbf{w}\|^2 \le c_0 \tag{3.38}$$

where c_0 is a constant and $L_\varepsilon(d, y)$ is an ε-insensitive loss function.

Two sets of non-negative relaxation variables $\{\xi_i\}_{i=1}^{N}$ and $\left\{\xi_i'\right\}_{i=1}^{N}$ are introduced to resolve the constrained optimization problem.

As defined below

$$d_i - \mathbf{w}^{\mathrm{T}} \mathbf{0}(x_i) \le \varepsilon + \xi_i; \quad i = 1, 2, \ldots, N \tag{3.39}$$

$$\mathbf{w}^{\mathrm{T}} \mathbf{0}(x_i) + d_i \le \varepsilon + \xi_i'; \quad i = 1, 2, \ldots, N \tag{3.40}$$

$$\xi_i \ge 0; \quad i = 1, 2, \ldots, N \tag{3.41}$$

$$\xi_i' \ge 0; \quad i = 1, 2, \ldots, N \tag{3.42}$$

Relaxation variables $\{\xi_i\}_{i=1}^{N}$ and $\left\{\xi_i'\right\}_{i=1}^{N}$ describe the ε-insensitive loss function defined in Formula (3.38), so this constrained optimization problem can be equivalent to minimizing the cost function, that is

$$\Phi\left(\mathbf{w}, \xi_i, \xi_i'\right) = C \sum_{i=1}^{N} \left(\xi_i + \xi_i'\right) + \frac{1}{2} \mathbf{w}^{\mathrm{T}} \mathbf{w} \tag{3.43}$$

It satisfies the constraints from Formulas (3.39)–(3.42).

When using the function combination term $\frac{1}{2}\mathbf{w}^{\mathrm{T}}\mathbf{w}$ of Formula (3.43), the inequality constraint of Formula (3.40) is not required. The constant C in Formula (3.43) is a parameter given by the user. At this point, we can define the Lagrange function as follows

$$\begin{aligned}L\left(\mathbf{w},\xi_i,\xi'_i,\alpha,\alpha',\gamma,\gamma'\right) &= \sum_{i=1}^{N}\left(\xi_i+\xi_i'\right)+\frac{1}{2}\mathbf{w}^T\mathbf{w}\\ &-\sum_{i=1}^{N}\alpha_i\left[\mathbf{w}^T\varphi(x_i)-d_i+\varepsilon+\xi_i\right]\\ &-\sum_{i=1}^{N}\alpha_i'\left[d_i-\mathbf{w}^T\varphi(x_i)+\varepsilon+\xi_i'\right]-\sum_{i=1}^{N}\left(\gamma_i\xi_i+\gamma_i'\xi_i'\right)\end{aligned} \tag{3.44}$$

where α_i and α_i' are Lagrange multipliers. The last term including γ_i and γ_i' on the right of formula (3.44) is to ensure that the optimal conditions of Lagrange multipliers α_i and α_i' become variable forms. As can be seen from Formula (3.44), $L\left(\mathbf{w},\mathbf{x}_\mathrm{i},\mathbf{x}_\mathrm{i}',\alpha,\mathbf{a}',\mathbf{g},\mathbf{g}'\right)$ requires the minimization of $\mathbf{w}$ and relaxation variables ξ_i and ξ_i' and the maximization of α_i and α_i' and γ_i and γ_i'.

To solve this optimization, there are

$$\mathbf{w}=\sum_{i=1}^{N}\left(\alpha_i-\alpha_i'\right)\mathbf{0}(x_i) \tag{3.45}$$

and

$$\gamma_i=C-\alpha_i \tag{3.46}$$

$$\gamma_i'=C-\alpha_i' \tag{3.47}$$

In order to construct the corresponding dual problem, Formulas (3.45)–(3.47) are substituted into Formula (3.44). After simplification, the convex function is obtained as follows

$$\begin{aligned}Q\left(\alpha_i,\alpha_i'\right) &= \sum_{i=1}^{N}d_i\left(\alpha_i-\alpha_i'\right)-\sum_{i=1}^{N}\varepsilon\left(\alpha_i+\alpha_i'\right)\\ &-\frac{1}{2}\sum_{i=1}^{N}\sum_{j=1}^{N}\left(\alpha_i-\alpha_i'\right)\left(\alpha_j-\alpha_j'\right)\mathbf{K}\left(x_i,x_j\right)\end{aligned} \tag{3.48}$$

where $\mathbf{K}\left(x_i,x_j\right)$ is the inner product kernel defined according to Mercer's theorem, $\mathbf{K}\left(x_i,x_j\right)=\mathbf{0}^T(x_i)\mathbf{0}\left(x_j\right)$.

The solution of the optimization problem is the maximization $Q(\alpha_i, \alpha_i')$ under a set of new constraints on Lagrange multipliers α_i and α_i' satisfying the relevant constant C, where C is included in the definition of function $\Phi(\mathbf{w}, \mathbf{x}_i, \mathbf{x}_i')$ in Formula (3.43).

The nonlinear regression dual problem using support vector machine is described as follows.

Given the training sample g, find Lagrange multipliers A and B so that the maximization objective function is

$$
\begin{aligned}
Q(\alpha_i, \alpha_i') = \sum_{i=1}^{N} d_i(\alpha_i - \alpha_i') - \sum_{i=1}^{N} \varepsilon(\alpha_i + \alpha_i') \\
-\frac{1}{2}\sum_{i=1}^{N}\sum_{j=1}^{N}(\alpha_i - \alpha_i')(x_j - x_j')\mathbf{K}(x_i, x_j)
\end{aligned}
\tag{3.49}
$$

It satisfies the following constraints

$$
\begin{aligned}
&\sum_{i=1}^{N}(\alpha_i - \alpha_i') = 0 \\
&0 \le \alpha_i \le C; \quad i = 1, 2, \ldots, N \\
&0 \le \alpha_i' \le C; \quad i = 1, 2, \ldots, N
\end{aligned}
$$

where C is the constant given by user.

In the Lagrange optimization problem, the constraint condition $\sum_{i=1}^{N}(\alpha_i - \alpha_i') = 0$ is generated by offsetting $b = w_0$ according to $\varphi_0(x) = 1$. Therefore, after obtaining the optimal values of α_i and α_i', the optimal value of weight vector $\mathbf{W}$ can be determined by Formula (3.45) for a given mapping $\mathbf{0}(x)$. It is noted that, like the solution of pattern recognition problem, in the extension of Formula (3.45), only some coefficients are nonzero. In particular, the data point corresponding to $\alpha_i \neq \alpha_i'$ is defined as the support vector of the machine.

The free parameters ε and C control the VC dimension of the following approximation functions

$$
F(\mathbf{x}, \mathbf{w}) = \mathbf{w}^{\mathrm{T}}\mathbf{x} = \sum_{i=1}^{N}(\alpha_i - \alpha_i')\mathbf{K}(x, x_i)
\tag{3.50}
$$

Both ε and C must be selected by the user. Conceptually, the selection of ε and C provides the same complexity control problem as the selection of parameter C in pattern recognition, but the complexity control in actual regression is a more difficult problem, which is caused by the following reasons:

(1) Parameters ε and C must be adjusted at the same time.
(2) In essence, regression is more difficult than pattern classification.

Finally, like the support vector machine for pattern recognition, the support vector machine for nonlinear regression can be realized by polynomial term learning machine, radial basis function network or two-layer perceptron.

3.2.3 Basic Algorithm of Support Vector Machine

The characteristics of SVM are as follows: Support vector is the nearest sample to hyperplane in all kinds of samples. Although the number of support vectors is small, it contains the information needed for classification. Because most of the training samples do not belong to support vectors, removing or reducing these samples has no effect on the classifier.

Because SVM has these characteristics, many related calculation methods have been produced for the theoretical basis and application of SVM method.

1. **SVM training algorithm for processing large data sets**

The optimization problem related to classification support vector machine can be written as

$$\max W(\mathbf{a}) = \sum_{i=1}^{l} \alpha_i - \frac{1}{2}\sum_{i=1}^{l}\sum_{j=1}^{l} \alpha_i \alpha_j y_i y_j \mathbf{K}(x_i, x_j) \tag{3.51}$$

$$\text{s.t.} \begin{cases} \sum_{i=1}^{l} \alpha_i y_i = 0 \\ 0 \le \alpha_i \le C; \quad i = 1, 2, \ldots, l \end{cases} \tag{3.52}$$

For regression problems, linear ε-insensitive loss problem can be expressed by the following formula

$$\max W(\mathbf{a}) = \sum_{i=1}^{l} \alpha_i y_i - \varepsilon \sum_{i=1}^{l} |\alpha_i| - \frac{1}{2}\sum_{i=1}^{l}\sum_{j=1}^{l} \alpha_i \alpha_j \mathbf{K}(x_i, x_j) \tag{3.53}$$

$$\text{s.t.} \begin{cases} \sum_{i=1}^{l} \alpha_i = 0 \\ -C \le \alpha_i \le C; \quad i = 1, 2, \ldots, l \end{cases} \tag{3.54}$$

SVM problem is actually about solving a quadratic equation. For this quadratic programming problem, the classical solutions include positive square set method, dual method and inner product algorithm. When the number of training samples increases, these algorithms will face the disaster of dimension, or the sample training cannot be carried out due to the limitation of computer memory. Because the kernel function matrix needs to be calculated and stored in the calculation process, when the number of sample points is large, it needs a lot of memory. For example, when the number of sample points exceeds 4000, storing the kernel function matrix requires up to 128 megabytes of memory. Secondly, support vector machine needs a lot of short matrix operations in the quadratic optimization process. In most cases, the optimization algorithm is the main part of the algorithm time.

In view of the slow speed of solving the traditional quadratic programming problem, the current training algorithm of support vector machine generally adopts cyclic iteration to solve the dual optimization problem. That is, the original problem is decomposed into several subproblems. According to some iterative strategy, the subproblem is solved repeatedly, and finally, the result converges to the optimal solution of the original problem. According to the division of subproblems and different iterative strategies, SVM training algorithms can be roughly divided into block algorithm and decomposition method.

(1) **Chunking algorithm**

In the support vector machine classification algorithm, the decision function obtained by the support vector machine method is only related to the support vector and has nothing to do with other sample data; that is, if only the support vector is taken as the training sample, the decision function obtained is consistent with the decision function obtained by all samples as the training sample. Therefore, if we know which are support vectors in advance, we can only retain the sample points corresponding to them and delete other sample points from the training set. The same decision function can be obtained by solving the dual problem with the remaining training set. This fact is very important for solving large-scale problems because the number of support vectors is generally small. Based on this idea, the chunking method is an optimization model for solving massive sample data; that is, the massive sample data set is divided into several small-scale sample sets. Train and learn various book subsets one by one in order. When learning each sample subset, we only need to make a new optimization calculation according to the support vector obtained from the previous sample subset and the current sample subset. However, we do not know which support vectors are in advance. The usual method is to find them by heuristic algorithm.

In other words, considering that removing the training samples with Lagrange multiplier equal to zero will not affect the solution of the original problem, select a part of the samples to form the working sample set for training, eliminate the non-support vectors and test the remaining samples with the training results, so as to combine the samples that do not meet the KKT (Karush–Kuhn–Tucker) conditions with the support vectors of this result into a new working sample set, then retraining. This is repeated until the optimal result is obtained.

It must be pointed out that the solution scale of the block method increases with the increase of the number of support vectors, so it is still difficult to realize the optimization calculation when the number of support vectors is very large.

The so-called KKT condition means that there is at least one nonlinear mapping for the mathematical model of nonlinear planning.

For the mathematical model of nonlinear programming

$$\min f(\mathbf{x})$$

$$\text{s.t. } g_i(\mathbf{x}) \geq 0 \quad (i = 1, 2, \ldots, m),$$

$$h_j(\mathbf{x}) = 0, \ (j = 1, 2, \ldots, l)$$

where $\mathbf{x} = (x_1, x_2, \ldots, x_n)^{\mathrm{T}} \in R^n$ is an n-dimensional vector, f, g_i and h_j are all mappings of $R^n \to R^l$ (i.e., the independent variable is a n-dimensional vector, and the dependent variable is a functional relationship of real numbers), and there is at least one nonlinear mapping.

Similar to linear programming, the solution satisfying the constraints is called feasible solution. If it is expressed as follows:

$$\chi = \left\{\mathbf{x} \middle| g_i(\mathbf{x}) \geq 0, \quad i = 1, 2, \ldots, m, h_j(\mathbf{x}) = 0, \quad j = 1, 2, \ldots, l\right\}$$

Then, χ is called feasible region. Therefore, the above nonlinear programming model can be abbreviated as

$$\min f(\mathbf{x})$$

$$\text{s.t. } \mathbf{x} \in \chi$$

KKT optimization condition means that the minimum point $\mathbf{x}^*$ of the above formula must meet the following conditions

$$\begin{cases} g_i(\mathbf{x}^*) \geq 0; \quad i = 1, 2, \ldots, m \\ h_j(\mathbf{x}^*) = 0; \quad j = 1, 2, \ldots, l \\ \mathbf{x}^* \in \chi \subset R^n \end{cases}$$

(2) **Decomposition method**

The decomposition method also transforms the large-scale quadratic programming problem into a series of small-scale quadratic programming solutions. The decomposition method selects q Lagrange multipliers α_i as the optimization variables ($q < N$, q is fixed), while the values of other α_i are fixed. Therefore, the subproblem solving scale of the decomposition method is not the same as that of the block method, and the solving scale will increase with the increase of the number of support vectors, but is fixed. The basic idea of decomposition is to divide the sequence number set $\{1, 2, \ldots, N\}$ of sample data into working set B and non-working set $\overline{B}$, and the size of working set B is q. In this way, the large-scale quadratic programming problem is transformed into a small-scale quadratic programming problem with only q optimization variables, $2 \times q$ linear inequality constraints and N equality constraints.

The key of the decomposition method is how to select the working set B in each iteration and the convergence of the algorithm.

Because the decomposition method transforms the large-scale quadratic programming problem into a series of small-scale quadratic programming solutions with the size of working set B fixed as q, it is also called fixed working sample set algorithm.

It must be pointed out that the size of the working sample set is fixed within the tolerable limit of the algorithm speed. In the iterative process, select an appropriate swap in and swap out strategy to exchange part of the remaining samples with the samples in the working sample set. Even if the number of support vectors exceeds the size of the working sample set, the size of the working sample set will not be changed, and only a part of the support vector is optimized.

2. **Sequential minimum optimization algorithm**

Sequential minimum optimization (SMO) is actually a special case of decomposition algorithm. The algorithm can avoid the instability and time-consuming problems of numerical solution in the case of multiple samples. At the same time, it does not need large matrix storage space. It is especially suitable for sparse samples, and the operation speed is very fast.

The sequential minimum optimization algorithm pushes the decomposition algorithm to the extreme, because each iteration only optimizes the minimum subset of two points (there are only two samples in working set B). The advantage of this algorithm is that the optimization problem of two data points can obtain an analytical solution, so the quadratic programming optimization algorithm is not a part of the algorithm. Its working set selection is not the traditional steepest descent method, but uses heuristics to find the optimized sample variables through two nested loops. Look for samples that violate KKT conditions in the outer loop, then select another sample in the inner loop for optimization and then recycle for the next optimization until all samples meet the optimization conditions.

Specifically, the condition $\sum_{i=1}^{l} \alpha_i y_i = 0$ needs to be implemented in the iteration, which means that the minimum number of multipliers that can be optimized in each step is 2. Whenever one multiplier is updated, at least another multiplier needs to be adjusted to keep the condition valid.

In each step, SMO selects two elements α_i and α_j for joint optimization, finds the optimal values of these two elements α_i and α_j when other parameters are fixed and updates the corresponding vector $\boldsymbol{\alpha}$. The selection of these two points is heuristic, and the optimization of multipliers of these two points can obtain analytical solutions. In the solving process, although more iterations are needed to converge, each iteration requires only a few operations, so the overall speed of the algorithm is improved by an order of magnitude, including the convergence time. At the same time, other features in the algorithm have no matrix operation, do not need to store the matrix in memory and are easy to be realized technically.

(1) **Two-point analytical solution**

Suppose that the two selected variables are α_1 and α_2, because α_3,α_4,..., α_l is fixed, the optimization problem of Formula (3.51) is equivalent to the following optimization problem

$$\max W(\boldsymbol{\alpha}) = \alpha_1 + \alpha_2 - \frac{1}{2}\left[\mathbf{K}(x_1, x_1)\alpha_1^2 + \mathbf{K}(x_2, x_2)\alpha_2^2\right] - y_1 y_2 \mathbf{K}(x_1, x_2)\alpha_1\alpha_2$$
$$- y_1\alpha_1 \sum_{i=1}^{l} y_i\alpha_i \mathbf{K}(x_i, x_1) - y_2\alpha_2 \sum_{i=1}^{l} y_i\alpha_i \mathbf{K}(x_i, x_2) \tag{3.55}$$

$$\text{s.t.}\begin{cases} \alpha_1 y_1 + \alpha_2 y_2 = -\sum\limits_{i=1}^{l} y_i\alpha_i = \text{const} \\ 0 \le \alpha_i \le C; \quad i = 1, 2, \ldots, l \end{cases} \tag{3.56}$$

Suppose the initial values of the optimization problem are m and N, and the optimal solutions are D and V. It can be seen that in order not to violate the linear constraint Q, the new value of the multiplier must be on a straight line, that is

$$\alpha_1^{\text{new}} y_1 + \alpha_2^{\text{new}} y_2 = \alpha_1^{\text{old}} y_1 + \alpha_2^{\text{old}} y_2 = \text{const} \tag{3.57}$$

This line is in the space of (α_1, α_2) and in the constraints of $0 \le \alpha_1 \le C$ and $0 \le \alpha_2 \le C$. The one-dimensional problem obtained by constraining the objective function to a straight line has an analytical solution.

The algorithm first calculates α_2^{new} and then uses it to calculate α_1^{new}. According to inequality constraints $0 \le \alpha_1 \le C, 0 \le \alpha_2 \le C$ and linear constraint $\sum_{i=1}^{l} \alpha_i y_i = 0$, a more-strict constraint on the feasible value of S is provided

$$U \le \alpha_2^{\text{new}} \le V \tag{3.58}$$

If $y_1 \ne y_2$, then

$$\begin{cases} U = \max\left(0, \alpha_2^{\text{old}} - \alpha_1^{\text{old}}\right) \\ V = \min\left(C, C - \alpha_1^{\text{old}} + \alpha_2^{\text{old}}\right) \end{cases} \tag{3.59}$$

If $y_1 = y_2$, then

$$\begin{cases} U = \max\left(0, \alpha_1^{\text{old}} + \alpha_2^{\text{old}} - C\right) \\ V = \min\left(C, \alpha_1^{\text{old}} + \alpha_2^{\text{old}}\right) \end{cases} \tag{3.60}$$

Use $f(x)$ to represent the current hypothesis determined by the values α and b at a specific stage of learning, that is

$$f(x_i) = \sum_{j=1}^{l} \alpha_j y_j \mathbf{K}\left(x_j, x_i\right) + b$$

And assume

$$E_i = f(x_i) - y_i = \sum_{j=1}^{l} \alpha_j y_j \mathbf{K}(x_j, x_i) + b - y_i; \quad i = 1, 2 \tag{3.61}$$

Here, E_{i} is the difference of functional target classification of the trained point on x_1 or x_2, even if the training point can be correctly classified, this value E_i may still be very large. The other quantity is the second derivative of the objective function on the diagonal, which can be expressed as $\mathbf{k}$; then, there is

$$\mathbf{k} = \mathbf{K}(x_1, x_1) + \mathbf{K}(x_2, x_2) - 2\mathbf{K}(x_1, x_2) = \|\mathbf{\Phi}(x_1) - \mathbf{\Phi}(x_2)\|^2 \tag{3.62}$$

where $\Phi(\cdot)$ represents the mapping from the original space to the feature space.

Theorem *When α_1 and α_2 are allowed to change, the maximum value of the objective function of Formula* (3.55) *can be obtained by calculating the following quantities*

$$\alpha_2^{\text{new,one}} = \alpha_1^{\text{old}} + \frac{y_2(E_1 - E_2)}{\mathbf{k}} \tag{3.63}$$

Clip it to implement constraint $U \le \alpha_2^{\text{new}} \le V$, that is

$$\alpha_2^{\text{new}} = \begin{cases} V; & \alpha_2^{\text{new,one}} > V \\ \alpha_2^{\text{new,one}}; & U \le \alpha_2^{\text{new,one}} \le V \\ U; & \alpha_2^{\text{new,one}} < U \end{cases} \tag{3.64}$$

wherein E_1 and E_2 are given by Formula (3.61), $\mathbf{k}$ is given by Formula (3.62), and U and V are given by Formula (3.59) or (3.60).

Further, α_1^{new} can be obtained from α_2^{new}, that is

$$\alpha_1^{\text{new}} = \alpha_1^{\text{old}} + y_1 y_2 \left(\alpha_2^{\text{old}} - \alpha_2^{\text{new}}\right) \tag{3.65}$$

Therefore, $\left(\alpha_1^{\text{new}}, \alpha_2^{\text{new}}\right)$ is the optimal solution of Formula (3.55).

(2) **Heuristic selection algorithm**

In order to improve the convergence speed, two of them can be selected as the subset optimization objective function according to the contribution of points in the solution process. If the amount of computation required to implement the selection strategy is less than that required by the saved iteration, it indicates that the benefit is obtained in the convergence process.

The selected stop condition can indicate which points are easier to make a greater contribution to convergence.

SMO uses two conditions to select two active points to ensure greater growth of the objective function in the optimization process. There are two separate heuristic algorithms to select the first point and the second point.

1) **The first selected heuristic algorithm**

The first point x_1 is selected from those points that most violate the KKT condition. The outer loop of the algorithm browses the data set to find the point that violates the KKT condition and updates it with any point. When such a point is found, the second point is selected by the heuristic method of the second selection to update the values of their multipliers. Then, the outer loop looks for the point violating the KKT condition again. In order to improve the chance of finding points that violate the KKT condition, the outer loop browses those points whose corresponding parameters meet $0 < \alpha_i < C$, which means that its value is not on the boundary of the feasible area. Only when these points meet the KKT condition at a specific tolerance level, the complete cyclic operation on the whole data set will end and then enter the next data set.

2) **The second selected heuristic algorithm**

The second point x_2 should be selected in such a way that the update on α_1 and α_2 causes greater changes and makes the dual target grow greatly. In order to find a good point without too much calculation, a fast heuristic method is to select the point of maximizing $|E_1 - E_2|$ as x_2. If E_1 is positive, SMO selects a sample x_2 with minimum error E_2. If E_1 is negative, SMO selects a sample x_2 with the maximum error E_2. Memory preserves the error of each point not on the boundary in the data set, which can reduce the amount of calculation. If this choice does not produce large growth on the dual target, the SMO will try each non-boundary point in turn. The loop that traverses each non-boundary point and the entire data set starts at a random position in their respective list, so that no bias is introduced at the beginning of either.

For regression, SMO algorithm can update α_1 and α_2 again from the optimization problem given in Formula (3.55). The programming of four independent problems is related to the symbols of two parameters. The interval of α_2 is given by the following formula

$$\begin{cases} U = \max\left(C_U^2, \alpha_1^{\text{old}} + \alpha_2^{\text{old}} - C_V^1\right) \\ V = \min\left(C_V^2, \alpha_1^{\text{old}} + \alpha_2^{\text{old}} - C_U^1\right) \end{cases} \tag{3.66}$$

Here, the parameters C_U^i and C_V^i of the four problems have different settings. See Table 3.1.

Use $f(x)$ to represent the current hypothesis determined by the values α and b at a specific stage of learning and make

Table 3.1 Parameter setting of C_U^i and C_V^i

Parameter	$\alpha_i \geq 0$	$\alpha_i < 0$
C_U^i	0	$-C$
C_V^i	C	0

$$E_i = f(x_i) - y_i = \sum_{j=1}^{l} \alpha_j \mathbf{K}(x_j, x_i) + b - y_i; \quad i = 1, 2 \tag{3.67}$$

This is the difference between the function output value on training point x_1 or x_2 and the target value.

Theorem *When* α_1 *and* α_2 *are allowed to change in a specific quadrant containing two values, the maximum value of the objective function of the optimization Formula* (3.55) *can be calculated by the following formula*

$$\alpha_2^{\text{new,one}} = \alpha_2^{\text{old}} + \frac{(E_1 - E_2) - \varepsilon(\text{sgn}(\alpha_2) - \text{sgn}(\alpha_1))}{\mathbf{k}} \tag{3.68}$$

Clip it to implement constraint $U \le \alpha_2^{\text{new}} \le V$

$$\alpha_2^{\text{new}} = \begin{cases} V; & \alpha_2^{\text{new,one}} > V \\ \alpha_2^{\text{new,one}}; & U \le \alpha_2^{\text{new,one}} \le V \\ U; & \alpha_2^{\text{new,one}} < U \end{cases} \tag{3.69}$$

Here, the value of the symbolic function sgn(·) is determined by the selected quadrant. E_1 and E_2 are given by Formula (3.61), **k** is given by Formula (3.62), and U and V are given by Formula (3.59). Further, α_1^{new} can be obtained from α_2^{new}

$$\alpha_1^{\text{new}} = \alpha_1^{\text{old}} + \alpha_2^{\text{old}} - \alpha_2^{\text{new}} \tag{3.70}$$

3.3 Fault Diagnosis of Traction Circuit of Metro Locomotive

The power electronic device inside the electric locomotive is the core component of the subway (or high-speed railway) locomotive. Most of the probability of locomotive operation faults are concentrated in power electronic devices. At the same time, relevant operation organizations need to master the fault-free operation cycle of locomotives in real time and must regularly predict and analyze the power electronic devices. Therefore, the real-time detection and fault diagnosis of power electronic circuits and their devices in the traction circuit of Metro locomotives naturally become the key technology necessary to ensure the safe operation of Metro (or high-speed railway) locomotives.

Because power electronic devices usually appear as power supply or actuator in traction system, they play an important and even decisive role in the reliability of the whole system. The power electronic device in Metro (or high-speed railway) locomotive is very special, because its transmission object is high-power DC motor or AC motor. For such a high-power transmission object, the dynamic effect produced when driving a huge mechanical system is very complex, so the possible fault types and properties of power electronic devices are also diverse.

Generally speaking, the fault of power electronic device generally refers to the fault of its main circuit, which can be divided into parametric fault and structural fault. Parametric fault refers to the fault caused by the deviation of circuit parameters (such as inductance and capacitance) from the normal value. It is usually diagnosed by parameter identification. Structural fault refers to the fault that changes the circuit topology due to short circuit, open circuit or loss of trigger signal of power electronic devices. Generally, it is difficult for operators to judge the fault element and/or location in a short time from failure to power failure. Even experienced personnel may be misdiagnosed due to external factors.

According to the current research situation, the fault diagnosis of power electronic devices is limited to individual specific circuits, so it is difficult to meet the needs of real-time diagnosis, and the universality is not strong, so it is more difficult to popularize and apply.

3.3.1 System Composition

This is a fault diagnosis system of Metro locomotive traction circuit based on waveform recognition. The core of the system is composed of embedded industrial computer and its recognition algorithm software. The embedded industrial control computer picks up the voltage input from both ends of the standard load, the terminal voltage of the standard load is input to the data input port of the most cutting-edge data collector in the embedded industrial control computer, the output end of the data collector is connected with the input end of the A/D converter, and the output end of the A/D converter is connected with the write port of the data buffer. The read outlet of the data buffer is connected with the read inlet of the CPU. The fault database module of the circuit under test, the record mark cache and the software package based on the waveform recognition algorithm all realize the two-way interaction of data and instructions with the CPU through the data bus, and the output of the CPU is connected with the display output through VGA.

On the hardware platform composed of the tested Metro locomotive traction circuit slot, analog standard load, data collector, A/D conversion, data cache, CPU, tested circuit fault database, record mark cache, waveform recognition algorithm software package and display output, the system realizes the diagnosis and location of Metro locomotive traction circuit fault through software. The software platform tests the traction circuit of the tested Metro locomotive under all working conditions by establishing the fault database of the traction circuit of the tested Metro locomotive. When running the software package based on waveform recognition algorithm, the system can read the waveform data in the data cache one by one, identify and match the elements in the fault waveform set, accurately identify the fault type and quality

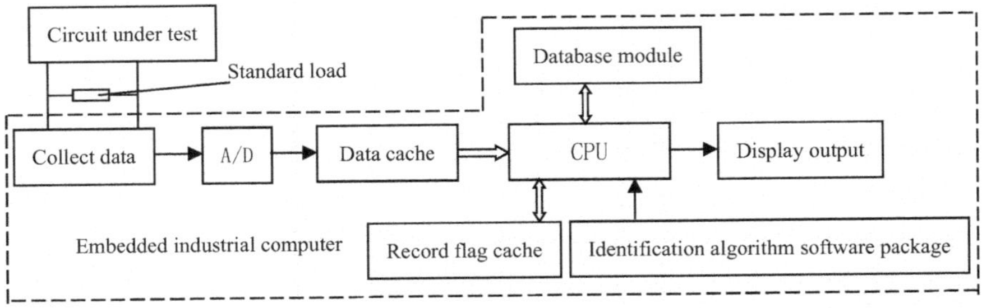

Fig. 3.8 Structure of locomotive traction circuit fault diagnosis system

of the traction circuit of the tested Metro locomotive within the determined similarity range and accurately locate the fault.

Among them, "analog standard load" is used for offline detection.

The fault diagnosis system of traction circuit of Metro locomotive based on waveform recognition includes traction circuit of Metro locomotive under test, standard load, embedded industrial computer and output load (or standard load) of traction circuit of Metro (or high-speed railway) locomotive under test, and embedded industrial computer picks up voltage input from both ends of standard load (see Fig. 3.8).

The embedded industrial computer includes data collector, A/D converter, data buffer, CPU, fault database module of the tested circuit, record mark cache, software package module based on waveform recognition algorithm and display output device. The terminal voltage of the standard load is input to the data input port of the data collector at the forefront of the embedded industrial computer, the output end of the data collector is connected with the input end of the A/D converter, the output end of the A/D converter is connected with the write port of the data buffer, and the read outlet of the data buffer is connected with the read inlet of the central processing unit. The fault database module of the circuit under test, the record mark cache and the software package module based on waveform recognition algorithm all realize the two-way interaction of data and instructions with the CPU through the data bus, and the output of the CPU is connected with the display output through VGA.

During offline detection, the standard load is used as the detection load of the traction circuit of the tested Metro locomotive, the traction circuit of the tested Metro locomotive is powered on according to the standard working environment, and the embedded industrial control computer detects and obtains the output voltage waveform of the traction circuit of the tested Metro locomotive from both ends of the standard load. The standard load is a semi-physical simulation load, which is made according to the similarity principle of mechanical inertia and motor transmission characteristics. The semi-physical simulation standard load can fully simulate the working state of the traction circuit of the tested Metro locomotive under train operation conditions. During online monitoring, the voltage at the output port of the main circuit is directly detected without adding a standard load. In the process of online detection, the "load" of the locomotive traction circuit is the locomotive electric drive system device.

The output voltage waveform is first collected by the data collector and then transferred to the data buffer through the A/D converter. After all the output waveforms collected by the data collector are completed, the CPU has called the software package based on waveform recognition algorithm and read the output waveform data from the data buffer for centralized processing.

Based on the waveform recognition algorithm, the software package module matches the output voltage waveform with the fault waveform in the fault database module of the tested circuit and calculates the one with the largest similarity within the allowable range of the determined similarity; that is, it is determined that the fault of the traction circuit of the tested Metro locomotive at this time belongs to the fault type and nature displayed in the record in the database and accurately points out the location of the fault point (including devices and lines).

The data collector is responsible for signal conditioning and filtering to obtain the measured voltage waveform signal with high signal-to-noise ratio. During the online detection of the operation site, the data collector has a parallel input interface to directly receive the output signal from the tested circuit of the locomotive.

The data buffer is used to separate data acquisition and waveform recognition operation; that is, the data acquisition and A/D conversion including signal conditioning and filtering are used as signal preprocessing, and waveform recognition is used as post-processing program. Therefore, it can greatly improve the speed of data acquisition and ensure the integrity of information collection under train operation conditions and then improve the accuracy of waveform recognition and fault diagnosis.

The data structure of the tested circuit fault database module includes field variables, that is, output waveform, fault type and nature and fault location. According to the circuit characteristics and fault detection parameters, curves and waveforms accumulated by the R & D institutions and manufacturers of the traction circuit of the tested Metro locomotive, the system implements sorting, digitization and eigenvalue extraction in advance and stores all fault characteristic parameters of the traction circuit of the tested Metro locomotive according to the above data structure. At the same time, the tested circuit fault database module has the automatic capacity expansion function of the recording space, which can add and enrich the records of the tested circuit fault database with the help of the expert system through the self-learning function in the process of using the software package.

The record mark cache is a buffer memory used for temporarily storing record marks in the identification process. Generally speaking, the faults of Metro locomotive traction circuit are diverse and complex. Using equal step sampling to identify and match the measured output waveform with all records in the database one by one will consume a lot of operation time, which is unrealistic and inapplicable. Therefore, the method of variable step size needs to be used for operation, that is, "coarse first and then fine". Every time a large step length coarse screening is performed, the records corresponding to the screening will be "marked"; that is, these record serial numbers will be stored in the record mark cache, so that the records marked with "mark" can be further screened when the next step length is reduced, so as to avoid unnecessary repeated calculation.

The waveform recognition algorithm software package module includes fault recognition module, marking module, compensation module and display module. The connection and signal processing relationship between these modules is the fault recognition module specifically recognizes the measured voltage waveform. Once the measured voltage waveform is considered to belong to a recorded fault, it notifies the marking module. Then, the marking module "marks" the record, that is, writes the record number into the record mark cache, and then proceeds to the next round of identification operation (i.e., fine selection and identification, and accurately determine and locate the fault). Otherwise (when there is no similar fault in the existing database), the fault identification module outputs the operation transfer instruction and transfers the operation task to the compensation module, and the compensation module determines the unknown signal characteristics. The fault identification module, marking module, compensation module and display module are connected in parallel. The display module can track the operation process and results of the first three in real time and output the results to the display output device according to the display output protocol [12].

3.3.2 *Identification Principle*

Taking the measured output voltage waveform as the template, when identifying for the first time, quickly roughly select all fault records in the database with large step sampling and "mark" the records with the highest similarity. Secondly, reduce the step length to resample and filter the "marked" records, then "mark" the new records with the highest similarity and update the records stored in the record mark cache. By analogy, until the only circuit with the highest similarity is found, the fault type and nature of the current circuit to be tested, the location of the fault and the device where the fault occurs can be confirmed; if no record has been found in the existing fault database, enter the function recovery program. The so-called compensation program is to identify the symmetry of the measured output voltage waveform itself (including its comparison with adjacent periodic waveforms and rotational symmetry) and then judge whether the measured waveform has no fault or new fault, and then, the new fault is added to the original fault database by using the expert system in the compensation program.

The working process of the software package module based on waveform recognition algorithm is further described as follows:

(1) Taking the measured output voltage waveform as the template, when identifying for the first time, quickly roughly select all fault records in the database with large step sampling and "mark" the records with the highest similarity.
(2) Reduce the step length, resample and filter the "marked" records, then "mark" the new record with the highest similarity and update the record mark cache.

(3) And so on until the only one with the highest similarity is found, so confirm the fault type and nature of the current circuit under test, as well as the location and device of the fault.
(4) Once there is a fault in the tested circuit, the above cycle results can confirm the fault nature and the probability of fault location > 98%. There is only a probability of < 2% that no record can be found in the existing fault database. At this time, enter the function compensation program.
(5) Startup of compensation program: identify the symmetry of the measured output voltage waveform (including its comparison with the adjacent periodic waveform and its rotation symmetry) and make the judgment of no fault or new fault. Once it is determined that there is an unknown type (property) fault, that is to add the new fault to the original fault database.
(6) Display output of conclusion.

3.3.3 Algorithm Steps

The specific steps of fault diagnosis method of traction circuit of Metro locomotive based on waveform recognition are as follows:

1. **Establish fault data set**

Establish the fault data set of traction circuit of tested Metro (or high-speed railway) locomotive

$$\Omega = \{X_i, a_i, s_i\}; \quad i = 1, 2, 3, \ldots, n \tag{3.71}$$

Here, X_i, a_i and s_i are fault waveform (event), fault type and nature (event type and nature) and fault location (event location), respectively.

It is established that the membership degree of any event falling into the set Ω is

$$\mu_{x \in \Omega}(x) \geq 98\% \tag{3.72}$$

where x is any unknown event (measured output waveform) that may belong to fault waveform set X_i.

2. **Data acquisition under all working conditions**

The test platform is used to test the traction circuit of the tested Metro (or high-speed railway) locomotive under all working conditions, and all the waveforms obtained from the test are stored in the data cache.

The test platform establishes a semi-physical simulation system for the traction circuit of the tested Metro locomotive. The test environment of the platform makes the traction circuit of the tested Metro locomotive work in the actual working condition of the locomotive traction system.

The test platform establishes the traction circuit fault data set of the tested Metro locomotive and constructs the traction circuit fault database of the tested Metro locomotive. The database includes fault waveform, fault type and nature and fault location information records.

The test platform tests the traction circuit of the tested Metro locomotive under all working conditions. First, all the waveforms obtained from the test are stored in the data cache, so as to ensure the rapidity of data acquisition and the integrity of information pickup.

3. **Rough identification process**

Run the waveform recognition algorithm software package, adopt the waveform recognition algorithm, read the waveform data in the data cache one by one as the matching template and identify and match the elements of the fault waveform set X_i. The identification and matching adopt the variable step method, that is, "coarse first and then fine": After each large step sampling coarse screening, the records corresponding to the screening will be "marked"; that is, these record serial numbers will be stored in the record flag cache, and then the "marked" records will be further screened by reducing the step size.

4. **Similarity calculation**

The similarity calculation process of each recognition process is as follows.

When $\Delta x = x_j - x_{ij} \leq \varepsilon$ (ε is the determined allowable error range, which can be set and changed through system initialization), it is considered that the output value of the sampling point j of the unknown event is equal to the output value of the sampling point j of the fault i n the fault set. At this time, add 1 to the counter (i.e., $m = m + 1$, and set the counter to zero before the identification operation, i.e., $m = 0$). Otherwise, it is considered that the two are not equal, and the count statement is not executed.

After the sampling in the waveform period, the recognition matching similarity is obtained as follows

$$p = \frac{m}{M} \tag{3.73}$$

where M is the total number of samples in the waveform period.

When $p = \mu(x)_{x\in\Omega} \geq 98\%$, it is considered that the measured event may belong to fault event. In the rough screening stage, the membership degree of possible fault types and properties is generally greater than 1. With the reduction of sampling step size, the unique fault type and nature will be determined and the fault (including devices and/or lines) will be located accurately.

When $p = \mu(x)_{x\in\Omega} < 98\%$, the compensation procedure must be entered: identify the symmetry of the measured output voltage waveform, including the comparison of adjacent periodic waveforms and rotational symmetry, make the judgment of no fault or new fault, define new types and properties of new faults in combination with the expert system and add the new faults to the original fault database.

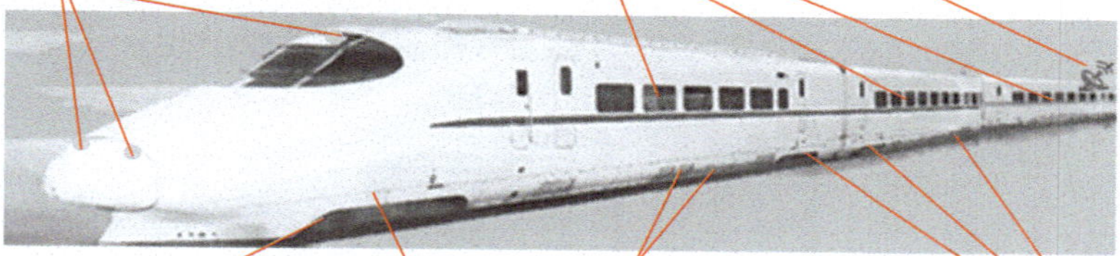

Fig. 3.9 Locomotive safety guarantee system

Finally, the output conclusion is displayed [13].

The fault identification and safety guarantee of the core power unit of the train finally need to be supported by system intelligent technology. The complete support system is shown in Fig. 3.9.

References

1. Li X, Zhang X, Wu H et al (2010) Intelligent identification device and method for abnormal state of locomotive engine and transmission mechanism. Chinese Invention Patent CN201010522633.0
2. Zhang X, Guan H, Ying J et al (2010) Micro ultrasonic sensor. Chinese Invention Patent CN201010109779.2
3. Chen A, Wu H, Du X et al (2006) Electric soft braking method for Metro locomotive. Chinese Invention Patent CN200610030330.0
4. Wang B, Qu D, Peng X (2005) Try out the basics of speech recognition. National Defense Industry Press, Beijing
5. Zhu M, Xion W (2002) Computer voice technology. Beihang University Press, Beijing
6. Mao H, Fan C, Song G (2000) Study on algorithms of low bit rate Hi-Fi audio coding based on wavelet transform and psychoacoustic model. Acta Electron Sin 28(1):26–29 (in Chinese)
7. Cristani M, Bicego M, Murino V (2004) On-line adaptive background modelling for audio surveillance. In: International conference on pattern recognition. IEEE, 399–402
8. Shen JL, Hung JW, Lee LS (1998) Robust entropy-based endpoint detection for speech recognition in noisy environments. In: The 5th international conference on spoken language processing, incorporating the 7th Australian International Speech Science and Technology Conference, Sydney, pp 232–235
9. Wang L, Li C (2010) An improved speech endpoint detection method based on adaptive band-partition spectral entropy. Comput Simul 27(12): 373.375+395 (in Chinese)
10. Zhang H (2019) Research on abnormal acoustic detection algorithm for train operation safety monitoring system in station. Dissertation, Tianjin University of Technology
11. Zhang X, Ying J (2011) Principles of automobiles intelligent technology. Shanghai Jiao Tong University Press, Shanghai

12. Chen A, Zhang X, Du X et al (2006) Fault diagnosis system for traction circuit of metro locomotive based on waveform recognition. Chinese Invention Patent CN200610030327.9
13. Chen A, Wu H, Du X et al (2006) Fault diagnosis method for traction circuit of metro locomotive based on waveform recognition. Chinese Invention Patent CN200610030328.3

Chapter 4
Artificial Intelligence Technology in the Operation and Management of Rail Transit System

In the twenty-first century, as a branch of the development of computer science, artificial intelligence (AI) technology began to rise and develop and has played a great role in human production and life. By studying the essence of human intelligence, people develop theories and technologies to simulate human intelligence and finally produce intelligent machines or systems that simulate human intelligence and respond in the same intelligent way. At present, artificial intelligence has been actively explored in the fields of theorem proving, medical diagnosis, intelligent automobile, voice assistant and so on. At present, AI has also been widely used in the operation and management system of rail transit.

4.1 Development Overview of AI

The development of AI can be understood from its formation reasons and technical aspects.

1. **Brief history of development**

In 1936, the British mathematician Alan Mathison Turing (1912–1954) proposed the Turing model in his paper "ideal computer" and the discussion that machines can think in his paper "can computers think" (Turing experiment) in 1950. Since then, the idea of artificial intelligence began to sprout, which laid the foundation for the birth of artificial intelligence.

The term "artificial intelligence" first appeared at the Dartmouth conference in 1956. Scientists such as John McCarthy, Marvin Minsky, Claude Shannon, Allen Newell and Herbert Simon gathered together and discuss the use of machines to imitate human learning and other intelligent academic problems. The meeting lasted for two months. Although there was no general consensus, it gave a name to the content discussed at the meeting: artificial intelligence. Since then, the new concept of artificial intelligence has become popular in the world, and 1956 has become

Y. Khan, *Principles of Intelligent Rail Transit*, Advances in High-speed Rail Technology,
https://doi.org/10.1007/978-981-19-6072-7_4

the first year of artificial intelligence. At that time, the main research directions and contents of artificial intelligence generally included game, translation, theorem proving and so on.

From 1980 to 1987, with the rapid development of theoretical research and computer software and hardware, the USA and Britain began to invest a lot of research funds in artificial intelligence. Subsequently, many technicians who studied artificial intelligence developed various AI practical systems tried to commercialize and put them into the market, so artificial intelligence began to stir up a wave.

2. **Technical layer of artificial intelligence**

The system of artificial intelligence is very huge. It involves not only natural science and engineering technology but also many disciplines of social science, including mathematics, computer science, software technology, topology, graph theory, graphics, logic, cognition, behavior, psychology, physiology, linguistics and so on.

The technical basis of artificial intelligence is mainly divided into three parts: Computer vision (visual perception), natural language processing and speech recognition.

In order for a machine to understand human behavior, it must first understand (including seeing and listening) external information. In this way, artificial intelligence can accurately execute people's instructions.

(1) **Computer vision (visual perception)**

The function of computer vision (visual perception) is to extract symbol and numerical information from image or video, analyze and calculate the information and carry out target recognition, detection and tracking at the same time. The images processed by computer vision (visual perception) are generally divided into static images and dynamic images. It is easy to recognize static images, just upload the collected images to the computer database for fuzzy comparison; it is troublesome to recognize dynamic images. All the information in the shooting scene needs to be sorted and classified and then processed and analyzed by intelligent devices.

In recent years, computer vision (visual sense) has also developed many industrial projects with the help of the concept and idea of artificial intelligence, such as face recognition, unlocking and payment function of mobile phone, application software (APP) for identifying animals and plants, electronic monitoring system, automatic control and processing of workshop production parts. As a technology to simulate and expand human visual ability, computer vision (visual perception) is the top priority for the development of artificial intelligence. Therefore, the research on computer vision technology will encounter more difficulties and challenges in the future.

(2) **Natural language processing**

Natural language processing is a technology that studies the effective communication between human and computer through natural language (also known as man–machine dialogue). With the development of artificial intelligence, the problems that computers have to deal with are becoming more and more complex, and the traditional programming language is obviously not practical. Therefore, in order to effectively solve this problem, computers need to learn human natural language by themselves. Through the "training algorithm" of computer simulating people's daily language communication habits, the computer can understand and use the languages commonly used by human beings, such as Chinese, English, Russian, French, Spanish, Arabic and so on. When people talk with the computer, the computer can quickly process people's requests, such as real-time translation, literature search and so on.

Natural language processing is a bridge (means, tools) between human and computer, but it is also a very complex step. Because natural language is not as rigorous as machine programming language and prefers oral form, and in real life, different people have different ways and habits of speaking, and even have local accents. When the computer receives it, it may not understand or even misinterpret its meaning, and execute it into wrong results, which brings unnecessary trouble to people. Therefore, in order to realize the barrier free communication between human and artificial intelligence in the future, the technology of natural language processing needs to be constantly updated and improved.

Of course, the natural language processing mentioned here mainly refers to a technology that people process natural language into text expression through intelligent algorithms.

(3) **Speech recognition**

Speech recognition is actually a process of converting speech signals into words or executing commands. The main method of speech recognition is pattern matching method. Firstly, the user's vocabulary is stored in the computer database, and then the similarity is matched with each template in the database. The one with the highest similarity is screened and output as the recognition result. The earliest speech recognition technology originated from Bell Labs. At present, people can see that speech recognition technology has been applied to all kinds of life service terminals and communication. Research on speech recognition technology is also the mainstream trend now. We should vigorously promote the application of artificial intelligence such as intelligent speech recognition, replace a large number of repetitive and fidgety artificial services and work contents, improve our work level and efficiency and develop toward a more advanced era.

3. **Application fields of artificial intelligence**

Artificial intelligence has a wide range of applications and can be put into almost all walks of life. Among them, the most striking is driverless technology and so on. The topic of driverless technology has been frequently discussed since 2016. Driverless technology also known as wheeled mobile robot. Its working principle is to automatically adjust the speed and direction of vehicles (including cars, trains and all kinds of vehicles) by sensing the current road conditions, weather and surrounding vehicles through intelligent control system and on-board sensors, so as to realize unmanned driving. The emergence of driverless technology can replace manual driving, reduce traffic accidents, reduce air pollution and so on.

The following describes in detail the application of computer vision (visual perception) technology in rail transit system.

4.2 Application of Face Recognition Technology in Traffic Management

The most typical example of AI application in rail transit is face recognition technology.

With the continuous progress of high-definition video technology, in order to improve the security level of social public environment, face recognition technology has been widely used in many aspects of social services and life (see Fig. 4.1).

At present, the method of face intelligent recognition has been widely used in the field of rail transit at home and abroad to realize the automatic identification of passenger identity. Therefore, it can provide convenient conditions for subway operation and maintenance, security and emergency.

Face recognition is a biometric recognition technology, which collects biometrics through non-contact distance, which is one of the important components of current AI technology.

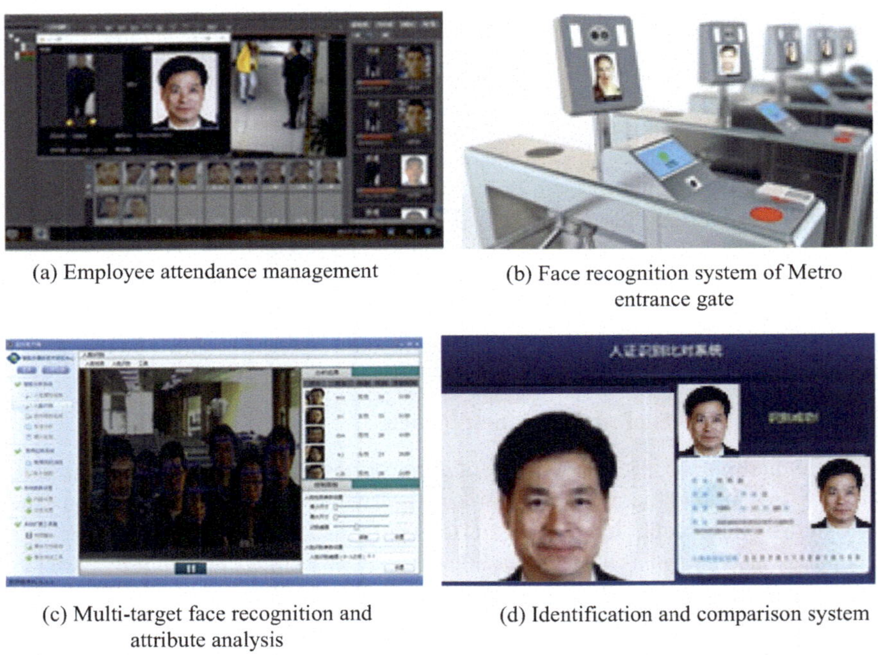

(a) Employee attendance management

(b) Face recognition system of Metro entrance gate

(c) Multi-target face recognition and attribute analysis

(d) Identification and comparison system

Fig. 4.1 Application example of face recognition technology

Face recognition technology is basically realized through three technical processes: face detection, feature extraction and face comparison [1].

4.2.1 Theoretical Basis of Face Recognition

The so-called face detection is to collect images of specific face parts in the environment, and then carry out necessary image preprocessing (Fig. 4.2).

1. **Face detection and preprocessing**

It is necessary to locate the specific location of the face in the acquisition of face images. Face detection usually uses the minimum circumscribed rectangle of the face to represent the size and position of the face (see Fig. 4.2) [2].

The basic image processing methods include image enhancement and filtering. The purpose of image enhancement is to improve the visual effect of the image and facilitate the observation, analysis and processing of the image by man or machine.

From the perspective of enhancement scope, image enhancement can be divided into two methods: spatial domain and frequency domain. The former directly processes each pixel of the image; the latter is to process the frequency spectrum of the image after Fourier transform, and then obtain the required image through inverse Fourier transform.

As for when to use what kind of enhancement method, it must be determined according to the environmental conditions of the scene object and the technical objectives of image processing and analysis.

Filtering can be divided into low-pass and high-pass. Low-pass can smooth the image and high-pass can sharpen the image.

(1) **Spatial domain enhancement**

Spatial domain enhancement algorithms, including gray-scale transformation, histogram correction, local statistics, image smoothing and image sharpening.

Fig. 4.2 Face detection

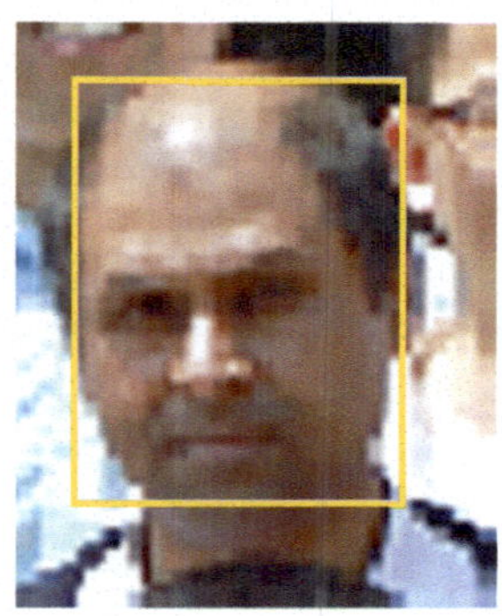

1) **Gray transformation**

Gray-scale transformation is used to adjust the gray dynamic range or contrast of the image. Specifically, the purpose of enhancing the image is achieved by modifying the pixel value. The modification is carried out by each pixel separately, so it is also called point processing operation.

Let the original pixel gray value of pixel coordinate (i, j) be $r(i, j)$, $r \in \left[a\ b\right]$, where a and b are the gray level of the original pixel.

Through the transformation function $T(\cdot)$, the pixel transformation result $S(i, j) = T(r(i, j))$ can be realized. $S(i, j) \in \left[a'\ b'\right]$. Here, a' and b' are the gray level of the transformed pixels. For example, $T(\cdot)$ adopts linear transformation, at this time

$$S(i, j) = a' + \frac{b' - a'}{b - a}(r(i, j) - a) \tag{4.1}$$

In the case of underexposure or overexposure, the gray scale of the image may be limited to a very small range. At this time, the image seen by people may be a blurred image without gray level. Linear transformation is used to stretch the gray level of pixels in the image, which can effectively improve the visual effect of the image.

The above is the linear transformation of the whole image. In addition, there are piecewise linear transformation and nonlinear transformation, Gamma (γ) correction transformation and curve transformation. Its essence is to adopt different expressions for the transformation function $T(\cdot)$.

2) **Histogram correction**

Histogram correction method is a method based on probability theory to enhance the image by modifying the histogram. Gray histogram reflects the relationship between each gray level and its occurrence frequency in digital image. It can describe the general picture of the image.

Histogram correction method includes histogram equalization and histogram specification.

First, Histogram Equalization

This is a method to obtain a new image with uniformly distributed gray histogram through some transformation of the original image.

Let $f(i, j)$ and $g(i, j)$, respectively, represent the original image gray after normalization of pixel coordinate (i, j) and the image gray after histogram correction. $f(i, j) \in [0, 255]$ and $g(i, j) \in [0, 255]$ represent the gray level of pixel coordinate from 0 to 255. For any $f(i, j)$ value, a $g(i, j)$ value can be generated, that is

$$g(i, j) = G(f(i, j)) \tag{4.2}$$

The transformation function $G(\cdot)$ in the formula (4.2) meets the following conditions: $G(\cdot)$ is a monotonic increasing function in $0 \le f(i, j) \le 255$, and the order of gray level from black to white is unchanged, and there is $0 \le g(i, j) \le 255$, that is, to ensure that the gray level of pixels after mapping and transformation is within the allowable range.

According to the probability theory, when the probability densities of $f(i, j)$ and $g(i, j)$ are $P_f(f)$ and $P_g(g)$, respectively, the distribution function $F_g(g)$ of random variable $g(i, j)$ has the following relationship

$$F_g(g) = \int_{-\infty}^{g} P_g(g)\mathrm{d}g = \int_{-\infty}^{f} P_f(f)\mathrm{d}f \tag{4.3}$$

Using the relationship that the density function is the derivative of the distribution function, the derivative of g on both sides of the equation is obtained

$$P_g(g) = \frac{\mathrm{d}}{\mathrm{d}g}\left[\int_{-\infty}^{f} P_f(f)\mathrm{d}f\right] = P_f(f)\frac{\mathrm{d}f}{\mathrm{d}g} = P_f(f)\frac{\mathrm{d}}{\mathrm{d}g}\left[G^{-1}(g)\right] \tag{4.4}$$

where $G^{-1}(g)$ represents the inverse operation of Eq. (4.2).

It can be seen that the probability density function of the output image can be obtained by controlling the probability density function of the gray level of the original image through the transformation function $G(\cdot)$, so as to improve the gray level of the original image, which is the mathematical basis of histogram correction technology.

Let the discrete functions of $f(i, j)$ and $g(i, j)$ be r_k and S_k, respectively, then the discrete expression of formula (4.2) is

$$S_k = T(r_k) = \left\lfloor 255 \sum_{l=1}^{k} \frac{n_l}{n} \right\rfloor \tag{4.5}$$

where $T(\cdot)$ is the transformation function, r_k represents the gray level of the original image, S_k represents the gray level of r_k after histogram equalization, $k = 1, 2, \ldots L$, L is the current statistical gray level, $L \in [0, 255]$, n is the total number of pixels in the image, and n_l is the number of occurrences of the lth gray level; $\lfloor \cdot \rfloor$ is rounded down, that is, take the largest integer less than the operation result.

In the actual operation, the mapping value after equalization corresponding to each gray level pixel point in the image to be equalized can be calculated according to formula (4.5), and the mapping value can replace the pixel value before the gray level pixel point equalization, so a brand-new gray level equalization image can be obtained, and then the image can be significantly enhanced.

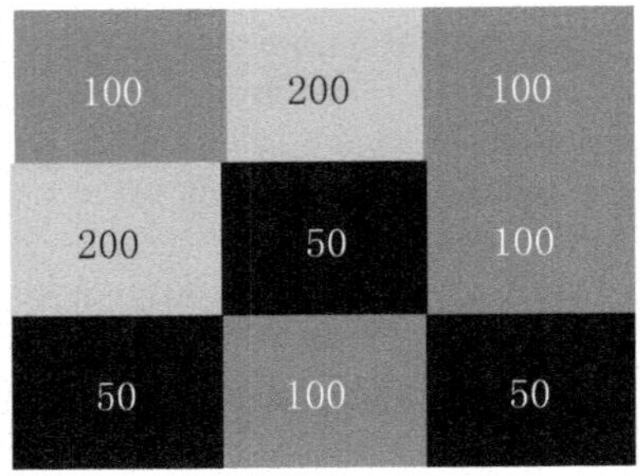

Fig. 4.3 Gray level before equalization

Example Take a simple picture with only 3◊3 pixels as an example to illustrate the algorithm of gray histogram equalization.

The gray level of 9 pixels in Fig. 4.3 is the value before equalization.

As can be seen from the figure, the occurrence times of pixel gray levels 50, 100 and 200 are 3, 4 and 2, respectively. The gray mapping value obtained from formula (4.5) is

$$\begin{cases} S_{50} = T(r_{50}) = 85 \\ S_{100} = T(r_{100}) = 198 \\ S_{200} = T(r_{200}) = 255 \end{cases} \tag{4.6}$$

Use the gray mapping value to modify the corresponding original gray value, that is, the results shown in Fig. 4.4 are obtained. The pixel gray level value transformed by this mapping will be able to generate a new image after gray histogram equalization.

Of course, in the process of black-and-white image enhancement, sometimes gray level interpolation is also included. The simplest interpolation method is nearest

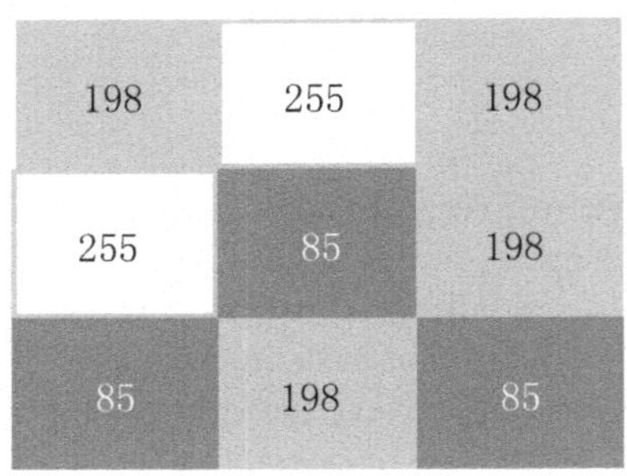

Fig. 4.4 Gray level after equalization

neighbor interpolation, that is, select the gray value of the input pixel closest to the mapped position as the interpolation result.

Second, Histogram Specification

In some cases, an image with a uniform histogram is not necessarily required, and sometimes an image with a specific histogram is required in order to enhance some gray levels in the image.

Histogram specification method is proposed for the above idea.

Histogram specification is an enhancement method to modify the original image by changing the gray histogram of the original image into a histogram with a specified shape. It can be seen that it is an effective extension of histogram equalization. It can be said that histogram equalization is a special case of histogram specification.

The basic idea of image histogram normalization algorithm is that the probability density of gray distribution of the original image and the image after histogram normalization correction is equal. Therefore, the probability density of gray distribution obtained by histogram equalization of the original image can be used as the probability density of gray distribution of the image; after histogram normalization correction, the histogram specified image is obtained by the inverse operation of the histogram equalization transformation function.

Suppose that $P_f(f)$ and $P_{\hat{z}}(\hat{z})$, respectively, represent the normalized probability density function of the original image gray distribution and the desired image gray distribution probability density function, and the possible histogram equalization transformation function of the desired image is

$$\hat{z} = Z(P_z(z)) = \int_{-\infty}^{z} P_z(z)\mathrm{d}z \tag{4.7}$$

where $P_z(z)$ is the probability density of gray distribution of histogram specification correction graph $z(i, j)$ to be achieved.

Firstly, the probability density $P_g(g)$ of image gray distribution obtained by histogram equalization of the original image according to formula (4.2);

Secondly, let $P_{\hat{z}}(\hat{z}) = P_g(g)$ to obtain the histogram equalization transformation function $Z(P_z(z))$;

Finally, replace $\hat{z}$ with g and then substitute it into the inverse transformation equation of formula (4.7), as follows

$$z = Z^{-1}(g) \tag{4.8}$$

Therefore, the gray image $z(i, j)$ of the histogram specification correction image is obtained.

It must be pointed out that the main difficulty of image enhancement using histogram specification method is to form a meaningful histogram, because after the image is specified by histogram, its enhancement effect should be conducive to human visual interpretation or machine recognition, which requires a certain degree of technical requirements.

3) **Local statistical algorithm**

One of the simplest ways to implement local statistical algorithms is to use mean and variance operations.

Assuming that the histogram with gray mean m and standard deviation σ of the original image is transformed into the histogram with mean $\hat{m}$ and standard deviation $\hat{\sigma}$, the gray transformation formula is as follows

$$f_{\text{out}} = \frac{\hat{\sigma}}{\sigma}(f_{in} - m) + \hat{m} \tag{4.9}$$

where f_{in} and f_{out} are the pixel gray values of the original image and the transformed image, respectively.

4) **Image smoothing**

For any original image, in any process of acquisition and transmission, it will be disturbed by various noises, which will worsen the image, degrade the quality, blur the image and annihilate the features, which will be detrimental to image analysis. In order to suppress noise and improve image quality, the processing is called image smoothing or denoising technology. This technical method can be carried out in spatial domain and frequency domain, respectively.

① **Local smoothing method**

Local smoothing method is a technology of smoothing directly in spatial domain.

It is assumed that the image is composed of many small blocks with constant gray level, there is a high spatial correlation between adjacent pixels and the noise is statistically independent. Therefore, the average gray value of each pixel in the neighborhood can be used to replace the original gray value of the pixel to realize image smoothing.

If $f(i, j)$ is a $M \times N$ neighborhood image and the smoothed image is $g(i, j)$, there is

$$g(i, j) = \frac{1}{M \times N} \sum_{m=1}^{M} \sum_{n=1}^{N} f(i + m, j + n) \tag{4.10}$$

where M and N are the size of smooth neighborhood centered on $f(i, j)$, respectively.

It must be pointed out that the advantage of this algorithm is simple operation, but its main disadvantage is that it will blur the image while reducing noise, especially at the edges and details. While the denoising ability is enhanced, the larger the neighborhood is, the more serious the ambiguity is.

In order to overcome the disadvantages of simple local average method, many local smoothing algorithms that preserve edges and details can be used. Their starting points are focused on how to choose: the size, shape and direction of the neighborhood, the number of points participating in the average, the weight coefficient of each point in the neighborhood and so on. For example, transfinite pixel smoothing method, median filter and gray nearest neighbor smoothing method are all improvements to the above algorithm.

② **Transfinite pixel smoothing method**

Compare the absolute value of the difference between $f(i, j)$ and the neighborhood average $g(i, j)$ with the selected threshold τ, and determine the final gray $\widehat{g}(i, j)$ of point (i, j) according to the comparison result, that is

$$\widehat{g}(i, j) = \begin{cases} g(i, j), (|f(i, j) - g(i, j)| > \tau) \\ f(i, j), (|f(i, j) - g(i, j)| \le \tau) \end{cases} \tag{4.11}$$

Compared with the local smoothing method, this algorithm is more effective in suppressing salt-and-pepper (also known as impulse noise) noise and protecting the details and textures with only small gray difference. However, with the increase of neighborhood, the denoising ability is enhanced, and the degree of blur will also increase. In order to suppress the salt-and-pepper noise, reduce the degree of blur and preserve the details, the median filter method can be used to enhance the image.

It can be said that the median filtering method is actually a variation of the transfinite pixel smoothing method, and its algorithm steps are as follows:

The first step is to coincide the pixel position with the template;
The second step is to read the gray value of each corresponding pixel under the template;
The third step is to arrange these gray values into a column from small to large;
Step 4: find the middle one of these values;
Step 5: assign this intermediate value to the pixel at the center of the template.

The noise elimination effect of median filter is related to the size of the template and the number of pixels involved in the operation. The too bright or dark areas in the image whose size are less than half of the template size will be eliminated after filtering.

③ **Gray near neighbor smoothing method**

The starting point of the algorithm is that in the $M \times N$ window, the gray values of pixels belonging to the same aggregate will be highly correlated. Therefore, the average gray level of K adjacent pixels closest to the gray level of the central pixel in the window can be used to replace the gray level value of the central pixel of the window. This is the gray level near neighbor average method. A smaller K value makes the noise variance decrease less, but it has a better effect of maintaining details; a larger K value makes the smoothing effect better, but it will blur the edge of the image. Experiments show that for 3×3 window, $K = 6$ is appropriate.

5) **Image sharpening**

In the process of image recognition, it is often necessary to highlight the edge and contour information, especially the blurred image caused by lens motion. Image sharpening is to enhance the edge or contour of the image.

Image smoothing blurs the image edge through the integral process, and image sharpening makes the image edge prominent and clear through the differential process. There are two commonly used image sharpening methods: gradient sharpening method and Laplacian enhancement operator algorithm.

① **Gradient sharpening method**

Gradient sharpening is the most commonly used method in image sharpening.

For image $f(i, j)$, the gray gradient at point (i, j) is defined as

$$\mathrm{grad}(i, j) = \begin{bmatrix} f_i' \\ f_j' \end{bmatrix} = \begin{bmatrix} \frac{\partial f(i,j)}{\partial i} \\ \frac{\partial f(i,j)}{\partial j} \end{bmatrix} \quad (4.12)$$

For discrete images, the concept of gradient size is often used, so the gradient size is often directly called "gradient", and the first-order partial derivative is expressed by the first-order difference approximation, that is

$$f_i' = f(i + 1, j) - f(i, j) \quad (4.13)$$

$$f_j' = f(i, j + 1) - f(i, j) \quad (4.14)$$

In order to simplify the operation, gradient operator algorithm is often used, that is

$$\mathrm{grad}(i, j) = \max\left(\left|f_i'\right|, \left|f_j'\right|\right) \quad (4.15)$$

Or

$$\mathrm{grad}(i, j) = \left|f_i'\right| + \left|f_j'\right| \quad (4.16)$$

② **Laplacian enhancement operator algorithm**

Laplacian operator is a linear second-order differential operator, that is

$$\nabla^2 f(i, j) = \frac{\partial^2 f(i, j)}{\partial i^2} + \frac{\partial^2 f(i, j)}{\partial j^2} \quad (4.17)$$

For discrete digital images, the second-order partial derivative can be approximated by the second-order difference. At this time, the expression of Laplacian operator is

$$\nabla^2 f(i,j) = f(i+1,j) + f(i-1,j) + f(i,j+1) + f(i,j-1) - 4f(i,j) \tag{4.18}$$

Laplacian enhancement operator is

$$g(i,j) = f(i,j) - \nabla^2 f(i,j) = 5f(i,j) - f(i+1,j) - f(i-1,j) - f(i,j+1) - f(i,j-1) \tag{4.19}$$

where $g(i, j)$ is the image gray level sharpened by Laplacian enhancement operator.

(2) **Frequency-domain enhancement algorithm**

Image smoothing can be carried out not only in the spatial domain but also in the frequency domain. Frequency-domain enhancement algorithms include low-pass filtering, high-pass filtering and homomorphic filtering enhancement.

Assuming that the original image is $f(i, j)$ and transformed into $F(u, v)$ by Fourier transform, frequency-domain enhancement is to select an appropriate filter $H(u, v)$ to process the spectral components of $F(u, v)$ and obtain the processed Fourier function $G(u, v)$, that is

$$G(u,v) = F(u,v) \cdot H(u,v) \tag{4.20}$$

Then, $G(u, v)$ is transformed by inverse Fourier transform to obtain the enhanced image $g(i, j)$.

1) **Low-pass filter enhancement algorithm: frequency-domain smoothing method**

Since the noise is mainly concentrated in the high-frequency part, in order to remove the noise and improve the image quality, the low-pass filter $H(u, v)$ is usually used to suppress the high-frequency component, pass through the low-frequency component, and then carry out the inverse Fourier transform to obtain the filtered image, so as to achieve the purpose of smoothing the image. There are four kinds of common low-pass filters $H(u, v)$ in frequency domain: ideal low-pass filter, Butterworth low-pass filter, exponential low-pass filter and trapezoidal low-pass filter.

① **Ideal low-pass filter**

Let the cut-off frequency of the ideal low-pass filter leaving the origin on the Fourier plane be D_0, then the transfer function of the ideal low-pass filter is

$$H(u,v) = \begin{cases} 1, & D(u,v) \le D_0 \\ 0, & D(u,v) > D_0 \end{cases} \tag{4.21}$$

where D_0 is a non-negative integer; $D(u, v)$ is the distance from point (u, v) to the origin of frequency-domain coordinates (the same below), that is

$$D(u,v) = \sqrt{u^2 + v^2} \tag{4.22}$$

Because the high-frequency component contains a large amount of edge information, the use of low-pass filter will lead to the loss of edge information and blur the image edge while removing noise. Therefore, it is necessary to find other more suitable filtering methods.

② **Butterworth low-pass filter**

The transfer function of n-order Butterworth filter is

$$H(u,v) = \frac{1}{1 + \left(\sqrt{2} - 1\right)\left[\frac{D(u,v)}{D_0}\right]^{2n}} \tag{4.23}$$

③ **Exponential low-pass filter**

Exponential low-pass filter is another smoothing filter commonly used in image processing, and its transfer function is

$$H(u,v) = \exp\left\{\left[\ln\left(\frac{1}{\sqrt{2}}\right)\right]\left[\frac{D(u,v)}{D_0}\right]^n\right\} \tag{4.24}$$

While using exponential low-pass filter to suppress noise, the blur degree of image edge is greater than that produced by Butterworth filter. However, exponential low-pass filter and Butterworth filter have no obvious ringing effect.

The common Gaussian filter is a kind of exponential low-pass filter. The basic idea of Gaussian filter is to convolute the Gaussian kernel function with the original signal to obtain the filtered output signal.

Let one-dimensional Gaussian function be

$$g(\tau,\sigma) = \frac{1}{\sqrt{2\pi}\sigma}\exp\left(-\frac{\tau^2}{2\sigma^2}\right) \tag{4.25}$$

Its first-order derivative is

$$g^{(1)}(\tau,\sigma) = \frac{-\tau}{\sqrt{2\pi}\sigma^3}\exp\left(-\frac{\tau^2}{2\sigma^2}\right) \tag{4.26}$$

Among them, $g^{(1)}(\tau,\sigma)$ is called Gaussian filter. The result $S(\tau,\sigma)$ of function $f(\tau)$ filtered by $g^{(1)}(\tau,\sigma)$ is

$$S(\tau,\sigma) = f(\tau) * g^{(1)}(\tau,\sigma) \tag{4.27}$$

where * is the symbol of convolution operation; σ is the standard variance of Gaussian function. The smoothing effect of Gaussian filter can be controlled by σ, that is, the smoothing degree of signal can be adjusted by changing the value of Gaussian standard variance σ, the greater the σ value, the better the smoothing effect.

Gaussian convolution is a linear image smoothing filtering process. Its convolution template is a linear template. The template weight can be calculated directly from the two-dimensional zero mean discrete Gaussian function. The expression of the two-dimensional zero mean discrete Gaussian function is

$$H(u, v) = \frac{1}{2\pi\sigma^2} \exp\left(-\frac{u^2 + v^2}{2\sigma^2}\right) \tag{4.28}$$

④ **Trapezoidal low-pass filter**

Trapezoidal low-pass filter is a compromise between ideal low-pass filter and complete smoothing filter, and its transfer function is

$$H(u, v) = \begin{cases} 1 & D(u, v) < D_0 \\ \frac{D(u,v)-D_1}{D_0-D_1}; & D_0 \le D(u, v) \le D_1 \\ 0 & D(u, v) > D_1 \end{cases} \tag{4.29}$$

Its performance is between ideal low-pass filter and exponential filter. The filtered image has certain blur and ringing effect.

It must be pointed out that what kind of low-pass filter is selected to enhance the image in the frequency domain needs to be determined according to the environmental conditions of the generated image.

2) **High-pass filter enhancement algorithm: frequency-domain sharpening method**

The edges and details of the image are mainly located in the high-frequency part, and the blur of the image is caused by the weak high-frequency component. Frequency-domain sharpening is to eliminate blur and highlight edges. High-pass filter can make high-frequency components pass through and low-frequency components weaken. After high-pass filter, the image with sharp edge can be obtained after inverse Fourier transform.

Common high-pass filters include ideal high-pass filter, Butterworth high-pass filter, exponential high-pass filter and trapezoidal high-pass filter.

① **Ideal high-pass filter**

二维理想高通滤波器的传递函数为.

The transfer function of two-dimensional ideal high-pass filter is

$$H(u, v) = \begin{cases} 0, D(u, v) \le D_0 \\ 1, D(u, v) > D_0 \end{cases} \tag{4.30}$$

② **Butterworth high-pass filter**

The transfer function of n-order Butterworth high-pass filter is defined as follows

$$H(u, v) = \frac{1}{1 + \left(\sqrt{2} - 1\right)\left[\frac{D_0}{D(u,v)}\right]^{2n}} \tag{4.31}$$

③ **Exponential high-pass filter**

The transfer function of the exponential high-pass filter is

$$H(u, v) = \exp\left\{\left[\ln\left(\frac{1}{\sqrt{2}}\right)\right]\left[\frac{D_0}{D(u, v)}\right]^n\right\} \tag{4.32}$$

④ **Trapezoidal high-pass filter**

The transfer function of trapezoidal high-pass filter is

$$H(u, v) = \begin{cases} 0 & D(u, v) < D_1 \\ \frac{D(u,v) - D_1}{D_0 - D_1}; & D_1 \leq D(u, v) \leq D_0 \\ 1 & D(u, v) > D_0 \end{cases} \tag{4.33}$$

The selection principle of the four high-pass filter functions is similar to that of the low-pass filter. Comparing the four high-pass filters, the ideal high-pass has obvious ringing effect; Butterworth high-pass filter has good effect, but the calculation is complex. Its advantage is that it allows a small amount of low-frequency to pass through, which is gradual, and the ringing effect is not obvious; the effect of exponential high-pass is worse than Butterworth, and the ringing effect is not obvious; trapezoidal high-pass will produce micro ringing effect, but the calculation is simple and commonly used. Generally speaking, whether in the image space domain or frequency domain, the use of high-frequency filtering will not only enhance the useful information but also enhance the noise. Therefore, it cannot be used arbitrarily.

(3) **Image edge detection**

The so-called image edge refers to those pixel sets in which the gray level of the surrounding pixels expressing the object changes step by step in the image. It exists between the target and background, target and target, region and region, primitive and primitive. The edge of an object is characterized by local discontinuity in the image. In essence, the edge of an object means the end of one region and the beginning of another. Image edge is a very important feature information in image analysis and visual perception detection.

Image edge has two characteristics: direction and amplitude. Generally, the pixels along the edge change gently; pixels that run perpendicular to the edge change dramatically. Therefore, these two characteristics of image edge can be used to judge the distribution and direction of edge pixels. The description of edge features generally includes:

① Edge point, that is, the gray value of pixels on both sides adjacent to the point is significantly different from it;
② The normal direction of the edge, that is, the direction where the gray level changes most violently;
③ Edge direction, that is, the direction perpendicular to the edge normal, that is, the tangent direction of the target boundary;
④ Edge position, that is, the coordinate position of the edge;
⑤ Edge intensity, that is, the change intensity of local gray level along the normal direction of the edge.

At present, the academia divides the object edge into two categories: first, it is called step edge, and the gray values of pixels on both sides are significantly different; second, the edge of the roof, that is, the measured pixels are located at the turning point of the gray value from increasing to decreasing. The edge features of different classes correspond to different physical states of the scene.

Image edge detection is the most basic processing method in all edge based image segmentation algorithms. It is one of the basic steps of image processing, image analysis, pattern recognition and machine vision detection. In general, people can regard the singular points and abrupt points in the signal as the edge points in the image, and the change of gray level near them can be reflected by the gradient of gray level distribution of adjacent pixels. Therefore, according to such image characteristics, many edge detection algorithms have been proposed, mainly based on differential method and combined with template method to extract image edge. Common edge detection algorithms include Roberts operator, Sobel operator, Prewitt operator and Laplace operator. Most of these methods take the neighborhood centered on the pixel to be processed as the basis of gray analysis in order to extract the image edge.

So far, there are thousands of image edge detection algorithms. As for the more classical algorithms, they mainly include threshold edge detection based on gray histogram, edge detection based on gradient, Laplacian operator and Canny edge operator.

1) **Threshold edge detection**

The so-called thresholding is to divide the gray level into two or more equal interval or unequal interval gray level intervals according to the subjective desire, and use the threshold to determine its region to obtain the boundary of the region.

The thresholding edge detection method based on gray histogram is to set the binary state of gray, use the threshold T of histogram to divide the histogram into two parts A and B belonging to two gray levels, respectively, according to the specified $H1$ and $H2$ gray levels and determine the gray mutation position of object image (A) and background (B). The thresholding method can be completed by two scans and one synthesis. The specific steps are as follows:

Step 1: Scan each row of the image $f(i, j)$ $(i = 0, 1, 2, \ldots, n - 1)$, compare each pixel point $f(i, j)$ in the scanned row with the pixel point $f(i, j - 1)$ of the adjacent column and generate the pixel histogram $g_1(i, j)$ of the middle image row according to the following formula (4.34).

$$g_1(i,j) = \begin{cases} H1, f(i,j) - f(i,j-1) \geq T \\ H2, f(i,j) - f(i,j-1) < T \end{cases} \tag{4.34}$$

Step 2: Scan each column of the image $f(i,j), (j = 0, 1, 2, \ldots, m-1)$, compare each pixel point $f(i,j)$ in the scanned column with the pixel point $f(i-1,j)$ in the adjacent row and generate the pixel histogram $g_2(i,j)$ of the middle image column according to the following formula:

$$g_2(i,j) = \begin{cases} H1, f(i,j) - f(i-1,j) \geq T \\ H2, f(i,j) - f(i-1,j) < T \end{cases} \tag{4.35}$$

Step 3: After combining $g_1(i,j)$ and $g_2(i,j)$, the boundary image $\hat{f}(i,j)$ of the object is obtained as follows

$$\hat{f}(i,j) = g_1(i,j) \cap g_2(i,j) = \begin{cases} A; g_1(i,j) = H1 \wedge g_2(i,j) = H1 \\ B; g_1(i,j) = H2 \vee g_2(i,j) = H2 \end{cases} \tag{4.36}$$

where $A \in [0, 255]$, $B \in [0, 255]$; when the image in A area takes the highest gray level 255 and the image in B area takes the lowest gray level 0, then

$$\hat{f}(i,j) = \begin{cases} 255; g_1(i,j) = H1 \wedge g_2(i,j) = H1 \\ 0; g_1(i,j) = H2 \vee g_2(i,j) = H2 \end{cases} \tag{4.37}$$

When the image in A area takes the lowest gray level 0 and the image in B area takes the highest gray level 255, then

$$\hat{f}(i,j) = \begin{cases} 0; g_1(i,j) = H1 \wedge g_2(i,j) = H1 \\ 255; g_1(i,j) = H2 \vee g_2(i,j) = H2 \end{cases} \tag{4.38}$$

2) **Edge detection based on gradient**

The so-called edge detection based on gradient algorithm is to distinguish edge points and non-edge points by using the gradient value of pixel points based on the characteristics of gray mutation of image edge. Among them, the selection of gradient threshold is very important. It is the key to ensure whether the real edge points can be extracted.

Edge detection based on gradient algorithm is more suitable for images with sharp edge gray value transition and low noise.

The most commonly used classical operators in edge detection based on gradient algorithm include Roberts operator, Sobel operator, Prewitt operator, Krisch operator and so on.

Fig. 4.5 Roberts operator template

1	0
0	−1

0	−1
1	0

① **Roberts operator**

Roberts is an operator that uses the local difference operator to find the edge, which uses the difference between the coordinates of two adjacent pixels in the diagonal direction.

For digital images $f(i, j)$, its Roberts gradient has two formulas

$$G_R(i, j) = |f(i, j) - f(i+1, j+1)| + |f(i+1, j) - f(i, j+1)| \tag{4.39}$$

Or

$$G_R(i, j) = \left\{[f(i, j) - f(i+1, j+1)]^2 + [f(i+1, j) - f(i, j+1)]^2\right\}^{\frac{1}{2}} \tag{4.40}$$

Among them, $f(i, j)$ is the input image with integer pixel coordinates, $G_R(i, j)$ is the processed gray value of $f(i, j)$ pixels and the square root operation makes the processing method similar to the process of human visual system. The template used in the calculation process is shown in Fig. 4.5. This is the template of 2◇2, which is represented by matrix, that is

$$\mathbf{H}_i = \begin{bmatrix} 1 & 0 \\ 0 & -1 \end{bmatrix}; \quad \mathbf{H}_j = \begin{bmatrix} 0 & -1 \\ 1 & 0 \end{bmatrix} \tag{4.41}$$

The gradients of $f(i, j)$ pixels in two orthogonal directions are the convolution of the image and the template, that is

$$G_i(i, j) = f(i, j) * \mathbf{H}_i \tag{4.42}$$

$$G_j(i, j) = f(i, j) * \mathbf{H}_j \tag{4.43}$$

It must be pointed out that Roberts edge detection operator has high positioning accuracy and good effect in horizontal and vertical directions, but it is sensitive to noise.

② **Sobel operator**

Sobel edge detection operator calculates the gradient in the 3◇3 sub domain of digital image, that is

Fig. 4.6 Sobel operator template

−1	−2	−1
0	0	0
1	2	1

−1	0	1
−2	0	2
−1	0	1

$$G_S(i, j) = \max\left[G_i(i, j), G_j(i, j)\right] \tag{4.44}$$

Among

$$\begin{aligned} G_i(i, j) = |&f(i-1, j+1) + 2f(i, j+1) + f(i+1, j+1) \\ &- f(i-1, j-1) - 2f(i, j-1) - f(i+1, j-1)| \end{aligned} \tag{4.45}$$

$$\begin{aligned} G_j(i, j) = |&f(i+1, j-1) + 2f(i+1, j) + f(i+1, j+1) \\ &-f(i-1, j-1) - 2f(i-1, j) - f(i-1, j+1)| \end{aligned} \tag{4.46}$$

Formulas (4.45) and (4.46) represent the differential (difference) of pixel coordinates in the direction of i and j axes, respectively.

Sobel edge detection operator is composed of two convolution kernels shown in Fig. 4.6. Each point in the image is convoluted with these two kernels. One convolution check usually has the largest vertical edge response, while the other convolution check has the largest horizontal edge response, and the maximum value of the two convolutions is used as the output value of the point.

After calculating the gradient through formula (4.44), a constant T can be set. When $G_s(i, j) > T$, mark the point as the boundary point, and its pixel value is set to 0 (or 255), and other pixel values are set to 255 (or 0). The operation result is an edge image. The best effect can be achieved by appropriately adjusting the size of constant T.

Sobel operator can usually deal better with images with gray gradual change and more noise.

③ **Prewitt operator**

Prewitt operator is a kind of edge detection template operator, which uses the gray difference between the upper and lower, left and right adjacent points of pixels to explore the edge points that reach the extreme value to realize detection and has a smoothing effect on noise. Because the gray value of edge point pixels is significantly different from that of adjacent point pixels, differential operator and template matching method are usually used to detect the edge of image in practical application.

Prewitt operator is defined as follows

$$G_P(i, j) = \max\left[G_i(i, j), G_j(i, j)\right] \tag{4.47}$$

Fig. 4.7 Prewitt operator template

1	1	1
0	0	0
−1	−1	−1

−1	0	1
−1	0	1
−1	0	1

Among

$$G_i(i, j) = |f(i-1, j-1) + f(i-1, j) + f(i-1, j+1) \\ -f(i+1, j-1) - f(i+1, j) - f(i+1, j+1)| \tag{4.48}$$

$$G_i(i, j) = |f(i-1, j+1) + f(i, j+1) + f(i+1, j+1) \\ -f(i-1, j-1) - f(i-1, j) - f(i+1, j+1)| \tag{4.49}$$

Like Sobel edge detection operator, formulas (4.48) and (4.49) also represent the differential (difference) of pixel coordinates in the direction of i and j axes, respectively.

Prewitt operator adopts the template, as shown in Fig. 4.7. Like the method of Sobel operator, each pixel in the image adopts two cores for convolution, and the maximum value is taken as the output.

After the gradient is calculated, a constant T can also be set. When $G_P(i, j) > T$, mark the point as the boundary point, and its pixel value is set to 0 (or 255), and other pixel values are set to 255 (or 0). Prewitt edge detection operator also obtains an edge image. Appropriately adjusting the size of constant T can achieve the best effect.

Prewitt operator can not only detect edge points, but also suppress the influence of noise, so it can process more noisy images better.

④ Krisch operator

Krisch is a direction operator, which uses a set of templates to calculate the difference values in different directions, takes the maximum value as the edge strength and takes the corresponding direction as the edge direction. The template of eight direction Krisch operator is shown in Fig. 4.8, and the included angle of each direction is 45 degrees.

Krisch edge detection operator consists of eight convolution kernels in Fig. 4.8. Each point in the image is convoluted with eight convolution kernels. Each convolution kernel makes the maximum response in a specific edge direction. The maximum value in all eight directions is output as the edge amplitude value. The serial number of the maximum response mask constitutes the coding of edge direction detection and thinning research.

−5	−5	−5	3	−5	−5	3	3	−5	3	3	3
3	0	3	3	0	−5	3	0	−5	3	0	−5
3	3	3	3	3	3	3	3	−5	3	−5	−5

3	3	3	3	3	3	−5	3	3	−5	−5	3
3	0	3	−5	0	3	−5	0	3	−5	0	3
−5	−5	−5	−5	−5	3	−5	3	3	3	3	3

Fig. 4.8 Krisch operator template

3) **Laplacian operator for edge detection**

Generally speaking, in view of the edge detection algorithm of the first derivative, when the first derivative is higher than a certain threshold, the point can be determined as an edge point. However, this alone may lead to too many edge points to be detected. To overcome the defects of this kind of algorithm, a better method can be adopted, that is to find the points corresponding to the local maximum of the gradient and identify them as edge points. The point corresponding to the local maximum of the first derivative is the zero intersection of the second derivative. In this way, the exact edge point can be found by finding the zero intersection of the second derivative of image intensity.

Laplacian operator is an alternative edge detection operator. Laplacian is an isotropic second-order differential operator with rotation invariance. The Laplace transform of a binary image function $f(i, j)$ is defined as

$$\nabla^2 f(i, j) = \frac{\partial^2 f(i, j)}{\partial i^2} + \frac{\partial^2 f(i, j)}{\partial j^2} \tag{4.50}$$

Among them, the digital second-order partial differential can be approximately expressed as

$$\begin{cases} \dfrac{\partial^2 f(i, j)}{\partial i^2} = f(i+1, j) + f(i-1, j) - 2f(i, j) \\ \dfrac{\partial^2 f(i, j)}{\partial j^2} = f(i, j+1) + f(i, j-1) - 2f(i, j) \end{cases} \tag{4.51}$$

The Laplace transform of Eq. (4.50) can be obtained by adding the two components expressed in Eq. (4.51), that is

Fig. 4.9 Laplacian operator template

0	1	0
1	−4	1
0	1	0

1	1	1
1	−8	1
1	1	1

$$\nabla^2 f(i, j) = f(i+1, j) + f(i-1, j) + f(i, j+1) + f(i, j-1) - 4f(i, j) \tag{4.52}$$

The commonly used Laplacian operator template is shown in Fig. 4.9.

It must be pointed out that in practical application, the original form of Laplacian operator is generally not used for edge detection. The main reasons are as follows: as a second derivative, Laplacian operator is very sensitive to noise; the amplitude of Laplacian operator produces bilateral peaks, which is an undesirable result in complex image segmentation.

At the same time, Laplacian operator cannot detect the direction of the edge. In order to solve this difficult problem, Gaussian filtering and Laplacian operator edge detection can be combined to form LoG (Laplacian of Gaussian) operator, that is, Gaussian–Laplacian operator, also commonly known as Mar–Hildreth operator. Firstly, Gaussian operator is used to smooth the image, and then Laplacian operator is used to detect the image edge according to the second-order differential zero crossing.

Gaussian–Laplacian operator usually has the following form

$$\nabla^2 G(r) = \frac{-1}{\pi\sigma^4}\left(1 - \frac{r^2}{2\sigma^2}\right)\exp\left(-\frac{r^2}{2\sigma^2}\right) \tag{4.53}$$

where σ is the variance and r is the radial distance from the origin, that is, $r^2 = i^2 + j^2$.

Gaussian–Laplace operator is a combination of two operators, which has both the smoothing characteristics of Gauss operator and the sharpening characteristics of Laplace operator. Smoothing and sharpening, integral and differential are two sides of a pair of contradictions. When they are unified, they become the best factor. Because the image contains noise, smoothing and integration can filter out these noise, and then sharpen and differentiate the edge detection after eliminating the noise, so a better effect can be obtained.

With the continuous emergence of new theories and algorithms, some new image edge detection methods appear accordingly. These new algorithms have a good suppression effect on noise, and can better detect edge details, such as Canny edge detection operator, mathematical morphology method, wavelet analysis method, neural network method, fuzzy operator method, multi-scale edge detection method and so on.

4) **Canny edge detection operator**

Canny operator is a kind of optimal edge detection operator, which has been widely used in many image processing fields.

The detection indicators of Canny edge detection operator include: Low misjudgment rate, which will identify as many actual edges in the image as possible; high positioning accuracy, which will make the marked edge as close as possible to the actual edge of the image; suppress false edges, which is a minimum response, that is, the edges in the image can only be identified once, and the possible image noise will not be identified as edges.

In order to meet these requirements, Canny uses the variational method, and the optimal detection is expressed by the sum of four exponential function terms, which is a method to find a specific function.

The basic idea of Canny edge detection operator is:

First, Noise Reduction

Any edge detection algorithm cannot work well on the unprocessed original data, so the first step is to convolute the original data with Gaussian convolution kernel, and the resulting image is slightly blurred compared with the original image. In this way, a single pixel noise has little effect on the Gaussian smoothed image.

Second, Find the Brightness Gradient in the Image

The edges in the image may point in different directions, so Canny algorithm uses four convolution kernels to detect the edges in horizontal, vertical and diagonal directions. The convolution between the original image and each convolution kernel is stored. For each point, we identify the maximum value at this point and the direction of the generated edge. In this way, we generate the brightness gradient map of each point in the image and the direction of the brightness gradient from the original image.

Third, Track the Edge in the Image

Through a large number of experiments, it is found that all image gradient histograms have similar statistical distribution rules, as shown in Fig. 4.10.

In the figure, the x axis represents the gradient value, the y axis represents the gradient number, a is the gradient mean and b is a statistic that can separate the image edge from the non-edge.

Fig. 4.10 Image gradient histogram distribution curve

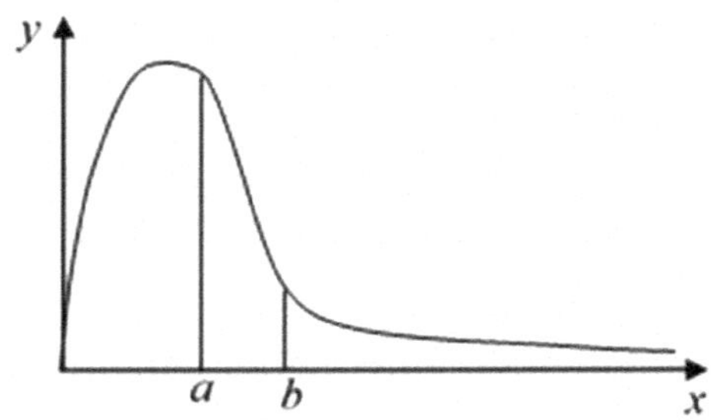

The peak part of the main body with low gradient value corresponds to the non-edge region in the image, while the gentle tail with high gradient value corresponds to the edge region. In probability theory, the gradient mean is a statistical feature that describes the location of the main gradient value concentration, and the standard deviation is a statistical feature that separates the main peak part from the gentle tail part. Therefore, using the statistical characteristics of gradient histogram and selecting b value as gradient threshold, the edge can be detected adaptively under the principle of global optimization.

However, there is no exact value to limit how much the brightness gradient is the edge, so Canny uses the hysteresis threshold.

Hysteresis-threshold requires two thresholds: high threshold and low threshold. It is assumed that the important edges in the image are continuous curves, so that the fuzzy part of the given curve can be tracked, and the noise pixels that do not form the curve can be avoided as edges. Therefore, starting from a large threshold, the real edges that people are more convinced of will be identified. Using the previously derived direction information, the whole edge can be tracked in the image from these real edges. When tracking, as long as a small threshold is used, the fuzzy part of the curve can be tracked until it returns to the starting point.

Once this process is completed, a binary image is obtained, which can confirm whether each point is an edge point. At the same time, by improving the subpixel precision edge, it can detect the zero crossing of the second-order directional derivative in the gradient direction.

The specific calculation steps of finding edges in the image are as follows:

Step 1 Calculate the Gradient Amplitude and Direction of Pixels

Calculate the gradient amplitude and gradient direction of pixels in the image $f(i, j)$.

The partial derivative in the row direction is

$$f_i(i, j) = \frac{\partial f(i, j)}{\partial i} = I(i + 1, j) - I(i - 1, j) \tag{4.54}$$

$$f_j(i, j) = \frac{\partial f(i, j)}{\partial j} = I(i, j + 1) - I(i, j - 1) \tag{4.55}$$

where $I(i, j)$ is the gray value of pixel (i, j), and so on.

The gradient amplitude $M(i, j)$ of pixel (i, j) can be calculated by the following formula

$$M(i, j) = \sqrt{f_i^2(i, j) + f_j^2(i, j)} \tag{4.56}$$

The gradient direction of pixel (i, j) is

$$\theta(i, j) = arctg\frac{f_j(i, j)}{f_i(i, j)} \tag{4.57}$$

Step 2 Suppress Non-local Maxima
For each pixel (i, j) with nonzero gradient intensity value, find two adjacent pixels along its own edge direction, that is, gradient direction $\theta(i, j)$. If the gradient amplitude $M(i, j)$ of the pixel is less than the corresponding value of adjacent pixels in the gradient direction, mark the point as a non-edge point and set its gradient value to zero; otherwise, it will be regarded as a candidate edge point.

Step 3 Select Adaptive Double Threshold
After the non-local maximum suppression of the image, the gradient is counted, and the gradient mean D and gradient variance σ are calculated. The sum of the two is taken as the high threshold T_h of edge detection, that is, $T_h = D + \sigma$; Take $0.4T_h$ as the low threshold T_l, that is, $T_l = 0.4 \times T_h$.

When the scene illumination is uneven, the image can be divided into several subgraphs. In each subgraph, the gradient histogram is used to select the adaptive threshold according to T_h and T_l. At the same time, in order to avoid the detected edge discontinuity, partial overlap of each sub image is allowed.

Step 4 Edge Connection
Firstly, the image after non-local maximum suppression is scanned. If the gradient intensity value of the pixel is greater than the high threshold, the pixel is marked as an edge point; if the gradient intensity value of a pixel is greater than the low threshold and less than the high threshold, the pixel is marked as a candidate edge point, otherwise it is a non-edge point.

Secondly, the image processed in the first step is scanned finally. If the candidate edge point is connected with the four neighborhood edge point, the candidate edge point is marked as an edge point and finally, the edge detection image is obtained.

Canny edge detection operator is different from other edge detection methods in that it uses two different thresholds to detect strong edges and weak edges, respectively, and the weak edges are included in the output image only when the weak edges are connected with the strong edges. Therefore, compared with other methods, this method is not easy to be "filled" by noise and is easier to detect the real weak edges [2].

5) **Mathematical morphology**

The theoretical basis of mathematical morphology and its language belong to the category of set theory. Mathematical morphology can simplify image data, maintain their basic shape characteristics and remove irrelevant structures. Mathematical morphology is composed of a group of morphological algebraic operators. Its basic operations include four: expansion, corrosion, opening and closing. Mathematical morphology is divided into two categories: binary morphology and gray morphology, which have their own characteristics in binary image and gray image. The morphology applied to binary images is called binary morphology, and the morphology applied to gray images is called gray morphology. Based on these basic operations, we can also deduce and combine various practical mathematical morphology algorithms, which can be used to analyze and process the image shape and structure, including image segmentation, feature extraction, boundary detection, image filtering, image enhancement and restoration. The algorithm of mathematical morphology has a natural parallel structure, which realizes the parallel of morphological analysis and processing algorithms, and greatly improves the speed of image analysis and processing.

The so-called structural element is a "probe" used to collect image information in mathematical morphology. When the probe moves continuously in the image, investigate the relationship between various parts of the image to understand the structural characteristics of the image. As the structural element of the probe, it can directly carry knowledge, such as shape, size, and even add gray and chroma information to detect and study the structural characteristics of the image.

The so-called expansion operation refers to the process of merging all background points in contact with an object into the object. The result of the process is to increase the area of the object by a corresponding number of points. If the object is round, its diameter will increase by two pixels after each expansion; if two objects are separated by less than three pixels in any direction at a point, they will be connected at that point. The expansion operation object in binary morphology is a set. The result of expanding the binary image with binary structural elements is to translate the structural elements and make the intersection of the two non-empty points form a new set. Gray morphology is the natural extension of binary mathematical morphology to gray image. The operation object of gray morphology is not a set, but an image function. In gray morphology, the intersection and union operations used in binary morphology will be replaced by maximum and minimum extremum operations, respectively. The expansion process of gray image can be directly calculated from the gray level function of image and structural elements. The calculation of expansion operation is carried out point by point. The calculation involves the gray value of the surrounding points and the value of structural elements. In fact, it is the sum of the gray values of the points in the local range and the corresponding points in the structural elements, and the maximum value is selected, so after the expansion operation, the edges are extended.

The so-called corrosion operation, simple corrosion, is a process of eliminating all boundary points of the object. The result is that the area of the remaining object along its periphery is one pixel smaller than that of the original object. If the object is round, its diameter will be reduced by two pixels after each corrosion. If the pixels connected in any direction at a certain point are less than three, the object will be divided into two objects at that point after one corrosion. The result of etching the binary image with binary structural elements is to translate the structural elements and make the binary structural elements include all the points of the binary image to form a new set. The corrosion process of gray image can also be calculated directly from the gray level function of image and structural elements. The corrosion operation is also carried out point by point. The operation result of a point is the difference between the gray value of the point in a local range and the corresponding point in the structural element, and the minimum value is selected. It can be seen that after corrosion operation, the gray value of relatively large points at the edge will be reduced, and the edge will shrink to the area with high gray value.

The so-called opening operation is a process of corrosion first and then expansion. It can eliminate small objects, separate objects in very small places and smooth the boundary of larger objects.

The so-called closure operation is an operation process that expands first and then corrodes. It has the function of filling small holes in objects, connecting adjacent objects and smoothing boundaries.

The specific process of edge detection based on multi-structural element mathematical morphology is as follows:

Step 1 Select Multiple Structural Elements

If the center point of the image is the coordinate origin (x_0, y_0), the square structure element matrix is

$$\mathbf{B}_i = \{f(x + x_0, y + y_0), \theta_i = i \times \alpha | -N \le x_0, y_0 \le N\} \tag{4.58}$$

Among them, $\mathbf{B}_i = \mathbf{B}_i(s, t), i = 0, 1, 2, ...4N - 1, N$ is the natural number, (s, t) is the coordinate of two-dimensional pixel matrix, $\alpha = 180°/4N$, θ_i is the direction angle formed by structural elements in the matrix, which is referred to as direction angle for short.

Step 2 Data Transformation of Image

The $B = B(s, t)$ structure element is used to perform the gray expansion operation $f \oplus B$ on the input image of the gray value $f = f(x, y)$ of the converted gray image at (x, y) point. The definition formula of the expansion operation is

$$(f \oplus B)(x, y) = \max\{f(s - x, t - y) + B(s, t) | (s - x, t - y) \in D_f \text{ and } (s, t) \in D_B\} \tag{4.59}$$

Then use the structural element of B to carry out gray corrosion operation $f \Theta B$ on the input image of f. The definition formula of corrosion operation is

$$(f \Theta B)(x, y) = \min\{f(x + s, y + t) - B(s, t) | (s + x, t + y) \in D_f \text{ and } (s, t) \in D_B\} \tag{4.60}$$

In the above two formulas, D_f and D_B are the definition fields of f and B, respectively.

After the above expansion and corrosion operations are combined, the morphological gradient $g_i(f)$ is obtained, and its operation formula is

$$g_i(f) = (f \oplus B) - (f \Theta B) \tag{4.61}$$

where i corresponds to the value of i in B_i

Step 3 Weighted Synthesis of Morphological Gradient
The morphological gradient $g_i(f)$ is synthesized after being weighted, so the synthesized morphological gradient $\hat{g}(f)$ is obtained as follows

$$\hat{g}(f) = \sum_{i=1}^{M} \omega_i g_i(f) \tag{4.62}$$

where $i = 1, 2, \ldots M$, M is the number of square structural elements, and ω_i is the weight of different directions in edge detection.

Step 4 Binarization of Morphological Gradient
The synthesized morphological gradient $\hat{g}(f)$ was binarized based on statistical principle.

First, calculate the frame average gray value $\overline{A}_{\text{gray}}$ of the gray image, and then calculate the average value main of the gray a_i of nine pixels in the region in the unit of 3×3 pixel small area, that is

$$\text{main} = \frac{1}{9}\sum_{i=1}^{9} a_i \tag{4.63}$$

Then, the average variance e_{A} of the gray level of the small area of the pixel is calculated one by one, that is

$$e_{\text{A}} = \sqrt{\sum_{i=1}^{9} (\text{main} - a_i)^2} \tag{4.64}$$

Finally, compare the gray average variance e_{A} of the pixel small area with the frame average gray value $\overline{A}_{\text{gray}}$ of the gray image one by one, and determine that the pixel small area takes 1 or 0 according to the following formula

$$F(x, y) = \begin{cases} 1 \ \overline{A}_{\text{gray}} \le n \cdot e_{\text{A}} \\ 0 \ \overline{A}_{\text{gray}} > n \cdot e_{\text{A}} \end{cases} \tag{4.65}$$

Among them, $F(x, y)$ is the binary image corresponding to the gray image $f(x, y)$, and n is the multiple, which is determined by experiment.

6) **Brief introduction of other algorithms**

Other algorithms mentioned here mainly include wavelet analysis, fuzzy operator, neural network method and multi-scale edge detection method.

① **Wavelet analysis**

From the perspective of signal processing, the image edge shows the singularity of the signal, and it has been proved mathematically that the Lipschitz index can be calculated from the trans-dimension maximum of wavelet transform. Therefore, as long as the modulus maximum of wavelet transform is detected, the edge can be detected. The application of multi-scale of wavelet can suppress noise in large scale, reliably identify edges and accurately locate in small scale.

② **Fuzzy operator method**

The generalized fuzzy operator is used for edge detection. Firstly, the generalized property set P is generated on the basis of the original input image, then the generalized fuzzy operator is used to generate P', the P' is inversely transformed to generate x' and finally, the edge image output can be obtained by simply closing the value of x'. The edge detected by generalized fuzzy operator has the advantages of small width and high signal-to-noise ratio.

③ **Neural network method**

Firstly, the input image is mapped into a neural network, then a certain prior knowledge (such as the original edge image) is input, and then the prior knowledge (image) is trained until the learning process converges or the user is satisfied. Because the edge extraction of neural network uses the existing knowledge of the original image, it recognizes the object from the macro and extracts the details from the micro, so it has a strong anti-noise ability.

④ **Multi-scale edge detection**

The so-called multi-scale edge detection is to effectively combine and use the edge detection operators of different scales to correctly detect the edges produced in an image. The idea is to suppress noise in large scale and identify edges reliably; precise positioning on a small scale. Firstly, several edge detection operators with different scales are used to detect the edge, and then their output results are combined to obtain the ideal edge.

2) **Image binarization processing and segmentation**

Image binarization is the basic method before image segmentation.

Generally speaking, the so-called binarization is to express the gray level of the image according to the two values of "1" or "0" through the comparison of the threshold, so that the image shows the black-and-white colors of the scene and the background.

Image segmentation is to divide the image into some meaningful areas, and then describe these areas, so as to extract the characteristics of some target area images and judge whether there are interesting target scenes in the image.

In other words, image segmentation is to decompose the image array into several non-overlapping regions. Some characteristics or features in each region are the same or close, while there are obvious differences in image features between different regions, that is, the characteristics in the same region change gently and relatively uniformly, while the characteristics at the boundary of the region change violently.

The consistent attribute criterion is P_0, when the image $f(i, j)$ meets the following conditions:

$$\begin{cases} f(i,j) = \bigcup_{k=1}^{n} f_k(i,j);\ f_k(i,j) \neq \phi \\ f_k(i,j) \cap f_l(i,j) = \phi \\ P[f_k(i,j)] = 1;\ P[f_k(i,j)] \subseteq P_0 \\ P[f_k(i,j)] = 0;\ P[f_k(i,j)] \not\subset P_0 \end{cases} \tag{4.66}$$

where ϕ is an empty set and $P[\cdot]$ is an attribute function; the process of dividing the image $f(i,j)$ into non-overlapping region sets $\{f_1(i,j), f_2(i,j), \ldots, f_n(i,j)\}$ is called segmentation.

The measurement criterion used for image segmentation is not unique; it is related to the application scene image and application purpose. For example, the scene image feature information used for image segmentation includes brightness, color, texture, structure, temperature, spectrum, motion, shape, position, gradient and model.

There are many methods to realize image segmentation. The following introduces some common image segmentation algorithms.

(1) Image binarization based on histogram Valley threshold

The so-called image binarization method based on histogram valley threshold is to segment the image according to the distribution characteristics of image histogram.

For example, when the gray histogram of the image is bimodal distribution, the segmentation is relatively easy. Just take its valley point as the threshold value to separate the object from the background, that is

$$f(i,j) = \begin{cases} 1;\ f(i,j) < \breve{z} \\ 0;\ f(i,j) \geq \breve{z} \end{cases} \tag{4.67}$$

where $\breve{z}$ is the threshold value of histogram valley point, which is the pixel gray level of its histogram valley point. See Fig. 4.11. This is a typical image gray histogram with bimodal distribution characteristics, where z is the pixel gray level and $h(z)$ is the pixel gray level distribution function.

There are many methods to realize image segmentation. The following introduces some common image segmentation algorithms.

In such gray histogram, when $z \in [0, 255]$, $\breve{z}$ shall meet the following relationship

Fig. 4.11 Image gray histogram

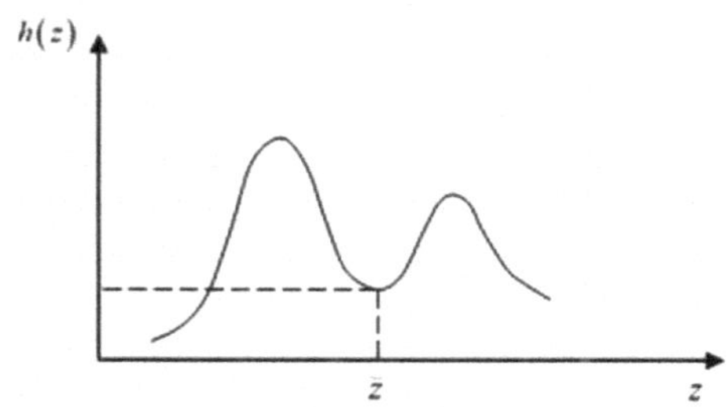

$$\left[h\left(\breve{z}\right) < h\left(\breve{z}-1\right)\right] \wedge \left[h\left(\breve{z}\right) < h\left(\breve{z}+1\right)\right] \tag{4.68}$$

However, if the histogram is not smooth, the valley will not be found. At this time, the histogram needs to be smoothed in advance. For example, for the discussion area $z \in [0, 255]$, after using the "5-point smoothing method", the smoothed image gray histogram can be obtained through the calculation of formula (4.69).

$$\begin{cases} h'(0) = \frac{1}{70}[69h(0) + 4h(1) - 6h(2) + 4h(3) - h(4)] \\ h'(1) = \frac{1}{35}[2h(0) + 27h(1) + 12h(2) - 8h(3) + 2h(4)] \\ h'(k) = \frac{1}{35}[-3h(k-2) + 12h(k-1) + 17h(k) + 12h(k+1) - 3h(k+2)] \\ h'(254) = \frac{1}{35}[2h(251) - 8h(252) + 12h(253) + 27h(254) + 2h(255)] \\ h'(255) = \frac{1}{70}[-h(251) + 4h(252) - 6h(253) + 4h(254) + 69h(255)] \end{cases} \tag{4.69}$$

where $k \in [3, 253]$

(2) **OTSU algorithm**

OTSU algorithm, also known as maximum interclass variance method, is one of the global threshold selection methods based on points in image segmentation. This method has been widely used because of its simple calculation, stability and effectiveness. It is still used as the standard algorithm for automatic selection of single threshold of gray image in the image processing toolbox of MATLAB.

OTSU algorithm divides the image gray histogram into two parts with the best threshold, so that the interclass variance of the two parts is maximized, that is, its separation is maximized.

If the gray level of the image is from 0 to 255, that is, $z \in [0, 255]$, there are n_z gray pixels of level z, and the total number of pixels is $N = \sum_{z=0}^{255} n_z$; the occurrence probability of gray pixels of level z is $P_z = n_z/N$.

If the gray threshold value is k, the image pixels are divided into two categories according to the gray level, that is

$$\begin{cases} C_0 = \{0, 1, \ldots, k\} \\ C_1 = \{k+1, k+2, \ldots, 255\} \end{cases} \tag{4.70}$$

The total average gray level of the image is

$$\mu = \sum_{z=0}^{255} z \cdot P_z \tag{4.71}$$

The average gray level of C_0 class is

$$\mu_0 = \sum_{z=0}^{k} z \cdot P_z = \mu(k) \tag{4.72}$$

The number of pixels of C_0 class is

$$N_0 = \sum_{z=0}^{k} n_z = N(k) \tag{4.73}$$

The average gray level of C_1 class is

$$\mu_1 = \sum_{z=k+1}^{255} z \cdot P_z = \mu - \mu(k) \tag{4.74}$$

The number of pixels of C_1 class is

$$N_1 = \sum_{z=k+1}^{255} n_z = N - N(k) \tag{4.75}$$

The image proportions of the two types of regions are

$$w_0 = \sum_{z=0}^{k} P_z = w(k) \tag{4.76}$$

and

$$w_1 = \sum_{z=k+1}^{255} P_z = 1 - w(k) \tag{4.77}$$

Therefore, the total average gray level of the image can be expressed as

$$\mu = w_0\mu_0 + w_1\mu_1 \tag{4.78}$$

The interclass variance can be expressed as

$$\sigma^2(k) = w_0 w_1 (\mu_0 - \mu_1)^2 \tag{4.79}$$

After combining formulas (2.72)–(2.79), the following formula can be obtained

$$\sigma^2(k) = \frac{w(k)[\mu(k) - \mu]^2}{1 - w(k)} \tag{4.80}$$

When $k \subseteq z$ changes from 0 to 255 Gy levels, the gray level $\overleftrightarrow{k}$ that makes $\sigma^2(k)$ get the maximum value will be found, which is called the optimal threshold value. At the same time, $\sigma^2(k)$ is called the objective selection function.

(3) **Image segmentation with minimum error**

If the percentage of target pixel points in the total pixel points of the image is θ, the percentage of background pixel points in the total pixel points of the image is $1 - \theta$.

If the gray distribution of target and background is $P_1(z)$ and $P_2(z)$, respectively, the mixing probability density of image gray is

$$P(z) = \theta P_1(z) + (1 - \theta)P_2(z) \tag{4.81}$$

When the selected threshold is T, the probability of target point and background being wrongly marked is

$$E_1(T) = \int_T^{\infty} P_1(z)\mathrm{d}z \tag{4.82}$$

and

$$E_2(T) = \int_{-\infty}^{T} P_2(z)\mathrm{d}z \tag{4.83}$$

Therefore, the total error probability is

$$E(T) = \theta E_1(T) + (1 - \theta)E_2(T) \tag{4.84}$$

Let $\frac{\partial E(T)}{\partial T} = 0$, then

$$-\theta P_1(T) + (1 - \theta)P_2(T) = 0 \tag{4.85}$$

and

$$\frac{(T - z_1)^2}{2\sigma_1^2} - \frac{(T - z_2)^2}{2\sigma_2^2} = \ln \frac{\theta\sigma_2}{(1 - \theta)\sigma_1} \tag{4.86}$$

Referring to Fig. 4.12, it is assumed that the gray histogram of the target and background follow the normal distribution, that is

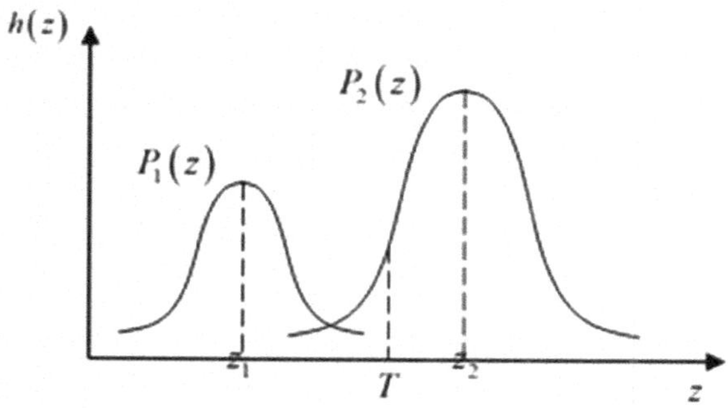

Fig. 4.12 Gray distribution of target and background

$$\begin{cases} P_1(z) \sim N(z_1, \sigma_1) \\ P_2(z) \sim N(z_2, \sigma_2) \end{cases} \tag{4.87}$$

When $\sigma_1 = \sigma_2 = \sigma$, the relationship is as follows

$$T = \frac{z_1 + z_2}{2} + \frac{\sigma^2}{z_2 - z_1} \ln \frac{\theta}{1 - \theta} \tag{4.88}$$

When $\theta = \frac{1}{2}$, it is a special case. At this time, the relationship is as follows

$$T = \frac{z_1 + z_2}{2} \tag{4.89}$$

This means that when the gray levels of the target and background are in normal distribution, the optimal threshold T (i.e., the threshold with minimum error) can be obtained according to formula (4.88) or (4.89); Otherwise, the optimal threshold T can be determined by formula (4.85).

3. **Color image enhancement**

Color image enhancement algorithm is developed on the basis of black-and-white image enhancement algorithm, but it is different from black-and-white image in description method.

The so-called color space is a mathematical description method of color, which is used to specify and generate the color of the scene and its image, and visualize the color of the scene and its image. The mathematical space formed by the mathematical model describing color is called color space. Different color mathematical models constitute different color space, also known as color space model. The color space model is usually represented by a three-dimensional model. Each color in the color space is specified by a set of coordinate parameters (three coordinate parameters) describing the position of the color in the color space.

A variety of color spaces can be adapted to different applications. For example, RGB space representing photoelectric devices such as digital cameras, CMYK space representing printing devices, YCbCr space representing television signals, HSV space representing human color perception characteristics and so on. "Color space model" and "color space" have the same concept. When describing the process of color image, there is no difference between them.

According to the principle of color synthesis

$$\mathbf{C} = \mathbf{R} + \mathbf{G} + \mathbf{B} = R\mathbf{r} + G\mathbf{g} + B\mathbf{b} \tag{4.90}$$

The color image **C** is the vector sum of the three primary colors of red **R** (wavelength 700 nm), green **G** (wavelength 546.1 nm) and blue **B** (wavelength 435.8 nm). In the formula, **r**, **g** and **b** are the unit vectors of the three primary colors of red **R**, green **G** and blue **B**, respectively, and R, G and B are the component coefficients of the three primary colors, that is, the brightness of the three primary colors, also known as luminous flux.

The vector $\left[R\ G\ B \right]^T$ of color image in RGB color space not only represents the color of red **R**, green **G** and blue **B** but also represents the brightness of the three primary colors. There is a great correlation between RGB three colors. In other words, different color effects can be formed through different values of $\left[R\ G\ B \right]^T$ three elements.

Hue is the feeling of the visual system to the wavelength of light wave. It is used to distinguish the name and type of color. It is an attribute that is easiest to distinguish color. Hue depends on the frequency (main wavelength) of the light wave in the visible spectrum. Colors without main wavelength are called achromatic colors, such as black, gray, white.

Saturation refers to the purity of color, which can be used to distinguish the degree of bright color. The more a color is infiltrated into other light components, the more unsaturated the color is. Fully saturated color refers to the color that is not penetrated into white light (i.e., single wavelength pure color light). For example, a spectral color composed of only a single wavelength is a fully saturated color.

Brightness is a physical property. The intensity of color can be described in different terms and methods, such as lightness, brightness and brightness. Lightness is a perceptual attribute, which is difficult to measure; Brightness is a physical property, which is easy to measure; Brightness is the sensing value of vision to brightness, which is between brightness and brightness. It can be calculated, but it is still different from the real perceptual characteristics. Lightness is the perceptual attribute of the visual system on how much light (or light reflection) a visible object emits. It is related to human perception. Because lightness is difficult to measure, the international lighting commission has defined a physical quantity that is easier to measure, called brightness. According to the definition of the international lighting commission, brightness is the radiation power weighted by the spectral sensitivity function reflecting visual characteristics, and reaches the peak at 555 nm, and its amplitude is directly proportional to the physical power. In this sense, brightness can be regarded as the intensity of light. The value of brightness is expressed by the intensity of light reflected or emitted per unit area, and the unit is candlelight / square meter (cd/m^2).

One Candela (abbreviated as *cd*) is equal to the luminous intensity in a given direction of a radiation light source with a frequency of 540 × 10^{12} Hz, and the radiation intensity of this light source in this direction is $\frac{1}{683}$ W/unit spherical angle. Brightness can also be expressed in lumens of luminous flux (abbreviated as lm). One lumen is equal to the luminous flux emitted by a uniform point light source of one candle in a unit solid angle. For light sources, the values of lm and cd/m^2 of the same brightness are equal in quantity. In fact, the specified brightness, i.e., white light, is often used as a reference, and it is normalized to 1 or 256 units.

HSV color space model has two important characteristics: first, V component is independent of image color information; second, H and S components are closely related to the way the human eye obtains color. These features make HSV model very suitable for image processing algorithms that perceive color characteristics with the help of human visual system. In HSV color space model, adding white to a solid color can change its color concentration, and adding black can change its color depth. At the same time, adding different proportions of white and black can finally get different colors.

Image enhancement refers to using pulse coupled neural network method to simulate the synchronization behavior of neurons related to features to show the connection model of pulse emission phenomenon, and enhance the images collected by the camera on the vehicle device or in the carriage in real time. Pulse coupled neural networks (PCNN) is a connection model that simulates the synchronization behavior of neurons related to characteristics to show the phenomenon of pulse emission. Therefore, it has a natural connection with the perception ability of the optic nerve system.

In the PCNN structure model applied to image processing, each pixel $f(i, j)$ of the image to be processed corresponds to each neuron N_{ij}. in the pixel coordinates, $i = 1, 2, 3, \ldots, j = 1, 2, 3, \ldots$. I_{ij} represents the pixel intensity value of the pixel point $f(i, j)$, in addition to receiving the external stimulation I_{ij}, each neuron N_{ij} also receives the feed input $F_{ij}(t)$ and connection input $L_{ij}(t)$ from other neurons in the internal network. Then, the internal behavior $U_{ij}(t)$ of the neuron N_{ij} is formed in the form of product coupling $F_{ij}(t)\left[1 + \beta L_{ij}(t)\right]$ through the neuron connection intensity β, and then the pulse signal output $Y_{ij}(t)$ (also known as ignition) of the neuron is excited or suppressed through the comparison of the dynamic threshold $\theta_{ij}(t)$ and $U_{ij}(t)$, t stands for time.

Since the luminance intensity difference of pixels on both sides of the edge in the conventional image is always larger than that of pixels adjacent to the space in the region, if PCNN is used for two-dimensional image processing, each neuron corresponds to the image pixels one by one, and its luminance intensity value is used as the external stimulation of neurons; the pixel clusters adjacent to the space and similar intensity can be ignited synchronously within PCNN; otherwise, they can be ignited asynchronously. In image enhancement, the image pixels corresponding to synchronous ignition present the same brightness intensity value, so as to smooth the image area; The image pixels corresponding to asynchronous ignition show different brightness intensity values, which increases the gradient of brightness intensity between image regions and then highlights the edge of the image, making the enhanced image brightness intensity distribution more hierarchical. In the standard PCNN model, the output is a binary image frame due to the function of hard limiting function. In order to make the PCNN output mapping function more effective in the processing of image overall contrast enhancement, based on the above human visual perception characteristics, the quasi-logarithmic mapping function is used to map the brightness intensity of the image to an appropriate visual range. The biggest advantage of this method is that it has a natural connection with the perception ability of the visual nervous system, so that the model can not only smooth the image area and highlight the image edge, but also significantly improve the visual effect of the color image and enhance the real effect of the image color. Of course, the enhancement of color image is also inseparable from the histogram equalization process. This will be covered in the next section.

Color space conversion refers to the color space conversion of the enhanced image, that is, the enhanced digital image is converted from RGB color space to HSV color space. The converted hue H, saturation S and brightness V are expressed as

$$V = \max(R, G, B) \tag{4.91}$$

$$S = \begin{cases} \frac{V - \min(R,G,B)}{V}; & \text{When } V \neq 0 \\ 0; & \text{Other} \end{cases} \tag{4.92}$$

$$H = \begin{cases} \frac{60^\circ (G-B)}{S}; & \text{When } V = R \\ \frac{180^\circ + 60^\circ (B-R)}{S}; & \text{When } V = G \\ \frac{240^\circ + 60^\circ (R-G)}{S}; & \text{When } V = B \end{cases} \tag{4.93}$$

In the process of calculating H, if $H < 0$, $\ddot{H} \leq H + 360^\circ$ is taken, and $\ddot{H}$ is the actual value of H.

RGB and HSV are two ways to describe the color space of an image. The space vector $\begin{bmatrix} H & S & V \end{bmatrix}^{\mathrm{T}}$ of the former not only represents the color of the three primary colors (R, G and B) but also represents the brightness of the three primary colors. There is a great correlation between the three colors of R, G and B. In other words, different color effects can be formed through different values of the three elements of $\begin{bmatrix} H & S & V \end{bmatrix}^{\mathrm{T}}$. The latter is a three-dimensional color space model including hue H, saturation S and brightness V, also known as hexagonal cone model. In this color space model, hue H is measured by angle, and the value range is from 0° to 360°. It is calculated counterclockwise from red. Red is 0°, green is 120°, blue is 240°, and their complementary colors: yellow is 60°, cyan is 180°, magenta is 300°; saturation S ranges from 0.0 to 1.0; brightness B ranges from 0.0 (black) to 1.0 (white). For example, pure red means $\begin{bmatrix} H & S & V \end{bmatrix}^{\mathrm{T}} = \begin{bmatrix} 0 & 1 & 1 \end{bmatrix}^{\mathrm{T}}$, while $S = 0$ means non-color. In this case, the hue H is undefined.

4. **Color-scale histogram equalization of color image**

The enhancement of color image is also inseparable from the process of histogram equalization.

The difference between color image equalization and black-and-white image equalization is that the former is brightness histogram equalization on the premise of unchanged hue, so it is also called color-scale histogram equalization.

Color-scale histogram equalization is actually the operation of gray-scale histogram equalization for one or more color channels of the image. Common methods are as follows:

(1) Count the histogram data of all RGB color channels and perform equalization operation, and then replace the color values of *R*, *G* and *B* channels, respectively, according to the mapping value obtained by equalization.
(2) The histogram data of *R*, *G* and *B* color channels are counted and equalized, respectively, and then the color values of *R*, *G* and *B* channels are replaced according to the mapping values of *R*, *G* and *B*.
(3) Calculate the brightness channel by brightness formula or RGB average value, count the histogram data of brightness channel and perform equalization operation and then replace the color values of *R*, *G* and *B* channels, respectively, according to the mapping value.

Example Use the conversion relationship between the brightness component $Y(x, y)$ in YCbCr color space and the brightness of three primary color components in RGB color space

$$Y(x, y) = 0.299 \times R(x, y) + 0.587 \times G(x, y) + 0.114 \times B(x, y) \tag{4.94}$$

Calculate the brightness $Y(x, y)$ of the true color image at point (x, y), where $R(x, y)$, $G(x, y)$ and $B(x, y)$ are the brightness of the three primary color components of the true color image at RGB color space point (x, y), respectively.

The luminance function $Y(x, y)$ of the color image is converted into the pixel coordinate function $g(i, j)$ through the gray level transformation function, and then the luminance color level statistical operation of the luminance formula is used to replace the color values of *R*, *G* and *B* channels, respectively, with the luminance color level mapping value. Finally, the enhancement effect of the color image can be obtained.

5. **Edge detection of color image**

Through research, it has been found that 10% of the edges of color images will be missed if gray-scale edge detection is adopted. This is because the edge of gray image is defined at the discontinuity of brightness function, and the precise definition of discontinuity will vary with the specific edge detection operator. For color edges, the edge can also be defined according to some discontinuity in the color space, and there are three alternative methods to define this discontinuity:

① A measurable distance is established in the color space, and the discontinuity of this distance is used to detect the existence of color edges.
② The edges of the three primary color components of the color image are calculated independently, and then the edges of the three primary color components are combined in some way.

③ It is allowed to have greater independence when calculating the edge on each component, but some consistency restrictions are added to each component to make use of the edge information on three components at the same time.

(1) **Color image edge detection based on gradient extremum**

Generally speaking, for black-and-white gray-scale images, gradient operator can not give a really ideal edge. The main reasons are as follows:

① Gradient operator is a multi-valued response operator.
② Gradient operator has no strict positioning function.
③ The gradient operator needs to classify the local eigenvalue, that is, the differential intensity value of each pixel to make the overall classification decision.

Therefore, we must seek a correction algorithm, which is to directly determine the attribution of boundary pixels in the smallest possible local area, and the result will not be affected by the uneven illumination of each area of the image.

For the edge detection of color image, for the convenience of description, take HSV color space model as an example.

In HSV color space, a gradient extremum operator window is designed to directly extract edge feature pixels in the window. In a small linear neighborhood, the color gradient of the pixel will have a maximum. Conversely, if the absolute gradient value of a pixel in a straight line direction of the image plane is the maximum value of the pixels in its neighborhood, the pixel may be a boundary pixel. Therefore, if all pixels with the maximum pixel gradient in the linear neighborhood in a certain direction can be extracted, the set will include all boundary point pixels (Fig. 4.13).

The specific steps are as follows:

Step 1 The RGB image is transformed into HSV image, and the original image is represented by $f(i, j)$, then the hue, saturation and brightness components of the corresponding HSV space are $H(i, j)$, $S(i, j)$ and $V(i, j)$, respectively.

Step 2 Design a 5×5 boundary detection operator window, see Fig. 2.40.

Let the window center point $A = (i, j)$ be the pixel to be detected, and $B = (x, y)$ be the eight adjacent pixels of the window center point.

The gradients of hue, saturation and brightness of the window center point $A = (i, j)$ in the θ direction are defined as

Fig. 4.13 Boundary detection window

	B	B	B	
	B	A	B	
	B	B	B	

$$H_\theta(i, j) = H(x, y) - H(i, j) \quad (4.95)$$

$$S_\theta(i, j) = S(x, y) - S(i, j) \quad (4.96)$$

$$V_\theta(i, j) = V(x, y) - V(i, j) \quad (4.97)$$

That is, the color information component of the center point pixel is subtracted from the color information component of the eight adjacent pixels of the window center point (i, j), where the subscript θ represents the detection direction of the window edge pixel, that is, the direction of the connecting line between the two pixels.

The final color gradient of the pixel is defined as

$$G_\theta(i, j) = H_\theta(i, j) + S_\theta(i, j) + V_\theta(i, j) \quad (4.98)$$

Step 3 In order to find the extreme value of the color gradient in the normal direction of the boundary, at least three color gradients need to be compared, that is, at least four pixels need to be compared. For example, select four consecutive pixels including the center point on a straight line passing through the center point of the window, and calculate the gradient of the center point and its adjacent three pixels, where the middle value is the color gradient of the center pixel of the window.

As shown in Fig. 4.14, among them, (a), (b), (c), (d), (E), (f), (g) and (h) are eight edge detection direction straight lines around the center of the window with angles of 0°, 45°, 90°, 135°, 180°, 225°, 270° and 315°: B-A-B-B.

Assuming that the distribution pixels along the above directions are set with subscript in turn, the order is: (x_0, y_0), (i, j), (x_1, y_1), (x_2, y_2).

Then, in any direction, the color gradients of the center point (i, j) and its three adjacent pixel points (x_0, y_0), (x_1, y_1) and (x_2, y_2) are defined as follows:

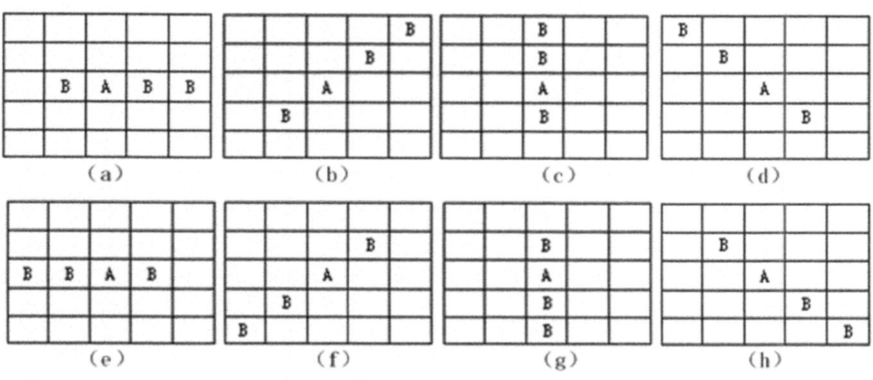

Fig. 4.14 Color gradient calculation window in eight directions

$$H_\theta(x_0, y_0) = H(i, j) - H(x_0, y_0) \tag{4.99}$$

$$H_\theta(i, j) = H(x_1, y_1) - H(i, j) \tag{4.100}$$

$$H_\theta(x_1, y_1) = H(x_2, y_2) - H(x_1, y_1) \tag{4.101}$$

$$S_\theta(x_0, y_0) = S(i, j) - S(x_0, y_0) \tag{4.102}$$

$$S_\theta(i, j) = S(x_1, y_1) - S(i, j) \tag{4.103}$$

$$S_\theta(x_1, y_1) = S(x_2, y_2) - S(x_1, y_1) \tag{4.104}$$

$$V_\theta(x_0, y_0) = V(i, j) - V(x_0, y_0) \tag{4.105}$$

$$V_\theta(i, j) = V(x_1, y_1) - V(i, j) \tag{4.106}$$

$$V_\theta(x_1, y_1) = V(x_2, y_2) - V(x_1, y_1) \tag{4.107}$$

The final color gradient of the above three pixel points (i, j), (x_0, y_0) and (x_1, y_1) is

$$G_\theta(x_0, y_0) = H_\theta(x_0, y_0) + S_\theta(x_0, y_0) + V_\theta(x_0, y_0) \tag{4.108}$$

$$G_\theta(i, j) = H_\theta(i, j) + S_\theta(i, j) + V_\theta(i, j) \tag{4.109}$$

$$G_\theta(x_1, y_1) = H_\theta(x_1, y_1) + S_\theta(x_1, y_1) + V_\theta(x_1, y_1) \tag{4.110}$$

If the window center pixel point (i, j) is the boundary pixel point, when $G_\theta(i, j) > 0$, the following two equations are established in one direction at the same time

$$\begin{cases} G_\theta(i, j) > |G_\theta(x_0, y_0)| \\ G_\theta(i, j) > |G_\theta(x_1, y_1)| \end{cases} \tag{4.111}$$

Combining the above two formulas, the gradient extreme value edge detection operator can be expressed as

$$\widehat{G}_\theta(i, j) = [G_\theta(i, j) - |G_\theta(x_0, y_0)|] \cap [G_\theta(i, j) - |G_\theta(x_1, y_1)|] \tag{4.112}$$

If $\widehat{G}_\theta(i, j) > 0$, $f(i, j)$ is the possible boundary pixel in the θ direction, where the θ direction is the normal direction of the edge. Scan all pixels of the image with the above edge detection window, and all pixels conforming to $\widehat{G}_\theta(i, j) > 0$ are boundary pixels. In the pixel set conforming to $\widehat{G}_\theta(i, j) > 0$, the boundary points and their edge point pixels will be included. It should be pointed out here that when comparing pixel color gradients, it is required that each pixel color gradient must be a gradient in the same direction, which aims to ensure the uniqueness of boundary response.

It must be pointed out that the color gradient of any point in the image is determined by the sum of the gradients of hue, saturation and brightness of the two pixel points. If the gradient value meets the above formula (4.112), the pixel point as an edge point has a direct impact on the image segmentation result, that is, the positioning of the edge. In other words, two adjacent edge points may be more than one unit of straight line distance after segmentation. In order to avoid the above situation, a determination principle for edge pixels is given: eight detection directions are designed in the operator window. The purpose is to ensure that if there are edge pixels in the window, when $\widehat{G}_\theta(i, j) > 0$, the pixel with smaller gray value is taken as the edge point among the two adjacent pixels. If $\widehat{G}_\theta(i, j) > 0$ is required in formula (2.112), the edge pixel can be determined in the center of the window.

(2) Practical method of color image edge detection

First, convert the original image from RGB space to HSV space from formula (4.91) to formula (4.93), and then perform edge detection by calculating the gradient of hue, saturation and brightness of the color image according to the detection steps.

In other words, the purpose of edge thinning is to narrow the edge with a certain width and keep its shape topology unchanged, which can greatly reduce the amount of edge data.

Hilditch algorithm and Sobel algorithm can be used for edge thinning.

1) Hilditch thinning algorithm

The algorithm is suitable for binary images with input images of 0 and 1. The area with pixel 1 is the part to be refined, and the area with pixel value 0 is the background. Hilditch thinning algorithm is described as follows:

If p is the detected pixel, $f(p)$ is the gray value of pixel p, and $n_i (i = 1, 2, \ldots, 8)$ is the 8 neighborhood pixels of p, the position of n_i is shown in Fig. 4.15.

Fig. 4.15 Pixel neighbourhood

n_4	n_3	n_2
n_5	p	n_1
n_6	n_7	n_8

It is assumed that set $I = \{1\}$ represents the subset of pixels to be refined, set $N = \{b|b - m \leq 0\}$ represents the subset of background pixels, and set $R = \{-m\}$ represents the pixels subtracted in $I = \{1\}$ during the m-th thinning.

The "cut/slice thin" conditions of image thinning are as follows:

$f(p) \in I$

$U(p) \geq 1, U(p) = \sum_{j=1}^{4} a_{2j-1}$ and $a_{2j-1} = \begin{cases} 1; f(n_i) \in N \\ 0; f(n_i) \notin N \end{cases}; i = 1, 2, \ldots, 8; j = 1, 2, 3, 4$

$V(p) \geq 2,\ V(p) = \sum_{j=1}^{8} \left(1 - a_j\right)$ and $a_j = \begin{cases} 1; j = 1, 3, 5, 7 \wedge f(n_i) \in N \\ 0; j = 2, 4, 6, 8 \wedge f(n_i) \notin N \end{cases}; i = 1, 2, \ldots, 8$

$W(p) \geq 1,\ W(p) = \sum_{i=1}^{8} a_i$ and $a_i = \begin{cases} 1; f(n_i) \in I \\ 0; f(n_i) \notin I \end{cases}$

$X(p) = 1,\ X(p) = \sum_{i=1}^{4} a_i$ and

$$a_i = \begin{cases} 1;\ f(n_{2i-1}) \in N \wedge \left[[f(n_{2i}) \in I \cup R] \vee \left[f(n_{2i+1}) \in I \cup R\right]\right] \\ 0;\ \text{Under other conditions} \end{cases}; i = 1, 2, 3, 4$$

$f(n_i) \notin R$ 或 $X[f(n_i)] = \sum_{i=2}^{3} a_{2i-1} = 1$ and

$$a_{2i-1} = \begin{cases} 1;\ f(n_{2i-1}) \in N \wedge \left[[f(n_{2i}) \in I \cup R] \vee \left[f(n_{2i+1}) \in I \cup R\right]\right] \\ 0;\ \text{Under other conditions} \end{cases}; i = 2, 3$$

2) **Sobel thinning algorithm**

Sobel thinning algorithm is suitable for edge images after edge detection.

The specific algorithm is as follows:

Step 1 Sobel algorithm is used to detect the edge of the detected edge image again, and the boundaries on both sides of the edge can be obtained.

Step 2 Subtracting the original edge image from the image processed by Sobel again will get an edge image with sharper edges.

Step 3 When the edge obtained by other methods is very fuzzy and weak, the Sobel algorithm can be repeated for two to three times until the edge is clear and sharp.

It must be pointed out that since the thinning process will weaken the signal intensity, the parts that have met the thinning requirements should retain the original results in the later thinning process without further processing.

4.2.2 Face Image Feature Extraction

Feature extraction involves a wide range, which is closely related to the characteristics of the physical form of the identified target, so there are a variety of different methods for feature extraction. Image features can be extracted from the global or local. The purpose of focusing on the local features of the image is to greatly reduce the amount of computation in the recognition process. This basic idea of recognition is especially suitable for the fast operation of aircraft to recognize the target object.

Feature extraction is a statistical pattern recognition method based on feature quantity. It mainly includes two steps: one is to extract the feature quantity that can represent the pattern, and the other is to determine the class of the target object to be recognized under the specific classification criteria.

1 **Decision theory and method**

Whether the feature is extracted from the whole or the part, the method of pattern recognition must come down to the problem of classification to implement decision-making and judgment. If N features have been extracted from the image, and the image can be divided into C categories. Then the N features of the image can be clearly classified, so that it can be further determined that the target waiting for detection should eventually belong to the specific type in C class.

Generally, the object waiting for recognition is called pattern, which is also called test sample. The pattern (test sample) constitutes the vector $\mathbf{X}$ in N-dimensional space, that is

$$\mathbf{x} = \left[x_1\ x_2\ \ldots\ x_N\right]^{\mathrm{T}} \tag{4.113}$$

Let C pattern categories $\omega^1, \omega^2, \cdots, \omega^c$ constitute the category set $\Omega = \{\omega^j\}$, $j = 1, 2, \ldots, c$;and N_j indicates the number of matching features of the j-th category.

Object image recognition is to judge whether $\mathbf{x}$ belongs to $\Omega = \{\omega^j\}$ and what kind of $\Omega = \{\omega^j\}$ it belongs to.

In this process, two problems should be solved: one is how to extract features, which requires the number N_j of features to be as small as possible and effective for classification judgment; second, assuming that there is a vector representing the pattern, how to determine which category it belongs to depends on the discriminant function. For the pattern category set $\Omega = \{\omega^j\}$, there should be $D_1(\mathbf{x}), D_2(\mathbf{x}), \cdots, D_c(\mathbf{x})$ and a total of C discriminant functions correspond to them.

The main program flow of image feature recognition and matching can be shown in Fig. 4.16.

(1) **Linear discriminant function**

The so-called linear discriminant function refers to that the discriminant function is a linear combination of all feature vectors of the image, that is

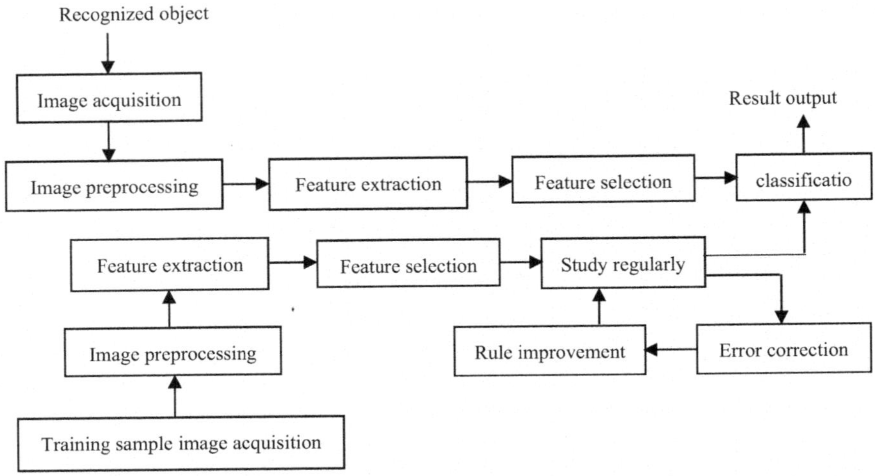

Fig. 4.16 Main program flow of image feature recognition

$$D_j(\mathbf{x}) = \sum_{i=1}^{N_j} \beta_{ji} x_i + \beta_{j0};\; j = 1, 2, \ldots, c \tag{4.114}$$

where $D_j(\mathbf{x})$ represents the discriminant function of the j-th category, β_{ji} represents the coefficient or weight of the i-th feature of the j-th category and β_{j0} is the constant (or threshold) of the j-th category.

The relationship between the two categories at the discrimination boundary is as follows

$$D_k(\mathbf{x}) - D_l(\mathbf{x}) = 0;\; k, l \in [1, 2, \ldots, c];\; k \neq l \tag{4.115}$$

Linear discriminant function is also called linear classifier. Any C class problem can be decomposed into $c - 1$ two-class identification problems. The method is to divide the pattern space into one class and other classes, and then proceed. When the linear discriminant function of each two categories in C class is designed separately, a total of $\frac{1}{2}c(c-1)$ linear discriminant functions will be used to realize all classification. See Fig. 4.17 for the classification process of multi-category problem by decomposing it into $c - 1$ two categories.

The equation is a straight line in two-dimensional space R^2, a plane in three-dimensional space R^3 and a hyperplane in n-dimensional space R^n.

In general, $D_k(\mathbf{x}) - D_l(\mathbf{x})$ is written in the following form

$$D_k(\mathbf{x}) - D_l(\mathbf{x}) = \sum_{i}^{N} (\beta_{ki} - \beta_{li}) x_i + (\beta_{k0} - \beta_{l0}) \tag{4.116}$$

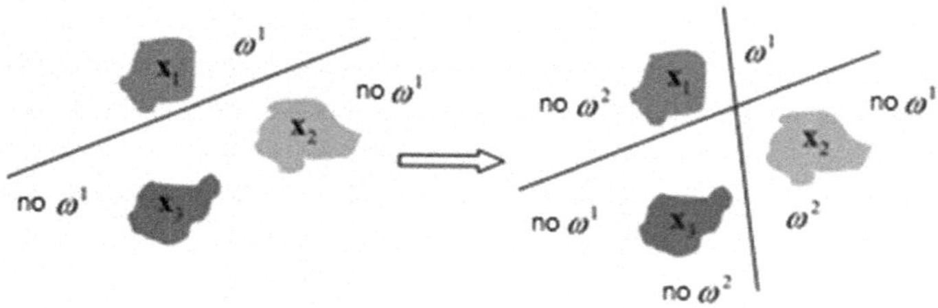

Fig. 4.17 Schematic diagram of multi-class problem decomposition into $c - 1$ two-class classification process

If $D_k(\mathbf{x}) - D_l(\mathbf{x}) > 0$, then $\mathbf{x} \in \omega^k$; If $D_k(\mathbf{x}) - D_l(\mathbf{x}) < 0$, then $\mathbf{x} \in \omega^l$.

It can be seen from the above that two-class linear classifier is the simplest and most basic decomposition method. The decision boundary of separating two-class is represented by $D_1(\mathbf{x}) - D_2(\mathbf{x}) = 0$. For any specific input image mode, the sizes of $D_1(\mathbf{x})$ and $D_2(\mathbf{x})$ must be determined.

For a function $D(\mathbf{x}) = D_1(\mathbf{x}) - D_2(\mathbf{x})$, when it is positive for the first type of mode $D(\mathbf{x})$ and negative for the second type of mode $D(\mathbf{x})$, the processing objects can be classified as long as a group of weight input modes corresponding to $D(\mathbf{x})$ are processed and their output symbols are judged. The principle block diagram of the classifier performing this operation is shown in Fig. 4.18.

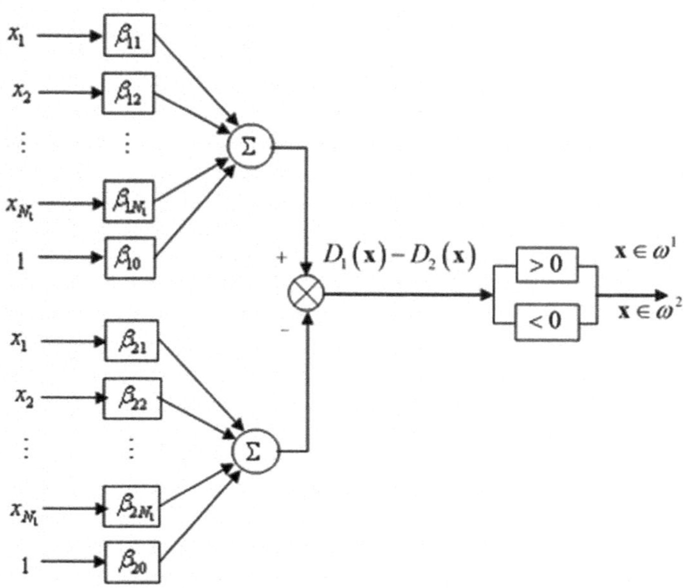

Fig. 4.18 Block diagram of classifier operation principle

In order to find the appropriate coefficients in the linear classifier and make the classification as error free as possible, the basic method is the experimental method. In the experiment, all coefficients can be set as 1 and sent to each mode. If the classification is wrong, the coefficients will be adjusted. This process is called the training or learning of linear classifier. For example, put the N-dimension features $\mathbf{x}$ and 1 together to form $\mathbf{y}$, that is

$$\mathbf{y} = \begin{bmatrix} x_1 \ x_2 \ \ldots \ x_N \ 1 \end{bmatrix}^{\mathrm{T}} \tag{4.117}$$

Let the combination coefficient be

$$\boldsymbol{\beta} = \begin{bmatrix} \beta_1 \ \beta_2 \ \ldots \ \beta_N \ \beta_{N+1} \end{bmatrix}^{\mathrm{T}} \tag{4.118}$$

Then the linear discriminant function can be rewritten as

$$D(\mathbf{x}) = \mathbf{y}^T \boldsymbol{\beta} \tag{4.119}$$

Considering that the eigenvectors belong to two different pattern classes, there are two training sets T_1 and T_2. When the two training sets are linearly separable, it means that the inner product of the test vector $\mathbf{y}$ and the weighted vector $\boldsymbol{\beta}$ has two states as follows

$$\mathbf{y}^T \boldsymbol{\beta} > 0; \ \mathbf{y} \in T_1 \tag{4.120}$$

$$\mathbf{y}^T \boldsymbol{\beta} < 0; \ \mathbf{y} \in T_2 \tag{4.121}$$

In the training process, when the output of the classifier cannot meet the conditions of Eqs. (4.120) and (4.121), the coefficients can be adjusted through the training of "error correction". Assuming the first mode $\mathbf{y}^T \boldsymbol{\beta}$ is not greater than zero, it indicates that the coefficient is not large enough, and the error can be corrected by increasing the coefficient, that is.

For $\mathbf{y} \in T_1$, if $\mathbf{y}^T \boldsymbol{\beta} \leq 0$, then

$$\boldsymbol{\beta}^* = \boldsymbol{\beta} + \alpha \mathbf{y} \tag{4.122}$$

Similarly, for $\mathbf{y} \in T_2$, if $\mathbf{y}^T \boldsymbol{\beta} > 0$, then

$$\boldsymbol{\beta}^* = \boldsymbol{\beta} - \alpha \mathbf{y} \tag{4.123}$$

The commonly used error correction methods are: fixed increment, absolute correction, partial correction rules and so on.

1) Fixed increment rule: select α as a constant non-negative number.

2) Absolute correction rule: take α as a minimum integer, which can make the value of $\mathbf{y}^T\boldsymbol{\beta}$ just greater than zero, that is

$$\alpha \geq \left\lfloor \frac{\left|\mathbf{y}^T\boldsymbol{\beta}\right|}{\mathbf{y}^T\mathbf{y}} \right\rfloor \tag{4.124}$$

3) Partial correction rule is

$$\alpha = \gamma \frac{\left|\mathbf{y}^T\boldsymbol{\beta}\right|}{\mathbf{y}^T\mathbf{y}}; \quad 0 < \gamma \leq 2 \tag{4.125}$$

(2) **Minimum distance classification function**

In the linear classifier, an important class is to take the mean vector of each category sample (input pattern and feature space) as the representative point of each category, and judge its category according to the minimum distance between the sample to be recognized and the representative point of each category. The so-called minimum distance classification function, that is, the minimum distance classifier, takes the closest distance between the unknown category image and one of the categories as the criterion for c class template. The decision surface is the vertical bisection plane of the mean line between the two categories.

It is assumed that the image has c class, including $\omega^1, \omega^2, \ldots, \omega^c$, respectively, that is, $\Omega = \{\omega^j\}, j = 1, 2, \ldots, c$; There are c reference vectors (i.e., training sample vectors) $\mathbf{R}_1, \mathbf{R}_2, \ldots, \mathbf{R}_c$; $\mathbf{R}_j$ $(j = 1, 2, \ldots, c)$ is associated with mode class ω^j. The minimum distance classification of $\mathbf{R}_j$ is to divide the input new mode $\mathbf{x}$ into ω^j category, and its classification criterion is to compare the distance between $\mathbf{x}$ and the reference model $\mathbf{R}_1, \mathbf{R}_2, \ldots, \mathbf{R}_c$. When $\mathbf{x}$ is closest to one of $\mathbf{R}_j$, $\mathbf{x}$ belongs to the closest category. The distance between $\mathbf{x}$ and $\mathbf{R}_j$ can be expressed as

$$\left|\mathbf{x} - \mathbf{R}_j\right| = \sqrt{\left[\mathbf{x} - \mathbf{R}_j\right]^T\left[\mathbf{x} - \mathbf{R}_j\right]} \tag{4.126}$$

Thus obtained

$$\left|\mathbf{x} - \mathbf{R}_j\right|^2 = \mathbf{x}^T\mathbf{x} - \left(\mathbf{x}^T\mathbf{R}_j + \mathbf{R}_j^T\mathbf{x} - \mathbf{R}_j^T\mathbf{R}_j\right) \tag{4.127}$$

It can be seen that $\mathbf{x}^T\mathbf{x}$ has nothing to do with $\mathbf{R}_j$. Therefore, the discriminant function at this time can be expressed as

$$D_j(\mathbf{x}) = \left(\mathbf{x}^T\mathbf{R}_j + \mathbf{R}_j^T\mathbf{x} - \mathbf{R}_j^T\mathbf{R}_j\right) \tag{4.128}$$

It can be seen that minimizing the distance $\left|\mathbf{x} - \mathbf{R}_j\right|$ is equivalent to maximizing $D_j(\mathbf{x})$. In other words, $D_j(\mathbf{x})$ gets the maximum value as long as $\mathbf{x} \in \omega^j$. Obviously, $D_j(\mathbf{x})$ is also a linear function, so the minimum distance classifier is also a linear classifier.

It must be pointed out that in the minimum distance classification, it is difficult to solve this situation when the point on the decision boundary is equidistant from the two adjacent classes. At this time, we must find new features and reclassify them.

The minimum distance classification function can also be expressed by decision region.

Suppose there are two kinds of problems ω^1 and ω^2, and their reference vectors correspond to $\mathbf{R}_1$ and $\mathbf{R}_2$, respectively, when

$$|\mathbf{x} - \mathbf{R}_1| < |\mathbf{x} - \mathbf{R}_2| \tag{4.129}$$

or

$$\sqrt{\sum_{i=1}^{N_1} (x_i - r_{1i})^2} < \sqrt{\sum_{i=1}^{N_2} (x_i - r_{2i})^2} \tag{4.130}$$

then $\mathbf{x} \in \omega^1$, otherwise $\mathbf{x} \in \omega^2$.

Where $r_{1i} \in \mathbf{R}_1$, that is $\mathbf{R}_1 = \left[r_{11}\ r_{12}\ \ldots\ r_{1N_1} \right]^{\mathrm{T}}$;

$r_{2i} \in \mathbf{R}_2$, that is $\mathbf{R}_2 = \left[r_{21}\ r_{22}\ \ldots\ r_{2N_2} \right]^{\mathrm{T}}$.

Make a bisector for the sample mean line of reference vectors $\mathbf{R}_1$ and $\mathbf{R}_2$. The $\mathbf{R}_1$ decision area and $\mathbf{R}_2$ decision area are respectively on the left and right of the bisector, which is called the middle decision line (surface).

For two kinds of cases, the boundary of the decision-making region is a line and the decision-making region is two planes. For three cases, the decision interface is hyperplane and the decision area is half space. The geometric meaning of the classification process for the two types of problems ω^1 and ω^2 is shown in Fig. 4.19. In the figure, μ_1 and μ_2 are the sample mean values of reference vectors $\mathbf{R}_1$ and $\mathbf{R}_2$, respectively.

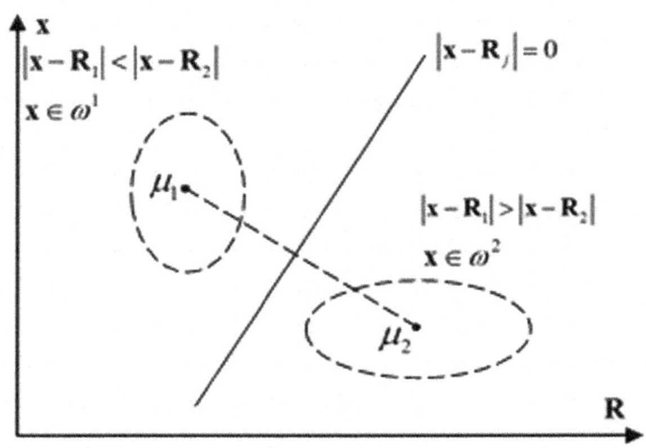

Fig. 4.19 Geometric significance of classification process of two types of problems

(3) **Nearest neighborhood classification function**

The nearest neighborhood classification function method (referred to as the nearest neighborhood classification method) is a method widely used in image recognition, and it is also one of the most important classification methods. In the minimum distance classification method, a standard vector is taken as the representative. Expand this kind of problem a little. A category cannot take only one representative. Expand the concept of minimum distance from the distance between a point and a point to the distance between a point and a group of points. This is the basic idea of nearest neighborhood classification.

Let c c reference vectors $\mathbf{R}_1, \mathbf{R}_2, \ldots, \mathbf{R}_c$,($\mathbf{R}_j$; $j = 1, 2, \ldots, c$) are the reference vector sets corresponding to classes $\omega^1, \omega^2, \cdots, \omega^c$, ($\omega^j$; $j = 1, 2, \ldots, c$), that is, the reference vector set $\mathbb{R}$ is

$$\mathbb{R} = \{\mathbf{R}_j\} \tag{4.131}$$

The reference direction quantum set $\mathbb{R}_j$ composed of each class is

$$\mathbb{R}_j = \{\mathbf{R}_j^k\}; k = 1, 2, \ldots, N_k \tag{4.132}$$

Then, the distance between the input feature vector $\mathbf{x}$ and the reference feature vector $\mathbf{R}_j$ can be expressed as

$$d(\mathbf{x}, \mathbf{R}_j) = \min_k |\mathbf{x} - \mathbf{R}_j^k| \tag{4.133}$$

That is, the distance between $\mathbf{x}$ and $\mathbf{R}_j$ is the smallest distance between $\mathbf{x}$ and each vector in the reference vector subset $\mathbb{R}_j$. The calculation method of the distance between spatial points is the same as that in the minimum distance classification method.

The decision boundary of image recognition using this discriminant function will be piecewise linear. Its geometric meaning is shown in Fig. 4.20. Among them, (a) and (b) are two different piecewise linear classification orders. The decision boundary of (a) is much shorter and simpler than that of (b), and the total operation time of the former is much less than that of the latter. It can be seen from the figure that the concept of piecewise linearity is relatively simple, and the piecewise linear boundary can represent a very complex curve. Even if it may have been a nonlinear boundary, this method can be approximately replaced by piecewise linearity. Of course, how to segment and the selection of segmentation order will directly affect the classification effect and operation time. However, on the premise of lack of prior knowledge, the characteristics and distribution law of test samples are often unknown, so the classification process is generally carried out in exploration.

In the two kinds of discrimination of $\mathbf{R}_1^1, \mathbf{R}_1^2 \in \omega^1$ and $\mathbf{R}_2^1, \mathbf{R}_2^2 \in \omega^2$, when the pattern $\mathbf{x}$ is input into the recognition system, the distance between it and each point of each class shall be calculated, respectively.

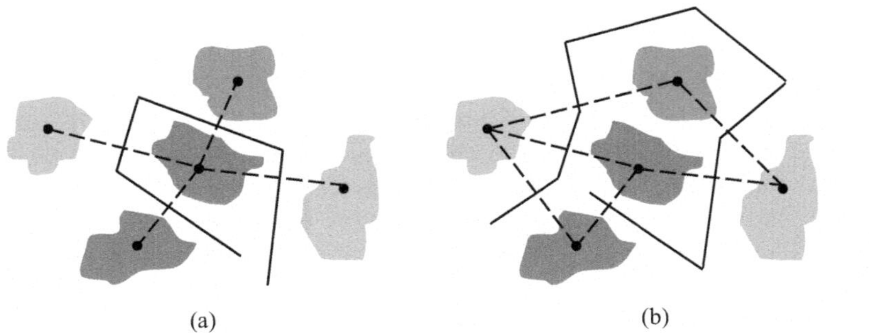

Fig. 4.20 Chematic diagram of geometric significance of piecewise linear classification. **a** Piecewise linear classification order 1, **b** piecewise linear classification order 2

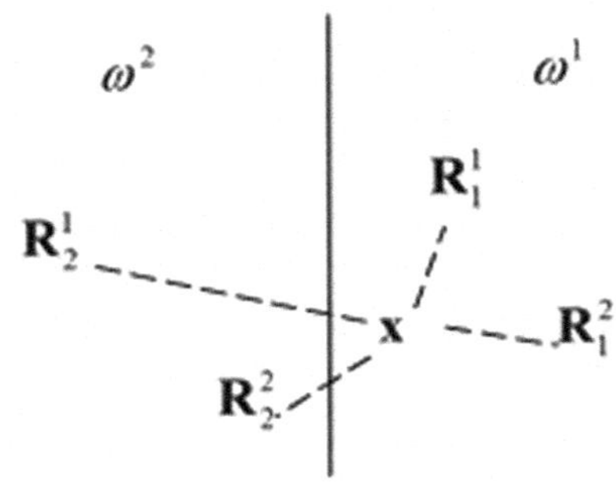

Fig. 4.21 Discrimination diagram of unknown modes in two types of discrimination

Referring to Fig. 4.21, since $\min_k |\mathbf{x} - \mathbf{R}_1^k|$ and $\min_k |\mathbf{x} - \mathbf{R}_2^k|$ obtain the minimum value when $k = 1$ and $k = 2$, respectively, and $\min_k |\mathbf{x} - \mathbf{R}_1^k| < \min_k |\mathbf{x} - \mathbf{R}_2^k|$, it can be determined that $\mathbf{x} \in \mathbf{R}_1^1$.

(4) **Nonlinear discriminant function**

Linear discriminant function is the basic method of pattern recognition. The outstanding advantage of this method is simple calculation. However, when the two types of sample distributions have multi-modal properties and interleave with each other, the simple linear discriminant function often leads to large classification errors. At this time, we need to find other solutions.

There are many ways to solve the above problems. One of them is to use the piecewise linear discriminant function method, which determines the hyperplane, which is composed of several hyperplane segments. Because its basic composition is still composed of hyperplanes, it is still simple compared with general hypercurves (such as Bayesian decision surface). Because it is composed of multiple hyperplanes, it can approach Hypersurfaces of various shapes and has strong adaptability.

Fig. 4.22 Schematic diagram of three discrimination effects

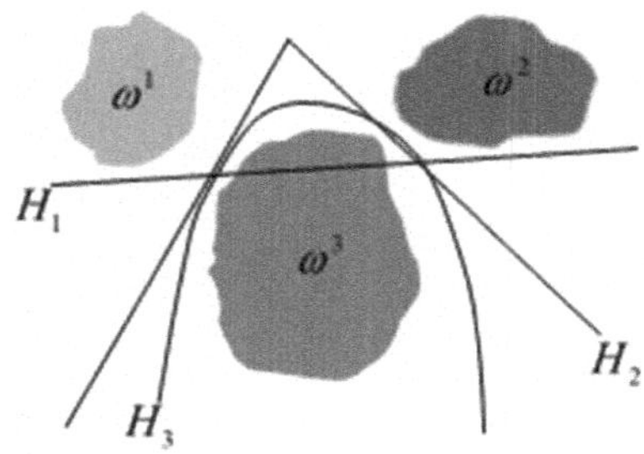

Referring to Fig. 4.22, for the three sample distributions ω^1, ω^2 and ω^3, the interfaces obtained by using linear discriminant function, piecewise linear discriminant function and quadratic nonlinear discriminant function are given, respectively. Among them, H_1, H_2 and H_3 are the decision surfaces generated by linear, piecewise linear and quadratic nonlinear discrimination, respectively. It is not difficult to see from the figure that the error rate of piecewise linear discriminant function is smaller than that of general linear discriminant function and larger than that of general nonlinear discriminant function.

According to the complexity of the problem, the discriminant function can be extended from linear to nonlinear. The nonlinear discriminant function (taking quadratic discriminant function as an example) can be written in the following form

$$D(\mathbf{x}) = \beta_0 + \sum_{k=1}^{N} \beta_k x_k + \sum_{k=1}^{N} \beta_{kk} x_k^2 + 2\sum_{k=1}^{N} \sum_{l=k+1}^{N} \beta_{kl} x_k x_l \tag{4.134}$$

This is a quadratic discriminant function. Usually, the decision boundary of the quadratic discriminant function is a superquadric surface.

In order to determine the discriminant function $D(\mathbf{x})$, M coefficients need to be determined, that is

$$M = \frac{1}{2} N(N+3) + 1 \tag{4.135}$$

This is a very complex calculation process. The decision surface of quadratic discriminant function includes hypersphere, hyperellipsoid, hyperhyperboloid and so on.

For two kinds of problems in which the sample distribution of one kind (assuming ω^1 type) is relatively compact and clustered, while the other kind of samples are evenly scattered around it, we can construct a quadratic discriminant function by the following method:

Calculate the sample mean matrix $\boldsymbol{\mu}_1$ and sample covariance matrix $\boldsymbol{\Sigma}_1$ of class ω^1 as follows

$$\boldsymbol{\mu}_1 = \frac{1}{N_1}\sum_{j=1}^{N_1} x_j \tag{4.136}$$

$$\boldsymbol{\Sigma}_1 = \frac{1}{N_1 - 1}\sum_{j=1}^{N_1} (\mathbf{x}_j - \boldsymbol{\mu}_1)(\mathbf{x}_j - \boldsymbol{\mu}_1)^T \tag{4.137}$$

where N_1 is the total number of sample records of reference class ω^1.

Define quadratic discriminant function

$$D(\mathbf{x}) = K^2 - (\mathbf{x} - \boldsymbol{\mu}_1)^T \boldsymbol{\Sigma}_1^{-1}(\mathbf{x} - \boldsymbol{\mu}_1) \tag{4.138}$$

The decision rules are: $D(\mathbf{x}) > 0, \mathbf{x} \in \omega^1$; $D(\mathbf{x}) < 0, \mathbf{x} \in \omega^2$.

$D(\mathbf{x}) = 0$ determines a hyperellipsoid in $\mathbf{x}$ space, and its size is controlled by K in formula (4.138). The larger K, the fewer sample points of the first type falling outside the hyperellipsoid. Therefore, the probability of being wrongly divided into the second type by the first type is smaller, but the probability of being wrongly divided into the first type by the second type will increase.

Of course, there are many other nonlinear discriminant functions, which will not be described here.

2. **Statistical classification method**

The classification method based on decision theory is carried out without noise interference. At this time, the measured features can indeed represent the pattern. If there is noise in feature extraction, the extracted features may not represent the pattern (image), so statistical method should be used for classification.

Using statistical methods for image feature extraction, learning and classification is one of the main methods of image recognition. One of the most basic contents of statistical methods is Bayesian analysis, including Bayesian decision-making method, Bayesian classification function, Bayesian estimation theory, Bayesian learning method, Bayesian distance and so on. The following focuses on: Bayesian rule, Bayesian classification with minimum error rate and Bayesian classification with minimum risk.

(1) **Bayesian rule**

In classical probability, Bayesian theorem has been familiar to everyone, that is

$$P(B_i|A) = \frac{P(B_i)P(A|B_i)}{\sum_{j=1}^{n} P(B_j)P(A|B_j)} \tag{4.139}$$

where $B_1, B_2, \ldots, B_n$ is n mutually incompatible events, $P(B_i)$ is the a priori probability of event B_i and $P(A|B_i)$ is the conditional probability of A under the condition that B_i has occurred. Bayesian theorem shows that given the prior probabilities $P(B_i)$ and conditional probability $P(A|B_i)$ of random events $B_1, B_2, \ldots, B_n$, the posterior probability of event B_i can be determined as $P(B_i|A)$ when event A occurs.

Bayesian formula is often used in classification problems. Suppose **x** represents the random variable of the state or feature of the event, which can represent the gray level or shape of the image, and ω_i represents the discrete random variable of the event category. The following formula can be used to classify the image according to the feature **x**

$$P(\omega_i|x) = \frac{P(\omega_i)P(x|\omega_i)}{\sum_{j=1}^{n} P(\omega_j)P(x|\omega_j)} \tag{4.140}$$

In the formula, $P(\omega_i)$ is the prior probability of event ω_i, which represents the preliminary rough understanding that the event belongs to ω_i, $P(x|\omega_i)$ represents the conditional probability that the event belongs to ω_i and has **x** state and $P(\omega_i|x)$ is the posterior probability of ω_i under **x** condition, which represents the possibility of judging that the event belongs to ω_i after observing the state of event **x**. It can be seen from Eq. (4.140) that as long as the prior probability of the category and the conditional probability of **x** are known, the posterior probability of the category can be obtained, and the statistical decision classification can be carried out by adding the minimum error probability or the minimum risk rule.

(2) **Bayesian classification with minimum error rate**

Suppose there are two categories, each of which is represented by two statistical parameters, that is

$$\begin{aligned} \omega_1 &: P(\omega_1),\ P(x|\omega_1) \\ \omega_2 &: P(\omega_2),\ P(x|\omega_2) \end{aligned}$$

where $P(\omega_1)$ and $P(\omega_2)$ are a priori probabilities, and $P(x|\omega_1)$ and $P(x|\omega_2)$ are conditional probability density functions. Under the influence of noise interference, each mode can only obtain the probability distribution of a certain type of mode.

The classification method using Bayesian rule is as follows:

If $P(\omega_1)P(x|\omega_1) > P(\omega_2)P(x|\omega_2)$, it is classified as $x \in \omega_1$;

Conversely, $P(\omega_1)P(x|\omega_1) < P(\omega_2)P(x|\omega_2)$, it is classified as $x \in \omega_2$.

In the above classification rules, product $P(\omega_i)P(x|\omega_i)$ plays the role of discriminant function. For convenience, in application, the logarithmic form of the product is usually taken as follows

$$\left.\begin{array}{l} \text{If}\ \ \lg[P(\omega_1)P(x|\omega_1)] > \lg[P(\omega_2)P(x|\omega_2)]\ \ \text{Then}\ \ x \in \omega_1 \\ \text{If}\ \ \lg[P(\omega_1)P(x|\omega_1)] < \lg[P(\omega_2)P(x|\omega_2)]\ \ \text{Then}\ \ x \in \omega_2 \end{array}\right\} \tag{4.141}$$

or

$$\left.\begin{array}{l} \text{If}\ \ \lg\frac{P(x|\omega_1)}{P(x|\omega_2)} > \lg\frac{P(\omega_2)}{P(\omega_1)}\ \ \text{Then}\ \ x \in \omega_1 \\ \text{If}\ \ \lg\frac{P(x|\omega_1)}{P(x|\omega_2)} < \lg\frac{P(\omega_2)}{P(\omega_1)}\ \ \text{Then}\ \ x \in \omega_2 \end{array}\right\} \tag{4.142}$$

For two kinds of classification problems, the interface is

$$\lg[P(\omega_1)P(x|\omega_1)] - \lg[P(\omega_2)P(x|\omega_2)] = 0 \tag{4.143}$$

Let the mean vector of the event eigenvector $\mathbf{x}$ be $\boldsymbol{\mu}_i$ and the covariance matrix be $\boldsymbol{\Sigma}_i$, $i = 1, 2$. If $P(\mathbf{x}|\omega_i)$ is a normal distribution, that is, $P(\mathbf{x}|\omega_i) \sim N(\boldsymbol{\mu}_i, \boldsymbol{\Sigma}_i)$, its decision interface can be obtained as

$$P(\mathbf{x}|\omega_i) = (2\pi)^{-\frac{N}{2}}|\boldsymbol{\Sigma}_i|^{-\frac{1}{2}}\exp\left[-\frac{1}{2}(\mathbf{x}-\boldsymbol{\mu}_i)^T\boldsymbol{\Sigma}_i^{-1}(\mathbf{x}-\boldsymbol{\mu}_i)\right] \tag{4.144}$$

where N is the number of rows (columns) of the covariance matrix $\boldsymbol{\Sigma}_i$, that is, $\boldsymbol{\Sigma}_i$ is the square matrix of $N \times N$.

If $x \in \omega_1$, the following formula can be obtained from formula (4.142) and formula (4.144)

$$-\frac{1}{2}\lg\left|\frac{\boldsymbol{\Sigma}_1}{\boldsymbol{\Sigma}_2}\right| - \frac{1}{2}\left[(\mathbf{x}-\boldsymbol{\mu}_1)^T\boldsymbol{\Sigma}_1^{-1}(\mathbf{x}-\boldsymbol{\mu}_1)\right] + \frac{1}{2}\left[(\mathbf{x}-\boldsymbol{\mu}_2)^T\boldsymbol{\Sigma}_2^{-1}(\mathbf{x}-\boldsymbol{\mu}_2)\right] > \lg\frac{P(\omega_2)}{P(\omega_1)} \tag{4.145}$$

At this time, the decision boundary between the two classes is quadratic.

If the two covariance matrices are the same, that is, $\boldsymbol{\Sigma}_1 = \boldsymbol{\Sigma}_2 = \boldsymbol{\Sigma}$, then

$$\left.\begin{array}{ll}\text{If}\ \ \mathbf{x}^T\boldsymbol{\Sigma}^{-1}(\boldsymbol{\mu}_1-\boldsymbol{\mu}_2)+\frac{1}{2}(\boldsymbol{\mu}_1+\boldsymbol{\mu}_2)^T\boldsymbol{\Sigma}^{-1}(\boldsymbol{\mu}_1-\boldsymbol{\mu}_2) > \lg\frac{P(\omega_2)}{P(\omega_1)} & \text{Then}\ \ \mathbf{x}\in\omega_1\\ \text{If}\ \ \mathbf{x}^T\boldsymbol{\Sigma}^{-1}(\boldsymbol{\mu}_1-\boldsymbol{\mu}_2)+\frac{1}{2}(\boldsymbol{\mu}_1+\boldsymbol{\mu}_2)^T\boldsymbol{\Sigma}^{-1}(\boldsymbol{\mu}_1-\boldsymbol{\mu}_2) < \lg\frac{P(\omega_2)}{P(\omega_1)} & \text{Then}\ \ \mathbf{x}\in\omega_2\end{array}\right\} \tag{4.146}$$

In other words, the boundary decision becomes linear, and the Bayesian classification function with the minimum error rate is the linear discriminant function. When $\boldsymbol{\Sigma} = \mathbf{I}$, the Bayesian classification function of minimum error rate is the minimum distance classification function.

Therefore, for the classification problem of two classes, if each class is normally distributed, but the covariance matrix is different, the boundary is a quadratic function; if N is very large, it is very troublesome to find $\boldsymbol{\Sigma}_i^{-1}$. If $\boldsymbol{\Sigma}_1 = \boldsymbol{\Sigma}_2 = \boldsymbol{\Sigma}$ is still assumed, it will make an error. For example, in image shape recognition, the problem of spherical ball may become a plane problem. In this case, it is necessary to find other effective methods to solve it.

(3) **Bayesian classification of minimum risk**

Let $\mathbf{x} = \{x_1, x_2, \ldots, x_N\} = \left[x_1\ x_2\ \ldots\ x_N\right]^T$ be a random variable, and there are c classes, that is, $\omega_1, \omega_2, \cdots, \omega_c$; At the same time, it is assumed that $P(\mathbf{x}|\omega_i)$ and $P(\omega_i)$ are known for each type of mode ω_i. Based on $P(\mathbf{x}|\omega_i)$ and $P(\omega_i)$, the success condition of a classifier is to complete the classification task under the condition of minimum misjudgment probability.

Define a decision function $D(\mathbf{x})$. When $D(\mathbf{x}) = d_i$, it means that hypothesis $\mathbf{x} \in \omega_i$ is accepted. The loss function $L(d_j, \omega_i)$ is introduced to represent the loss caused by classifying the input mode $\mathbf{x}$ from ω_i to ω_j (the decision made is d_j).

Due to the introduction of the concept of "loss", when considering the loss caused by misjudgment, we can not only classify according to the size of a posteriori probability but must consider whether the classification result minimizes the loss. The conditional expected loss in the case of decision d_i is

$$R(d_i|\mathbf{x}) = \sum_{j=1}^{c} L(d_i, \omega_j) P(\omega_j|\mathbf{x}) \tag{4.147}$$

where $R(d_i|\mathbf{x})$ is also called conditional risk.

Define the expected risk R as

$$R = \int R(D(\mathbf{x})|\mathbf{x}) P(\mathbf{x}) \, \mathrm{d}\mathbf{x} \tag{4.148}$$

where $\mathrm{d}\mathbf{x}$ is the volume element in the feature space, and the integration is carried out in the whole feature space.

Select the appropriate decision $D(\mathbf{x}) = d_i; i = 1, 2, \ldots, c$ to minimize the expected risk R. If the conditional risk is minimized when taking each decision, the expected risk must be minimized when making decisions on all $\mathbf{x}$. This decision is called minimum risk Bayesian decision, that is, if $R(d_k|\mathbf{x}) = \min_i\{R(d_i|\mathbf{x})\}$, then $D(\mathbf{x}) = d_k$, $\mathbf{x} \in \omega_k$.

3. **Discrimination similarity of feature classification in image recognition process**

With the feature vector, the test target can be recognized by establishing the similarity between the known training sample feature vector and the test target (sample) feature vector. How to establish feature vector similarity is the so-called classification principle. The classification principle can be determined according to prior knowledge or multiple experiments and field observations in advance.

Taking the nearest neighborhood algorithm as an example, among many classification criteria, this is one of the most important methods of non-parametric classification.

Let, the category set for judging similarity is $\Omega = \{\omega^j\}$, $j = 1, 2, \ldots, c$; N_j is the number of matching samples indicating the j-th category.

Then, the discriminant function of the nearest neighborhood class ω^j can be specified as

$$G_j(\mathbf{x}, \mathbf{y}) = \min_j \left\{ \left\| \mathbf{T}_y - \mathbf{T}_{xk}^j \right\| \right\} = \min_j \left\{ \sqrt{\left[\mathbf{T}_y - \mathbf{T}_{xk}^j\right]^{\mathrm{T}} \left[\mathbf{T}_y - \mathbf{T}_{xk}^j\right]} \right\}; \quad k = 1, 2, \ldots, N_j \tag{4.149}$$

where $\mathbf{x}$ and $\mathbf{y}$ are training samples and test target samples, respectively.

If $\mathbf{y}$ makes $G_j(\mathbf{x}, \mathbf{y})$ take the minimum value in the case of the k sample, it indicates that the test feature vector of the test target is the training sample feature vector when it is closest to class ω^j, that is, the target at a certain time in the sample set $\mathbf{y}$ is $y \in \omega^j$; in other words, the confidence of the target to be measured that is accurately identified at the moment reaches the highest.

4.3 Practical Algorithms for Face Recognition

In the feature extraction process of face recognition, the first is to extract the face features of each located face through spatial transformation, dimension reduction, machine learning and other methods and use the form of feature vector to represent the face features. That is, a face image uniquely corresponds to a feature vector, which can uniquely correspond to the original face image through transformation. In other words, after face feature extraction, the corresponding relationship between face image and its feature vector can be established.

At present, the mainstream face recognition methods can be roughly divided into: methods based on geometric feature matching, methods based on feature face, methods based on beam graph matching, methods based on neural network, methods based on support vector machine (SVM) and methods based on hidden Markov model (HMM), and so on.

4.3.1 Face Recognition Algorithm Based on Geometric Feature Matching

Since all human faces are composed of eyes, nose, mouth and other organs, although the size and shape of these organs will vary according to different human faces, their shape and distribution structure have certain regularity, so their geometric description can be used as an important feature of human face recognition [3].

The flow of face recognition algorithm based on geometric features is shown in Fig. 4.23. Among them, image preprocessing is a series of operations and processing, such as image graying, image filtering, histogram equalization and so on.

The following focuses on face location and feature extraction algorithms.

1. **Location method of facial organs**

In the process of face recognition, in order to find the parts with prominent features on the face, eyes, nose and mouth are generally selected as the main features of recognition. Therefore, eyes, nose and mouth need to be located. The most important one is the positioning of human eyes, because the characteristics of human eyes are relatively stable and are not easy to be affected by the changes of light or expression, and the accurate positioning of eyes plays an important reference role in the positioning of nose and mouth.

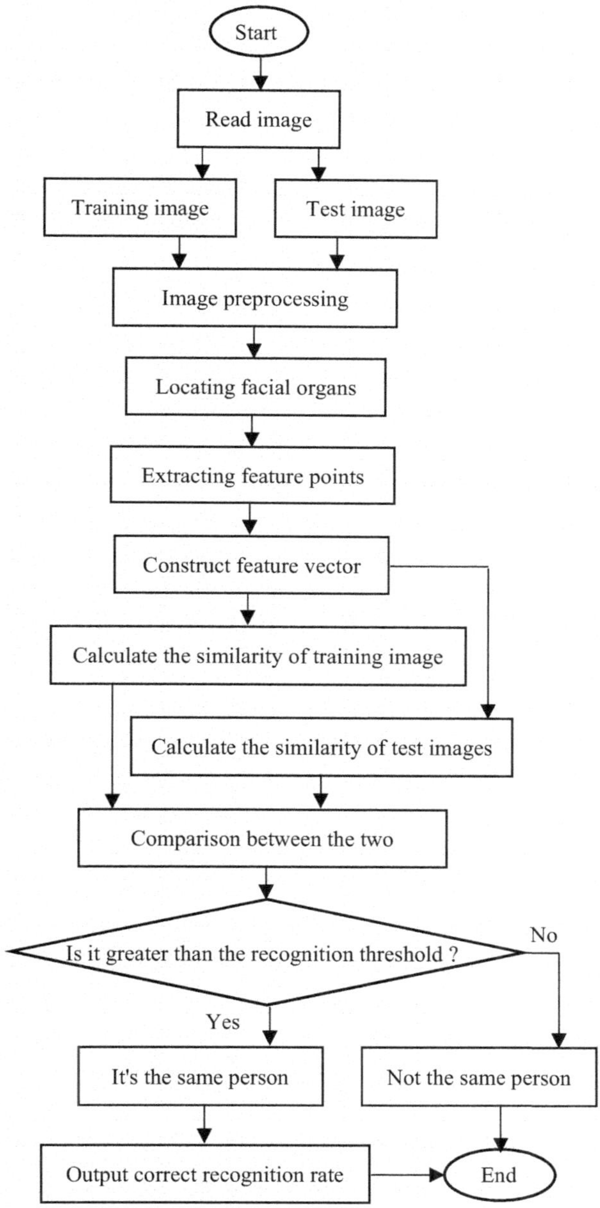

Fig. 4.23 Flow chart of face recognition algorithm based on geometric features

(1) **Common methods of eye location**

The general methods of eye location include gray projection method, template matching method, Hough transform method, neural network algorithm and so on.

1) **Gray projection method**

The gray projection method determines the position of human eyes according to the distribution information of peaks / troughs in the image and the projection of the horizontal and vertical directions of the face image. This method has fast localization speed, but it has poor robustness to the changes of face and pose.

2) **Template matching method**

Using the template matching method, the eye image template in the database can be used to locate the eye directly. However, the template method requires a large amount of calculation and is sensitive to the scale and illumination of the image.

3) **Hough transform method**

Hough transform method is based on edge detection and detects the circular features of eyelid or pupil through template for eye location. However, Hough transform method needs a lot of preprocessing and a large amount of calculation.

The basic principle of Hough transform is to transform the curve (including straight line) in the image space into the parameter space and determine the description parameters of the curve by detecting the extreme points in the parameter space, so as to extract the regular curve in the image. Hough transform can describe the shapes of line, circle, parabola and ellipse in the detected image with a certain functional relationship. It has been successfully applied in many fields such as image analysis and pattern recognition.

4) **Neural network method**

The eye location method based on neural network takes the pixels in the search image window as the input of neural network. If the image window contains an eye image, the output value of the neural network is large, so the position of the eye can be determined.

(2) **Eye positioning process**

Generally speaking, the eyebrows and eyes of the face have fixed image corner-point features:

1) The edge of the white part of the eye is bright, its gray value is also high and there are obvious sudden changes in the gray at the intersection of pupil and white part of the eye and between eye and skin, that is, the corner-point information is rich;
2) Pupil and eyebrow are the places with low gray value in this area, while the gray change frequency of eyebrow area is low, and there is no obvious mutation, that is, the corner-point information is relatively lacking. The general process of eye positioning is shown in Fig. 4.24.

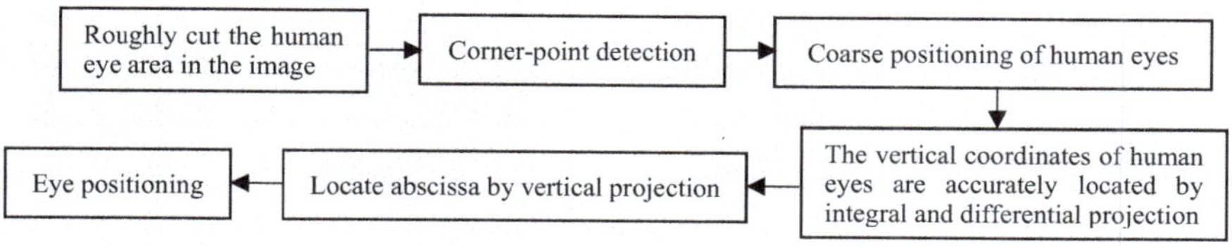

Fig. 4.24 Eye location process in face image

Fig. 4.25 Movement of face detection window

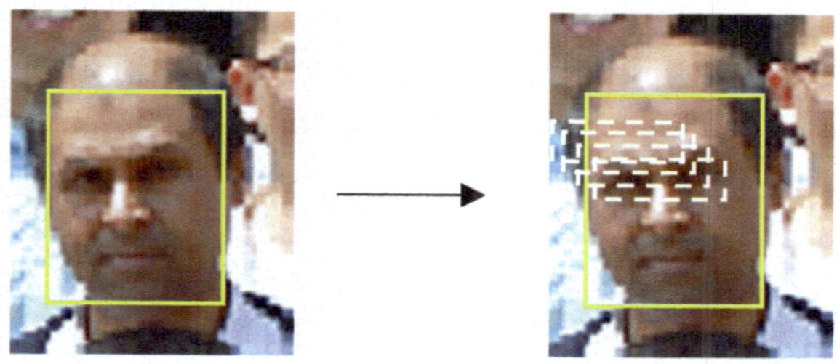

2. **Corner-point detection algorithm**

For the roughly intercepted human eye area, select the window and move continuously in the image (see Fig. 4.25).

If the gray value of the image is represented by $I(x, y)$ and the horizontal and vertical displacement of the window in the image is $[u, v]$, the relative gray change value $E(u, v)$ generated when the window moves $[u, v]$ is

$$E(u, v) = \sum_{x,y} w(x, y)[I(x + u, y + v) - I(x, y)]^2 \tag{4.150}$$

Transform Eq. (4.150) into:

$$E(u, v) = \begin{bmatrix} u & v \end{bmatrix} \begin{bmatrix} I_x^2 & I_x I_y \\ I_x I_y & I_y^2 \end{bmatrix} \begin{bmatrix} u \\ v \end{bmatrix} \tag{4.151}$$

where $I_x I_y$ is the first-order directional derivative of $I(x, y)$, and I_x^2 and I_y^2 are the second-order directional derivatives of $I(x, y)$.

Therefore, for the smaller offset value $[u, v]$, the window can be set to the matrix M of 2×2 size, and the following expression can be obtained approximately, that is

$$E(u, v) \approx \begin{bmatrix} u & v \end{bmatrix} M \begin{bmatrix} u \\ v \end{bmatrix} \tag{4.152}$$

Among them, $M = \sum_{x,y} w(x, y) \begin{bmatrix} I_x^2 & I_x I_y \\ I_x I_y & I_y^2 \end{bmatrix}$

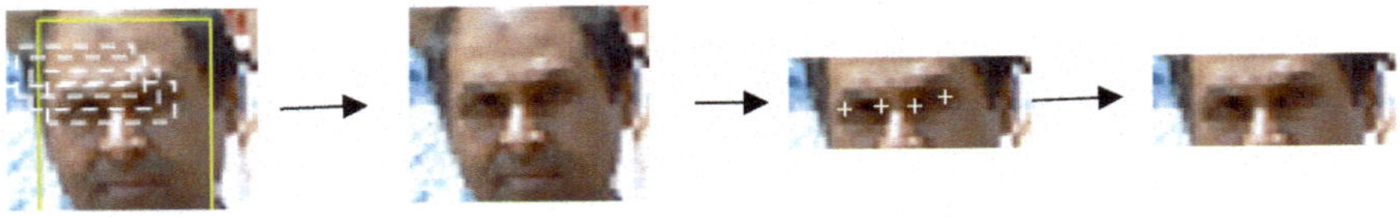

Fig. 4.26 Example of coarse positioning of human eye detection

Calculate the eigenvalues λ_1 and λ_2 of M, so you can create a metric corner function R, that is

$$R = (\lambda_1 \lambda_2) - k(\lambda_1 + \lambda_2)^2 \tag{4.153}$$

where k is the coefficient, which can be taken as 0.04.

For the judgment of image corners, the corner function value R needs to be compared with the set threshold. The existence of image corners can be judged only when the function value R is greater than the set threshold. The selection of threshold often needs to be determined according to the actual situation, because the image corners are redundant.

3 **Coarse positioning method of eye area**

After obtaining the minimum eigenvalue by using the local M-matrix, find the largest eigenvalue in the neighborhood around each pixel point and finally compare it with the minimum distance between the set corners (that is, compare it with the threshold value) to judge whether it is a corner. Figure 4.26 shows an example of coarse positioning of eye corners.

4. **Accurate eye location method based on integral and differential projection**

Assuming that the pixel value at (x, y) in the image is represented by $L(x, y)$, and the horizontal integral projection in the image area $[y_1, y_2]$ and the vertical integral projection in the area $[x_1, x_2]$ are $H(y)$ and $V(x)$, respectively, there is

$$H(y) = \sum_{x=x_1}^{x_2} L(x, y) \tag{4.154}$$

$$V(x) = \sum_{y=y_1}^{y_2} L(x, y) \tag{4.155}$$

Obviously, horizontal projection is to accumulate the gray values of all pixels in a row and then display them, while vertical projection is to accumulate the gray values of all pixels in a column and then display them.

It can be seen that the changes of $H(y)$ and $V(x)$ values, respectively, reflect the changes of gray values of pixels in a row or a column of the image. Therefore, the features in the image can be extracted by analyzing the value of the integral projection function. In the face image, compared with the surrounding area, the human eye area is darker, that is, the gray value is lower. Using this feature, the integral projection method is often used to locate the eyes. Through the horizontal and vertical projection of the face image, the eye area is located according to the distribution of peaks and troughs.

Let the difference between the gray value of $L(x, y)$ at (x, y) point and the adjacent pixel point be

$$\Delta L(x+1, y) = L(x+1, y) - L(x, y) \tag{4.156}$$

And the absolute value of the difference in the same line is accumulated, and then the accumulated value $S(y)$ is

$$S(y) = \sum_{x=x_1}^{x_2} |\Delta L(x+1, y)| \tag{4.157}$$

Because the gray value of the eye is smaller than the surrounding gray value, the integral projection value at the eye is usually smaller. Once the integral value at the eye is greater than or equal to the integral value at the eyebrow, the eye position cannot be accurately located at this time, and even misjudgment will occur.

Therefore, it is necessary to consider the gray change law of the eye area in the horizontal direction (from the skin to the pupil, and from the pupil to the skin), while the gray value of the eyebrow area does not have this regular distribution, and the gray value of the eyebrow is relatively concentrated, so only adding the gray difference in the horizontal direction for comparison will produce a large numerical difference. Then it can be used as another basis for locating the eyes: the greater the change rate is, the greater the difference cumulative value is, and it can be determined as the eyes; otherwise, the area is the eyebrows. In other words, the formula combining integral projection and differential projection can be expressed as follows

$$P(y) = k_1 H(y) - k_2 S(y) > \delta \tag{4.158}$$

where, k_1 and k_2 are the weights corresponding to $H(y)$ and $S(y)$, respectively, and they are taken from (0, 1) interval; δ is the reference threshold, which generally takes the gray mean value of the image area.

When formula (4.154) holds, the value of y is the ordinate value of the eye.

Integral projection and differential projection are complementary to a certain extent. If the two are combined, $P(y)$ not only considers the gray distribution of the image in a certain direction but also reflects the gray change in that direction. Considering the two comprehensively, it has stronger adaptability. As shown in Fig. 4.27, the determined position of the ordinate y of the human eye has been obtained. In the same way, the determined position of the abscissa x of the human eye can also be obtained (see Fig. 4.28) [2].

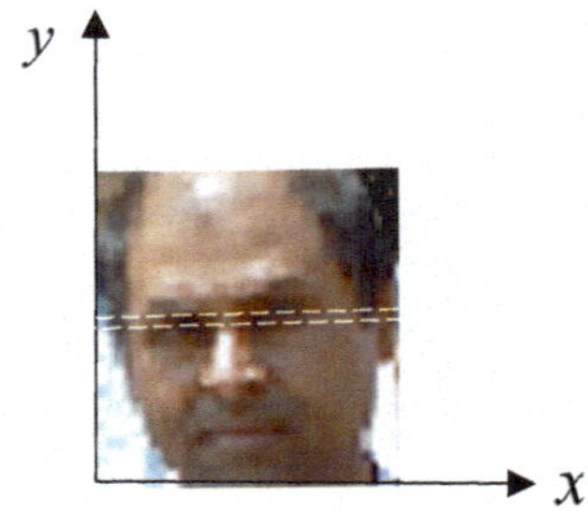

Fig. 4.27 Precise positioning of y coordinate of human eye

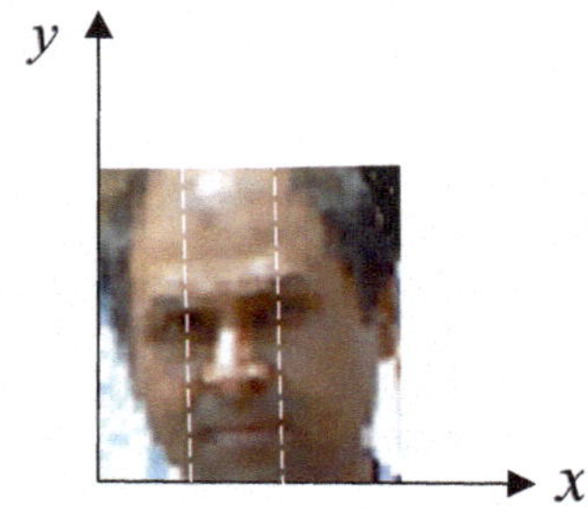

Fig. 4.28 Precise positioning of x coordinate of human eye

The method based on geometric feature matching is to extract the geometric features from the face image for face recognition. Due to the large amount of calculation and low recognition accuracy, this algorithm has not been popularized and applied on a large scale.

4.3.2 Recognition Algorithm Based on Feature Face

The method based on feature face is to map the face image to low-dimensional space by using principal component analysis (PCA) method to realize the recognition result. This method can not only improve the accuracy of the algorithm but also reduce the computational complexity [4].

Principal component analysis is an optimal orthogonal transformation based on the statistical characteristics of targets. The transformation algorithm can make the new components generated after transformation orthogonal or uncorrelated, represent the original vector with some new components to minimize the mean square error and has the characteristics of making the transformation vector more stable and energy more concentrated. Therefore, it has a very important application in feature extraction.

Principal component analysis is a classical technique of multi-variate analysis. It was introduced into the study of biological theory by Pearson in 1901. Karhunen expressed it in the form of probability theory in 1974. Loéve then developed and improved this theory. Therefore, PCA is sometimes called Karhunen–Loéve transformation.

Due to the small amount of calculation, this transformation can be well used for real-time processing. Therefore, using PCA for feature extraction is a good scheme for dimension reduction [5].

1. **Principle of principal component analysis**

As we all know, the attributes contained in data sets can range from more than a dozen to hundreds. With the development of information collection means, data sets with up to thousands of attributes have long been commonplace.

Although each attribute of the data set provides certain information, the amount and importance of information provided by these attributes are different, and in many cases, there are different degrees of correlation between attributes, resulting in a certain overlap of information provided by these attributes.

In other words, statistics show that abnormal behaviors often focus on a small number of attributes. If the algorithm is applied to all attributes, it will not only consume time, increase the complexity of calculation but also affect the correctness of data classification. Therefore, people hope to extract the main attributes from the data attributes and analyze the problem with fewer uncorrelated new variables.

Principal component analysis can just meet this requirement. It can deal with high-dimensional data well, so that low-dimensional data can describe high-dimensional original data in the sense of Least Squares error.

(1) **K-L transformation**

The K-L transformation is expressed as follows

$$\mathbf{y} = \mathbf{A}^T(\mathbf{x} - \boldsymbol{\mu}_x) \tag{4.159}$$

The inverse K-L transformation is expressed as follows

$$\mathbf{x} = \mathbf{A}\mathbf{y} + \boldsymbol{\mu}_x \tag{4.160}$$

Among them, $n \times n$ transformation matrix $\mathbf{A} = \begin{bmatrix} \mathbf{a}_1 \ \mathbf{a}_2 \cdots \mathbf{a}_n \end{bmatrix}$ meets the following characteristic:

1) $\mathbf{A}^T\mathbf{A} = \mathbf{A}\mathbf{A}^T = \mathbf{I}$, is orthogonal matrix, where $\mathbf{I}$ is identity matrix; The column vector $\mathbf{a}_i$ satisfies the standard orthogonality, that is

$$\mathbf{a}_i \cdot \mathbf{a}_j = \mathbf{a}_i^T\mathbf{a}_j = \begin{cases} 1 & i = j \\ 0 & i \neq j \end{cases} \tag{4.161}$$

2) $\mathbf{A}^T\boldsymbol{\Sigma}_x\mathbf{A} = \boldsymbol{\Lambda}$, where $\boldsymbol{\Lambda} = \mathrm{diag}\begin{bmatrix} \lambda_1 \ \lambda_2 \cdots \lambda_n \end{bmatrix}$, $\lambda_1 \geq \lambda_2 \geq \cdots \geq \lambda_n$ is the eigenvalue of $\boldsymbol{\Sigma}_x$ and arranged from large to small.

The column vector $\mathbf{a}_i$ is the eigenvector of $\boldsymbol{\Sigma}_x$, that is

$$\mathbf{\Sigma}_x \mathbf{a}_i = \lambda_n \mathbf{a}_i \tag{4.162}$$

where the subscript i represents the statistical characteristic quantity corresponding to the $\mathbf{x}$ vector.

Obviously, it is easy to obtain the mean vector and covariance matrix of transformed $\mathbf{y}$ as follows

$$\boldsymbol{\mu}_y = E\{\mathbf{y}\} = E\{\mathbf{A}(\mathbf{x} - \boldsymbol{\mu}_x)\} = \mathbf{A}(E\{\mathbf{x}\} - \boldsymbol{\mu}_x) = \mathbf{0} \tag{4.163}$$

$$\mathbf{\Sigma}_y = E\left\{(\mathbf{y} - \boldsymbol{\mu}_y)(\mathbf{y} - \boldsymbol{\mu}_y)^T\right\} = E\left\{\mathbf{A}^T(\mathbf{x} - \boldsymbol{\mu}_x)(\mathbf{x} - \boldsymbol{\mu}_x)^T\right\} = \mathbf{A}^T\mathbf{\Sigma}_x\mathbf{A} = \mathbf{\Lambda} \tag{4.164}$$

(2) **Properties of K-L transformation**

1) **Minimization of Squares error in dimensionality reduction estimation**

Recover $\mathbf{x}$ with component $y_i = \mathbf{a}_i \cdot (\mathbf{x} - \boldsymbol{\mu}_x) = \mathbf{a}_i^T(\boldsymbol{\mu} - \boldsymbol{\mu}_x)$ $(i = 1, 2, \cdots, n)$ of formula (4.159) as follows

$$\mathbf{x} = \sum_{i=1}^{n} y_i \mathbf{a}_i + \boldsymbol{\mu}_x \tag{4.165}$$

If the y_i component of $1 \le i \le m$ is retained and $\mathbf{x}$ is estimated by replacing the term of y_i in $m < i \le n$ with some constant components u_i, $\hat{\mathbf{x}}$ is estimated to be

$$\hat{\mathbf{x}} = \sum_{i=1}^{m} y_i \mathbf{a}_i + \sum_{i=m+1}^{n} u_i \mathbf{a}_i + \boldsymbol{\mu}_x \tag{4.166}$$

The resulting error is

$$\Delta\mathbf{x} = \mathbf{x} - \hat{\mathbf{x}} = \sum_{i=m+1}^{n} (y_i - u_i)\mathbf{a}_i \tag{4.167}$$

To find the squares error

$$\varepsilon^2 = E\{|\Delta\mathbf{x}|^2\} = E\{\Delta\mathbf{x}^T\Delta\mathbf{x}\} = E\left\{\sum_{i=m+1}^{n} (y_i - u_i)^2 \mathbf{a}_i^T \mathbf{a}_i\right\} = \sum_{i=m+1}^{n} E\{(y_i - u_i)^2\} \tag{4.168}$$

Minimize this error

$$\frac{\partial \varepsilon^2}{\partial c_i} = -2E\{y_i - u_i\} = -2E\{y_i\} - u_i = 0 \tag{4.169}$$

Since $E\{y_i\} = 0$, resulting in $u_i = 0$, the estimation formula (4.166) is simplified to

$$\hat{\mathbf{x}} = \sum_{i=1}^{m} y_i \mathbf{a}_i + \boldsymbol{\mu}_x \tag{4.170}$$

$y_1, y_2, \ldots, y_m$ is called the principal component, while $y_{m+1}, y_{m+2}, \ldots, y_n$ discarded is called the non-principal component.

Substitute $u_i = 0$ into the squares error formula (4.168) to obtain the following formula

$$\begin{aligned}\varepsilon^2 &= \sum_{i=k+1}^{n} E\{y_i^2\} = \sum_{i=k+1}^{n} E\left\{\left[\mathbf{a}_i^T(\mathbf{x}-\boldsymbol{\mu}_x)\right]^2\right\} \\ &= \sum_{i=k+1}^{n} \mathbf{a}_i^T E\left\{(\mathbf{x}-\boldsymbol{\mu}_x)(\mathbf{x}-\boldsymbol{\mu}_x)^T \mathbf{a}_i\right\} = \sum_{i=k+1}^{n} \mathbf{a}_i^T \boldsymbol{\Sigma}_x \mathbf{a}_i\end{aligned} \tag{4.171}$$

Let $\lambda_1 \geq \lambda_2 \geq \cdots \geq \lambda_n$ be the eigenvalue of $\boldsymbol{\Sigma}_x$. Under the constraint of $|\mathbf{a}| = 1$, from the quadratic optimization, $\mathbf{a}^T \boldsymbol{\Sigma}_x \mathbf{a}$ minimizing $\mathbf{a}$ is the eigenvector $\mathbf{a}_n$ corresponding to the smallest eigenvalue λ_n, which satisfies $\boldsymbol{\Sigma}_x \mathbf{a}_n = \lambda_n \mathbf{a}_n$.

If further dimensionality reduction is needed, the $n - 1$ component is discarded. By analogy, the basis vector corresponding to the discarded $n - m$ components is the eigenvector corresponding to the smaller $n - m$ eigenvalues. At this time, the relationship formula is as follows

$$\varepsilon^2 = \sum_{i=m+1}^{n} \mathbf{a}_i^T \lambda_i \mathbf{a}_i = \sum_{i=m+1}^{n} \lambda_i \tag{4.172}$$

After calculation, the minimum value is taken.

It can be seen that after K-L transformation, the estimated recovered data by discarding the components corresponding to small eigenvalues is optimal in the sense of squares error, and the squares error is the sum of discarded eigenvalues. Although some components are removed, the reduced dimension subspace established according to this principle can still restore the original information with the remaining features, so as to maintain the optimal Squares error level.

2) **Component variance maximization**

Consider projecting the deviation between vector $\mathbf{x}$ and mean $\boldsymbol{\mu}_x$ onto an axis. The direction of the axis is represented by vector $\mathbf{a}$, and its component is

$$y = \mathbf{a} \cdot (\mathbf{x} - \boldsymbol{\mu}_x) = \mathbf{a}^T(\mathbf{x} - \boldsymbol{\mu}_x) \tag{4.173}$$

We hope that the new component after projection should reflect the original information as much as possible, that is, the variation range of y should be as large as possible. The variation range reflected by the variance of y is as follows

$$\text{var}y = E\{(y - E\{y\})^2\} = E\{y^2\} = \mathbf{a}^T E\left\{(\mathbf{x} - \boldsymbol{\mu}_x)(\mathbf{x} - \boldsymbol{\mu}_x)^T \mathbf{a}\right\} = \mathbf{a}^T \boldsymbol{\Sigma}_x \mathbf{a} \tag{4.174}$$

Among them, $\lambda_1 \geq \lambda_2 \geq \cdots \geq \lambda_n$ is still set as the eigenvalue of $\boldsymbol{\Sigma}_x$.

Under the constraint of $|\mathbf{a}| = 1$, according to the quadratic optimization, $\mathbf{a} = \mathbf{a}_1$ that maximizes $\mathbf{a}^T \boldsymbol{\Sigma}_x \mathbf{a}$ is the eigenvector corresponding to the eigenvalue λ_1, satisfying $\boldsymbol{\Sigma}_x \mathbf{a}_1 = \lambda_1 \mathbf{a}_1$.

By projecting $\mathbf{x}$ onto the $n-1$-dimensional subspace perpendicular to $\mathbf{a}_1$, we can get a similar problem in $n-1$-dimension. At this time, the projection direction with the largest variance should be $\mathbf{a} = \mathbf{a}_2$. If further projection is needed, and so on, until the number of projections reaches m, the same definition of K-L transformation can be obtained.

It can be seen that K-L transform can best reflect the dimensionality reduction information of high-dimensional original information and project it onto the dimensionality reduction space of principal components, so as to maximize the variance of each component.

2. **Kernel principal component analysis**

The above PCA only involves the calculation in the input (data) space. When we consider another form of PCA: this calculation is carried out on the feature space, and its relationship with the input space is nonlinear, which is called linear indivisible problem. At this time, the kernel principal component analysis method should be applied.

Although there is a nonlinear relationship between input space and feature space, which is the nonlinearity of kernel PCA, unlike other forms of nonlinear PCA, the implementation of kernel PCA can still rely on linear algebra. Therefore, we can regard nuclear PCA as a natural extension of general PCA.

The so-called kernel principal component analysis is to extend PCA to kernel PCA by introducing kernel function. Firstly, the input space is mapped to the high-dimensional feature space, and then the principal component analysis is carried out in the feature space. The inner product kernel function of KPCA is based on Mercer's theorem, and the feature space used is the feature space defined in the inner product kernel. KPCA directly originates from PCA. The only difference is that it uses different application spaces.

(1) **Kernel function**

The so-called kernel function is a function $K(\mathbf{x}, \mathbf{y})$ satisfying a specific condition introduced in the process of studying the nonlinear mapping ϕ from a space to a high-dimensional space, which is called kernel function; where $\mathbf{x}$ and $\mathbf{y}$ are vectors in the original feature space D. Moreover, it can be proved that there is a space E, $K(\mathbf{x}, \mathbf{y}) \in E$, and $K(\mathbf{x}, \mathbf{y})$ is the dot product (inner product) of $\phi(\mathbf{x})$ and $\phi(\mathbf{y})$. The kernel function is also a symmetric function.

A specific condition mentioned here refers to the Mercer theorem condition. When the introduced kernel function meets the Mercer theorem condition, the above conclusion naturally holds.

Mercer Theorem An inner product description that can guarantee the symmetric function $K(u, v)$ in a characteristic space, that is, the sufficient and necessary condition for expanding the positive coefficient $\alpha_k > 0$ into $K(u, v) = \sum_{k=1}^{\infty} \alpha_k \phi_k(u)\phi_k(v)$ is that

$$\int g^2(u)\mathrm{d}u < \infty \tag{4.175}$$

For all $g \neq 0$, under the following conditions

$$\iint K(u, v)g(u)g(v)\mathrm{d}u\mathrm{d}v > 0 \tag{4.176}$$

The formula holds.

Using different functions as kernel function $K(u, v)$, we can construct learning machines to realize different types of nonlinear decision surfaces in input space.

(2) **Types of kernel function**

There are three types of kernel functions: polynomial kernel function, radial basis function and Sigmoid function.

Type 1, polynomial kernel function

By the following formula

$$K(x, x_i) = [(\mathbf{x} \cdot \mathbf{x}_i) + C]^q \tag{4.177}$$

A q order polynomial classification function is obtained. Where C is the undetermined parameter.

Type 2: radial basis function

By the following formula

$$K(x, x_i) = \exp\left[-\frac{\|\mathbf{x} - \mathbf{x}_i\|^2}{\sigma^2}\right] \tag{4.178}$$

the radial basis function is determined.

The important difference between the classification function obtained from formula (4.178) and the traditional radial basis function is that each basis function center corresponds to a support vector, and the support vector and output weight are automatically determined by the algorithm.

Type 3: Sigmoid function

The so-called Sigmoid function, also known as the hyperbolic tangent function in mathematics, is a kind of nonlinear action function of neurons.

$$K(x, x_i) = \left[\gamma \tanh(\mathbf{x} \cdot \mathbf{x}_i) + \eta\right] \tag{4.179}$$

Take the above formula (4.179) as the kernel function, where γ and η are undetermined parameters.

This is the learning algorithm of artificial neural network (BP algorithm). The learning of neural network is based on a set of samples. It includes input and output (which can be expressed by expected output). The number of input and output neurons corresponds to the number of components of input and output. The initial weight and threshold of the neural network are given arbitrarily. Learning is to gradually adjust the weight and threshold to make the actual output of the network consistent with the expected output.

(3) **Kernel principal component**

Let vector $\phi(\mathbf{x}_i)$ represent the image derived from the input vector $\mathbf{x}_i$ by transforming $\mathbf{x}_i \in R^N$ space into the feature space of high-dimensional space R^M ($\phi : R^N \to R^M$) under nonlinear mapping ϕ, where N and M represent the dimensions of original space R^N and high-dimensional space R^M after nonlinear mapping, respectively.

Given a set of samples $\{\mathbf{x}_i\}_{i=1}^N$, we have a set of corresponding eigenvectors $\{\phi(\mathbf{x}_i)\}_{i=1}^N$, so we can define a correlation matrix $\widetilde{\mathbf{R}}$ of $M \times M$ in the eigenspace, and

$$\widetilde{\mathbf{R}} = \frac{1}{N}\sum_{i=1}^{N} \phi(\mathbf{x}_i)\phi^T(\mathbf{x}_i) \tag{4.180}$$

Like ordinary PCA, the first thing to do is to ensure that the feature vector set $\{\phi(\mathbf{x}_i)\}_{i=1}^N$ has a zero mean, that is

$$\frac{1}{N}\sum_{i=1}^{N} \phi(\mathbf{x}_i) = 0 \tag{4.181}$$

Assuming that the eigenvectors have gathered in the center, there is the following relationship

$$\widetilde{\mathbf{R}}\tilde{\mathbf{q}} = \tilde{\lambda}\tilde{\mathbf{q}} \tag{4.182}$$

where $\tilde{\lambda}$ is the eigenvalue of $\tilde{\mathbf{q}}$ and $\tilde{\mathbf{q}}$ is its corresponding eigenvector.

Since all eigenvectors satisfying formula (4.182) fall in the space generated by eigenvector set $\{\phi(\mathbf{x}_i)\}_{i=1}^N$ when $\tilde{\lambda} \neq \mathbf{0}$, there is a set of corresponding coefficients $\{\alpha_i\}_{i=1}^N$, which can be written as

$$\tilde{\mathbf{q}} = \sum_{j=1}^{N} \alpha_j \phi(\mathbf{x}_j) \tag{4.183}$$

Substitute formulas (4.180) and (4.183) into formula (4.182) to obtain the following formula

$$\sum_{i=1}^{N}\sum_{j=1}^{N}\alpha_j\phi(\mathbf{x}_i)K\left(\mathbf{x}_i,\mathbf{x}_j\right)=N\tilde{\lambda}\sum_{j=1}^{N}\alpha_j\phi\left(\mathbf{x}_j\right) \tag{4.184}$$

where $K\left(\mathbf{x}_i,\mathbf{x}_j\right)$ is the inner product kernel, and the eigenvector passed is defined by the following formula

$$K\left(\mathbf{x}_i,\mathbf{x}_j\right)=\phi^T(\mathbf{x}_i)\phi\left(\mathbf{x}_j\right) \tag{4.185}$$

The transpose vector $\phi^T(\mathbf{x}_k)$ is used to multiply the two sides of the equal sign of formula (4.184) from the left to obtain the following formula

$$\sum_{i=1}^{N}\sum_{j=1}^{N}\alpha_j K(\mathbf{x}_k,\mathbf{x}_i)K\left(\mathbf{x}_i,\mathbf{x}_j\right)=N\tilde{\lambda}\sum_{j=1}^{N}\alpha_j K\left(\mathbf{x}_k,\mathbf{x}_j\right);\ k=1,2,\ldots,N \tag{4.186}$$

Among them, $K(\mathbf{x}_k,\mathbf{x}_i)=\phi^T(\mathbf{x}_k)\phi(\mathbf{x}_i)$; $K\left(\mathbf{x}_k,\mathbf{x}_j\right)=\phi^T(\mathbf{x}_k)\phi\left(\mathbf{x}_j\right)$

Two matrix definitions need to be introduced here:

Definition 1 $N\times N$ matrix $\mathbf{K}$ is a kernel matrix, and its *ij*-th element is $K\left(\mathbf{x}_i,\mathbf{x}_j\right)$;

Definition 2 $N\times 1$ vector $\boldsymbol{\alpha}$, whose *j*-th element is α_j.

Therefore, formula (4.186) can be written in the following compact matrix form

$$\mathbf{K}^2\boldsymbol{\alpha}=N\tilde{\boldsymbol{\lambda}}\mathbf{K}\alpha \tag{4.187}$$

where the square $\mathbf{K}^2$ of the matrix represents the self-multiplication of $\mathbf{K}$.

Because both ends of formula (4.187) have $\mathbf{K}$, all solutions around the eigenvalue problem can also be expressed as a simpler eigenvalue problem as follows

$$\mathbf{K}\alpha=N\tilde{\boldsymbol{\lambda}}\boldsymbol{\alpha} \tag{4.188}$$

Let $\lambda_1\geq\lambda_2\geq\cdots\geq\lambda_N$ represent the eigenvalue of the kernel matrix $\mathbf{K}$, that is

$$\lambda_j=N\tilde{\lambda}_j;\ j=1,2,\ldots,N \tag{4.189}$$

where $\tilde{\lambda}_j$ is the *j*-th eigenvalue of the correlation matrix $\tilde{\mathbf{R}}$.

Thus, formula (4.188) is changed into the following standard form

$$\mathbf{K}\alpha=\boldsymbol{\lambda}\alpha \tag{4.190}$$

Among them, the coefficient vector $\boldsymbol{\alpha}$ plays the role of the eigenvector corresponding to the eigenvalue $\boldsymbol{\lambda}$ of the kernel matrix $\mathbf{K}$. The coefficient vector $\boldsymbol{\alpha}$

is normalized because it has been required to normalize the eigenvector $\tilde{\mathbf{q}}$ of the correlation matrix $\tilde{\mathbf{R}}$ to the unit length, which is represented by the following formula

$$\tilde{\mathbf{q}}_k^T \tilde{\mathbf{q}}_k = 1 \tag{4.191}$$

where $k = 1, 2, \ldots, p$, and it is assumed that the eigenvalues are arranged in descending order, and λ_p is the minimum nonzero value of the eigenvalues of the kernel matrix $\mathbf{K}$.

The normalization conditions equivalent to formula (4.191) can be obtained by using formulas (4.183) and (4.190), as follows:

$$\boldsymbol{\alpha}_k^T \boldsymbol{\alpha} = \frac{1}{\lambda_k};\ k = 1, 2, \ldots, p \tag{4.192}$$

In order to extract the principal component, it is necessary to calculate the projection of the feature vector $\tilde{\mathbf{q}}_k$ on the feature space, as follows

$$\tilde{\mathbf{q}}_k^T \phi(\mathbf{x}) = \sum_{j=1}^{N} \alpha_{k,j} \phi^T(\mathbf{x}_j)\phi(\mathbf{x}) = \sum_{j=1}^{N} \alpha_{k,j} K(\mathbf{x}_j, \mathbf{x});\ k = 1, 2, \ldots, p \tag{4.193}$$

where vector $\mathbf{x}$ is the "test point", that is, the target vector to be tested, and $\alpha_{k,j}$ is the j-th coefficient of eigenvector $\boldsymbol{\alpha}_k$ corresponding to the k eigenvalue of matrix $\mathbf{K}$. The projection of formula (4.193) is defined on the nonlinear principal component in M dimensional feature space R^M.

The idea of kernel principal component and its analysis is shown in Fig. 4.29, in which the feature space is transformed into $\phi(\mathbf{x})$, which is nonlinear related to the input space. The coordinate planes shown in (a) and (b) in the figure are called input space coordinate plane and feature space coordinate plane, respectively. In (a), the nonlinear contour cluster corresponding to the linear contour in the feature space on the input space is displayed. In (b), the uniformly arranged dotted line indicates that the projection on the main eigenvector is a cluster of constant linear isolines, and the eigenvector is represented by dotted arrow; at the same time, it shows that the transformation $\phi(\mathbf{x})$ adopts the following method: in the feature space, the images induced by data points gather along the feature vector.

It must be pointed out that the feature vector is not drawn in the input space here because it may not exist.

3. **Face recognition based on principal component analysis**

The image recognition steps implemented by using the principle and method of principal component analysis generally include: importing the system training sample set, calculating the characteristic value and characteristic vector of the training sample, importing the test sample set, calculating the characteristic vector of the image to be tested, classification and recognition, etc.

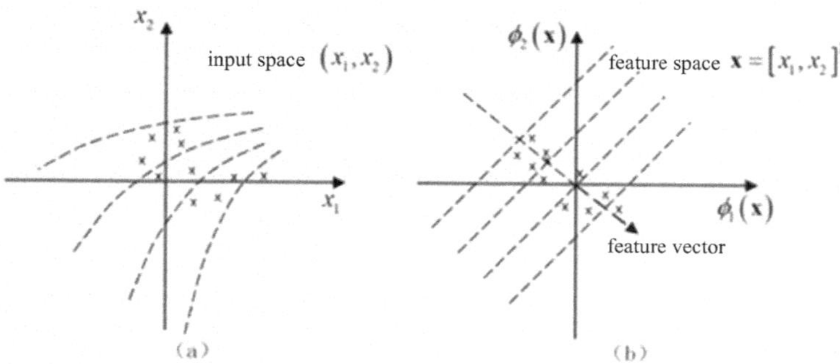

Fig. 4.29 Geometric significance of nonlinear mapping of kernel principal components

Taking face recognition as an example, its essence is the matching of two-dimensional projection images of three-dimensional plastic objects. The difficulty lies in: the uncertainty of face plastic deformation (such as expression); diversity of face patterns (such as beard, hairstyle, glasses, makeup, etc.); uncertainty in the process of image acquisition (such as light intensity, light source direction, etc.).

Face recognition is mainly based on the features on the face, that is, based on the relatively different features between different individuals, while for the same person, there are relatively stable features. Due to the complex changes of human face, generally speaking, its feature description and feature extraction are more difficult.

The identification process can be divided into the following steps.

(1) **Face image preprocessing**

Geometric normalization and gray normalization are generally needed before feature extraction and classification of face images. Geometric normalization refers to transforming the face in the image to the same position and size according to the face location results. Gray normalization refers to the processing of image such as illumination compensation. Illumination compensation can overcome the influence of illumination change to a certain extent and improve the recognition rate.

(2) **Import face training samples**

The so-called importing face training samples is to read each two-dimensional face image data and convert it into one-dimensional vectors. For face images with different expressions, a certain number of images are selected to form a training set [6].

(3) **Calculation of eigenvalues and eigenvectors of training samples**

The calculation steps are as follows:

1) Calculate the average of all kinds of faces;
2) Standardize the training samples of corresponding categories;
3) The training samples constitute the normalization vector of the corresponding category;

4) Rearrange the elements in the normalized vector from large to small to generate a new feature vector;
5) All kinds of eigenvectors are combined into an eigenmatrix.

(4) **The training samples are linearly transformed and projected into the feature space**

The calculation steps are as follows:

1) Because the feature vector corresponding to the larger eigenvalue contains more face feature information, the vector space composed of the feature vectors corresponding to the first few larger eigenvalues can be selected to approximately represent the main information of the face image.
2) Several images in the image database can be projected to this feature space to obtain the projection vector and its corresponding feature space.
3) The inter class covariance matrix is used to represent the faces of the corresponding categories.

(5) **Import test samples**

Select every test sample and import it into the test system one by one.

(6) **Calculate the feature vector of the sample image to be tested**

According to the above steps (3) and (4), calculate the test sample image eigenvalue and its eigenvector.

At this time, after each face image is projected into the feature subspace, it corresponds to a point in the subspace. Similarly, any point in the subspace also corresponds to a face image.

(7) **Face recognition**

Compare all the test images projected into the subspace with the trained images one by one, and then determine the category of the samples to be recognized.

However, this method has poor robustness to illumination and angle changes of face images [7].

4.3.3 *Face Recognition Algorithm Based on "Elastic Beam Graph" Matching*

The main idea of elastic graph matching algorithm is to select some special points as feature points; then, the filter is used to filter the pixel information at the feature point, and the filtered wavelet coefficients are used as the corresponding eigenvalues of the feature point; then the eigenvalues are stored in the data structure called face image. These specially selected feature points and the corresponding eigenvalues are the feature information of the face image; during recognition, the position of feature points in the face image to be recognized is located according to the unified feature

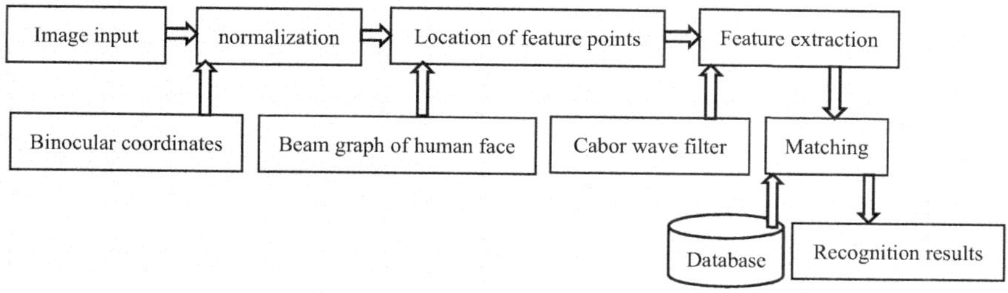

Fig. 4.30 Structure of face recognition system based on elastic beam graph matching

point location criterion, and then the pixel values on these feature points are also filtered to obtain the measured feature values and generate the face image; calculate the similarity between the feature vector of the measured face image and the feature vector of the existing face image in the database, and then get the similarity of the face image, and then get the recognition result [8].

1. **Face recognition system based on elastic beam graph matching**

The elastic beam graph matching algorithm uses the label graph to represent the face image, and the nodes of the label graph are represented by a set of two-dimensional Gabor wavelet-transform coefficients describing the local features of the face. The edge of the label graph is represented by the measurement information describing the corresponding position of two adjacent nodes.

Through the matching between the label images of different face images to realize the connection between the local features of the corresponding parts of the face, so as to compare and classify the solutions of face images. Then the best matching is made for each node position in the graph.

The structure of face recognition system based on elastic beam graph matching is shown in Fig. 4.30.

2. **Image preprocessing**

Because the face image is easily affected by the disturbance of illumination, expression and posture in the extraction process, the face image needs to be normalized before recognition. This is a preprocessing method. This method usually takes the eye coordinates as the reference point and normalizes the face image through geometric affine transformations such as translation, rotation, contraction / amplification and so on. The position and distance between eyes are the basis of face image normalization.

The basic steps of image preprocessing are as follows.

(1) **Geometric normalization**

Locate the eye to a predetermined coordinate and shrink / enlarge the image to a fixed size. Through geometric affine transformations such as translation, rotation, contraction / amplification, the face image can be geometrically normalized. The affine transformation expression is

$$\begin{bmatrix} x & y & 1 \end{bmatrix} = \begin{bmatrix} u & v & 1 \end{bmatrix} \begin{bmatrix} a_{11} & a_{12} & 0 \\ a_{21} & a_{22} & 0 \\ a_{31} & a_{32} & 1 \end{bmatrix} \tag{4.194}$$

where (u, v) represents the coordinates of the pixels in the input image and (x, y) represents the coordinates of the pixels in the output image.

Expand the above formula to obtain the following formula

$$x = a_{11}u + a_{21}v + a_{31} \tag{4.195}$$

$$y = a_{12}u + a_{22}v + a_{32} \tag{4.196}$$

Translation transformation is to add Δu and Δv to the coordinates of all pixels in the image, and its transformation expression is

$$\begin{bmatrix} x & y & 1 \end{bmatrix} = \begin{bmatrix} u & v & 1 \end{bmatrix} \begin{bmatrix} 1 & 0 & 0 \\ 0 & 1 & 0 \\ \Delta u & \Delta v & 1 \end{bmatrix} \tag{4.197}$$

Rotate the points of all pixels in the image counterclockwise around the coordinate origin by θ angle, so it is transformed into the following formula

$$\begin{bmatrix} x & y & 1 \end{bmatrix} = \begin{bmatrix} u & v & 1 \end{bmatrix} \begin{bmatrix} \cos\theta & \sin\theta & 0 \\ -\sin\theta & \cos\theta & 0 \\ 0 & 0 & 1 \end{bmatrix} \tag{4.198}$$

The so-called reduction / amplification transformation is to enlarge or reduce the image according to the given proportion r. When $r > 1$, the image is enlarged; when $0 < r < 1$, the image is reduced, and its transformation expression is

$$\begin{bmatrix} x & y & 1 \end{bmatrix} = \begin{bmatrix} u & v & 1 \end{bmatrix} \begin{bmatrix} r & 0 & 0 \\ 0 & r & 0 \\ 0 & 0 & 1 \end{bmatrix} \tag{4.199}$$

(2) **Gray level interpolation**

In the space of the image after spatial transformation, the gray value of each pixel should be equal to the pixel value of the corresponding position of the image before transformation. However, in practice, after geometric transformation, some pixels of the image will be squeezed together or dispersed, which will make some pixels of the image after transformation correspond to the non-integer coordinate position of the image before transformation. This requires interpolation to calculate the gray value of these pixels. The commonly used methods are nearest neighbor interpolation, bilinear interpolation and bicubic interpolation.

Nearest neighbor interpolation is the simplest interpolation method. The output pixel gray value is the gray value of the expected nearest pixel in the input image. The amount of operation of this method is very small, but the gray value of the transformed image has obvious discontinuity, which will enlarge the high-frequency component in the image and produce obvious blocky effect.

In the so-called bilinear interpolation, the gray value of the output pixel is the average value of the sampling points of the pixel in the 2×2 adjacent area in the input image, and the gray values of the surrounding four adjacent pixels are used for linear interpolation in the vertical and horizontal directions. Compared with the nearest neighbor interpolation method, this method has a slight increase in the amount of calculation, and there is no obvious discontinuity in the gray value of the transformed image, but bilinear interpolation has the property of low-pass filtering, which will lead to the partial loss of high-frequency component information and the blurred image contour.

The cubic interpolation function is best approximated by the cubic interpolation function. The size of its interpolation neighborhood is 4×4, which uses the gray value of 16 adjacent pixels in the calculation. Compared with the first two interpolation methods, this method has the largest amount of calculation, but it can overcome the shortcomings of the first two interpolation methods and has high calculation accuracy.

(3) **Gray normalization**

Through gray-scale transformation, the gray-scale distribution parameters of different images are uniformly adjusted to a predetermined value, which is called gray-scale normalization. Gray normalization is usually to adjust the mean and mean square deviation of image gray distribution to 0 and 1, respectively.

Suppose that the gray value distribution of an image with size $M \times N$ can be expressed in the form of matrix $I_{i \times j}$, $1 \leq i \leq M$ and $1 \leq j \leq N$, and the value of each element of the matrix is the pixel value of the point in the image, then the mean value and mean square deviation of the probability density function of the gray value distribution of the image are

$$\frac{1}{\mu} = \frac{1}{MN} \sum_{i=1}^{M} \sum_{j=1}^{N} I_{i \times j} \tag{4.200}$$

$$\sigma = \sqrt{\frac{1}{MN} \sum_{i=1}^{M} \sum_{j=1}^{N} \left(I_{i \times j} - \mu\right)^2} \tag{4.201}$$

(4) **Gray normalization**

Gray normalization is to improve the image quality through image smoothing, histogram equalization, gray transformation and other image processing methods and unify it to a given level.

The purpose of image smoothing is to suppress noise and improve image quality, which can be carried out in spatial domain and frequency domain. Common methods include neighborhood averaging, spatial filtering and median filtering.

Neighborhood average method is a local spatial domain processing method. It uses the gray average value of each pixel in the pixel neighborhood to replace the original gray value of the pixel to realize image smoothing. Because the noise in the image belongs to high-frequency component, the spatial filtering method uses low-pass filtering to remove the noise and realize image smoothing.

The median filtering method replaces the original gray value of the pixel with the median of the gray value of each pixel in the pixel neighborhood.

The gray histogram reflects the relationship between each gray level in the image and the frequency of pixels with this gray level, which can be expressed as follows

$$P(r_k) = \frac{n_k}{\mathbb{N}} \tag{4.202}$$

where r_k represents the k-th gray level, n_k is the number of pixels of the k-th gray level and $\mathbb{N}$ is the total number of pixels of an image.

Gray histogram is an important statistical feature of image and can be considered as the approximation of image gray probability density function. Histogram equalization is to convert the gray distribution of the image into uniform distribution. For the image with low contrast, its gray histogram distribution is concentrated in a small range. After equalization, the probability of all gray levels of the image is the same. At this time, the entropy of the image is the largest, that is, the amount of information contained in the image is the largest.

r and s, respectively, represent the normalized original image gray and the image gray after histogram equalization. $T(r)$ is the transformation function, then any r in interval [0, 1] corresponds to a s after transformation, $s = T(r)$.

$T(r)$ shall meet two conditions: In interval [0, 1], s is a monotonically increasing function; in interval [0, 1], $r = T^{-1}(r)$ inverse transformation also exists and is a monotonically increasing function.

s is a monotonic increasing function, which ensures the order of gray level from black to white; $r = T^{-1}(r)$ inverse transformation ensures that the gray level of the mapped pixel is within the allowable range.

After histogram equalization, the details of the image are clearer, and the distribution of gray levels in the histogram is more balanced.

3. **Gabor wavelet transform**

In terms of representing non-stationary signals, Fourier transform is difficult to accurately describe the local short-time characteristics of signals. Gaussian function is introduced into Fourier transform to strengthen its representation ability of short-time signals or narrow-band signals by windowing Fourier transform. Therefore, wavelet transform is also known as short-time Fourier transform or windowed Fourier transform.

Wavelet transform has excellent localization function in time domain and frequency domain when processing signals. It is not only applied to speech signal processing but also applied to face recognition technology based on elastic beam matching algorithm.

(1) **Gabor wavelet filter**

One-dimensional wavelet filter is

$$J_j(x) = \int I(x)\psi_j(x - x')\mathrm{d}x \tag{4.203}$$

where $J_j(x)$ represents a small gray value around a given pixel $X(x, y)$ in the face image $I(x)$.

The difference between wavelet transform and Fourier transform is that Fourier transform decomposes the time-domain signal into the sum of several accurate frequency components, while wavelet transform represents it as the sum of time-domain components describing several sub-bands. Therefore, wavelet has good locality and is suitable for analyzing high-frequency signal components such as local features of human face.

(2) **Two-dimensional Gabor wavelet filter**

Two-dimensional Gabor wavelet transform is a powerful tool for multi-scale representation and analysis of images. As the only Gabor function that can obtain the lower bound of the joint uncertainty relationship in spatial and frequency domain, it is often used as wavelet basis function to analyze images. Wavelet transform represents or approximates a signal by the convolution of a set of filter functions and a given signal.

The kernel function of two-dimensional Gabor filter can be expressed as

$$\psi_j(x) = \frac{\|\mathbf{k}_j\|^2}{\sigma^2}\exp\left(-\frac{\|\mathbf{k}_j\|^2\|\mathbf{x}\|^2}{2\sigma^2}\right)\left[\exp(ik_jx) - \exp\left(-\frac{\sigma^2}{2}\right)\right] \tag{4.204}$$

Gabor filter is

$$J_i(\mathbf{x}) = \int I(\mathbf{x}')\psi_j(\mathbf{x} - \mathbf{x}')\mathrm{d}^2\mathbf{x}' \tag{4.205}$$

where the wave vector is

$$\mathbf{k}_j = \begin{bmatrix} k_{jx} \\ k_{jy} \end{bmatrix} = \begin{bmatrix} k_v \cos\varphi_\mu \\ k_v \sin\varphi_\mu \end{bmatrix} \tag{4.206}$$

where $k_v = 2^{\frac{v+2}{2}}\pi$, $\varphi_\mu = \mu\frac{\pi}{8}$, $\vec{x}$ is the image coordinate of the given position; $\mathbf{k}_j$ is the center frequency vector of the filter; φ_μ reflects the directional selectivity of the filter.

According to the characteristics of visual field, visual cortical cells are divided into simple cells, complex cells and super complex cells. The research on the information processing mechanism of mammalian visual cortex shows that the visual response of most simple cells in the visual cortex can be simulated by a set of self-similar two-dimensional wavelets. When the gray value of image pixels is directly used for face recognition, the pattern features are easily affected by facial expression, lighting conditions and various geometric transformations, and it is difficult to achieve high recognition accuracy. Two-dimensional wavelet transform can capture the local structure information corresponding to spatial position, spatial frequency and direction selectivity, which is suitable for representing face images.

4 **Feature point positioning**

In the face recognition system, whether the feature points can be located accurately affects the performance of the recognition system to a great extent. The positioning methods for feature points include snake model method, variable template method, elastic beam graph matching technology and so on. Among them, elastic beam graph matching method locates feature points through displacement estimation based on phase prediction and graph matching technology. The specific process is divided into two steps: rough estimation and accurate estimation.

(1) **Rough estimate**

When locating feature points, the displacement estimation based on phase prediction is needed to make a rough estimation. According to the empirical knowledge provided by the face beam image established during preprocessing for the location algorithm, the approximate location of feature points is estimated by using the distance statistical characteristics.

During preprocessing, the left eye corner coordinates of each face image are fixed to a certain pixel position. Since the normalized pixel size of the image is exactly the same, the distance from other feature points to left eye pixels can be estimated according to the empirical value, and then the positions of all other feature points can be estimated.

(2) **Accurate estimation**

After the rough estimation is completed, it needs to be combined with the graph matching technology for accurate estimation. The selected pixels in a certain range around the rough estimation points can be selected; first, the values of these pixels are filtered to obtain the eigenvalues, and then the similarity between these eigenvalues and the template eigenvalues in the face database can be calculated, respectively. Finally, the point with the highest similarity can be regarded as the accurate estimation position of the feature points.

5. **Feature extraction**

After locating the feature points, extract the features. By Gabor wavelet filtering the pixel values on the located feature points, the obtained wavelet coefficients are stored in the face image structure database.

Feature extraction using wavelet filter requires a lot of computation and is very time-consuming. This is because the Gabor wavelet filter adopts convolution operation for feature extraction. At the same time, the feature vector extracted from Gabor filter is a vector of up to 40 dimensions in eight directions and five wavelengths, so the amount of calculation is huge.

However, the problem of too much convolution calculation between filter function and image gray can be solved by multi-channel fast filter. The multi-channel fast filter can be equivalent to expanding the image feature components according to the frequency-domain distribution. After entering a specific set of parameters, these expanded multi-channel fast filters are transformed into a discrete Gabor filter.

In other words, extracting features with Gabor filter requires convolution calculation for each pixel in the image, which will produce the so-called dimension disaster problem and seriously affect the speed of the recognition algorithm. To solve the problem of high dimension of eigenvector, the dimension reduction method is generally used. In other words, first process the collected samples and then extract the features, but in this way, some important feature information may be lost; or select some feature points with rich face information, only transform these feature points and discard the feature points that are not helpful for recognition, but this method must have high requirements for the positioning accuracy of feature points.

6. **Matching the degree of similarity**

In the elastic beam graph matching algorithm, comparing the similarity of feature points is to compare the similarity of feature vectors on feature points. The similarity calculation formula is

$$S_D\left(J, J', \vec{d}\right) = \frac{\sum_{n=1}^{N} a_n a'_n \cos\left[\varphi_n - \left(\varphi'_n + \vec{d}\vec{k}_n\right)\right]}{\sqrt{\sum\limits_{n=1}^{N} a_n^2 \sum_{n=1}^{N} a_n'^2}} \tag{4.207}$$

where $\vec{d}$ represents the position deviation vector of J and J'; $\vec{k}_n$ represents the unit vector of the n-th Gabor wavelet coefficient. The formula compensates for the adverse effect of small phase deviation on similarity and comprehensively reflects the similarity between the two eigenvalues by using the amplitude information of eigenvalues.

4.3.4 Face Recognition Algorithm Based on Neural Network

Compared with the traditional face recognition technology (such as local binary pattern recognition technology, feature recognition technology based on Gabor and scale invariance, etc.), deep convolution neural network has the ability of effective automatic feature extraction. Using deep convolution neural network for face recognition can effectively overcome the serious changes of face posture, expression, occlusion and so on.

Convolutional neural network is a deep learning neural network model, which has the characteristics of weight sharing, local connection and so on. This makes the convolutional neural network have certain invariance properties of parallel movement, contraction/amplification and rotation and can be widely used in image classification, face recognition and other scenes [9].

1. **Composition of convolutional neural network**

Convolutional neural network is generally composed of convolution layer, pooling layer and full connection layer. This belongs to feedforward neural network, which is trained by back propagation algorithm.

(1) **Convolutions layer**

The convolution layer is used to learn the feature representation of the input data, and the convolution kernel constituting the convolution layer is used to calculate different feature maps.

Convolution layer is an indispensable part of convolution neural network. The convolution layer consists of some convolution kernels of uniform size, and the parameters of each convolution kernel are learned through network training. Generally, the calculation in convolution layer is called convolution calculation. The main function of convolution calculation is to extract the feature information in the input image. The ability of convolution operation to extract features is very obvious, and with the deepening of the number of convolution layers, the feature information extracted by convolution layer is more and more abstract. The more abstract the feature information is, the more distinguishable the feature information is, and it is easier to solve the problems in practical application.

Convolution is the most common operation in life. We can use convolution to simulate most scenes in life. For example, when we watch a video online, the video gets stuck because of the network. If this situation is described by convolution, it can be expressed as the result of the convolution operation between the smooth playback of the video and a convolution kernel representing the network. It can be expressed as

$$Y(m, n) = X(m, n) * H(m, n) \tag{4.208}$$

where $X(m, n)$ represents normal video playback, $H(m, n)$ represents network pause convolution kernel and $Y(m, n)$ represents the result of video pause.

If $H(m, n)$ can be estimated from $Y(m, n)$ in some way, measures can be taken to recover $X(m, n)$ and finally solve the phenomenon of video pause.

Convolution operation is widely used in the field of image processing. Using convolution neural network can effectively extract the feature information in the image, so as to achieve the effect of image recognition.

The specific formula of using convolution to operate an image is

$$y_{nm} = f\left(\sum_{j=0}^{J-1}\sum_{i=0}^{I-1} x_{m+i,n+j} w_{ij} + b\right), (0 \le m \le M, 0 \le n \le N) \tag{4.209}$$

Among them, (M, N) represents the size of the input image, w_{ij} represents the size of the convolution kernel and b represents the offset size in the convolution layer. In order to enable the convolution neural network to learn the nonlinear characteristic information in the image, an inactive function is generally added after the convolution operation, and $f(\cdot)$ in formula (4.209) is the activation function.

The convolution layer in convolution neural network is generally composed of multiple convolution kernels. Each convolution kernel will convolute with the image, and then each convolution core will output a feature map, that is, the feature information extracted by convolution check image. With the increase of the number of convolution kernels in the convolution layer, the output feature map of the convolution layer also increases, but this is not to say that the more convolution kernels, the more feature information will be extracted, and there is no inevitable relationship between the two. With the increase of convolution kernel, the amount of irrelevant feature information extracted by convolution layer may also increase, which is bound to affect the image recognition by convolution neural network. Therefore, there are certain requirements for the number of convolution kernels, not the more, the better, which is closely related to the size of the input image and the size of the data set.

(2) **Pooling layer**

The pool layer is also called the lower sampling layer. It is used to reduce the parameter value and dimension of eigenvector output from convolution layer. Through the improvement of parameters and dimensions, more abstract features are obtained through the convolution layer and pooling layer.

The feature information after convolution operation of convolution layer is still very large, which will not only reduce the computational performance but also produce the phenomenon of over fitting. Therefore, it is necessary to use some features to replace the overall features in order to obtain a simple effect. This operation is called pooling. The so-called pooling means that while reducing the feature dimension, it can not only extract representative feature information but also make the processed feature map have a greater perception field of vision.

The pooled image still has the invariance of parallel movement. Pooling operation will blur the specific position of features, and the same features can still be generated after the image moves in parallel. The pooling operation can further abstract the characteristics of a local area. An element in the pooling corresponds to an area

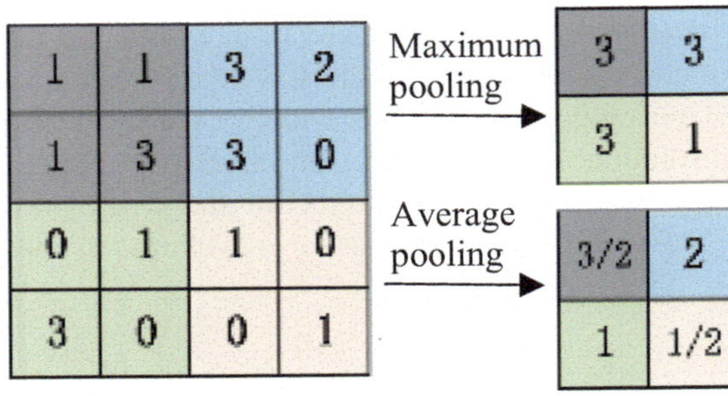

Fig. 4.31 Schematic diagram of two kinds of pooling

in the input data. The pooling operation can reduce the number of parameters and reduce the image dimension.

Common pooling operations include maximum pooling and average pooling. The former is to select the maximum value for each small area as the pooling result; the latter is to select the average value as the pooling result. Figure 4.31 shows the schematic diagram of two kinds of pooling effects, and the pooling step is 2.

(1) **Full connected layer**

Before the concept of local perception was put forward, all feedforward neural networks were connected in the form of full connection. The full connection layer can gather the characteristic information obtained from convolution kernel and pooling kernel. In addition, the full connection layer can also simplify the parameter model and reduce the amount of neuron data and training parameters of feedforward neural network to a certain extent. In order to use the back propagation algorithm to train the neural network, the full connection layer requires the image to have a fixed input size. Since the convolution layer is developed on the basis of the full connection layer, the full connection layer can also be represented by a special convolution layer. In the full connection layer, it can be considered that the sensory field of each neuron is the whole image. The more nodes in the hidden layer of the full connection layer, the stronger the model fitting ability, but it will also bring over fitting due to parameter redundancy. In order to solve the influence of over fitting on the training effect, regularization technology is generally used between the full connection layers. Some neurons can also be randomly discarded in the whole connection layer to solve the problem of over fitting in the training process.

The calculation methods of full connection layer and convolution layer are different. It is not as complex as convolution calculation. In the full connection layer, the full connection layer directly multiplies the input data and vector.

Assuming that the input of the full connection layer is x, the input dimension is D, the output is y and the output dimension is V, the full connection layer calculation can be expressed as

$$y = Wx \tag{4.210}$$

where W is the weight matrix of $V * D$.

Due to the use of weight sharing and local connection in convolution layer, the number of parameters and calculation of convolution layer are successfully reduced, so that the number of parameters of full connection layer is several times that of convolution layer. In convolution neural network, convolution layer mainly extracts the features of the input image, while the main function of full connection layer is to integrate the feature information extracted by convolution layer. Therefore, when constructing convolutional neural network, to improve the feature extraction ability of the network mainly lies in the design method of convolution layer. In order to reduce the amount of parameters and calculation, the key lies in the design skills of full connection layer.

(2) **Activation function**

In order to enable the convolutional neural network to simulate more complex functional relationships, the activation function is usually added behind the convolution layer or full connection layer, so that the convolutional neural network can learn more complex functional relationships. In addition, adding the activation function to the convolutional neural network can add nonlinear elements to the network, so that the convolutional neural network can simulate any nonlinear function and improve the ability of convolutional neural network to solve practical problems.

Common activation functions include: ReLU activation function, Sigmoid activation function and Tanh activation function.

1) The mathematical expression of the ReLU activation function is

$$f(x) = \max(0, x) \tag{4.211}$$

ReLU activation function is an unsaturated activation function, also known as modified linear element. Compared with other commonly used activation functions, ReLU activation function has stronger expression ability for linear functions in the network; for the nonlinear function in the network, the ReLU activation function can keep the convergence speed of the model in a relatively stable state. Therefore, ReLU activation function is also the most frequently used activation function of convolutional neural network.

2) The mathematical expression of Sigmoid activation function is

$$f(x) = \frac{1}{1 + e^{-x}} \tag{4.212}$$

Sigmoid activation function is an S-shaped curve, so Sigmoid activation function is also called S-shaped function. Because the Sigmoid activation function is continuous and monotonous between (0, 1) and the output range of the function is limited, the Sigmoid activation function is often used in binary classification problems.

3) Tanh 激活函数的数学表达式为.
4) The mathematical expression of Tanh activation function is

$$f(x) = \frac{e^x - e^{-x}}{e^x + e^{-x}} \tag{4.213}$$

Tanh activation function is a hyperbola, so it is also called hyperbolic tangent function. The advantage of Tanh activation function is that it works well when used in scenes with obvious difference in feature information, and will expand the feature effect in the training process [10].

2. **Steps of face recognition based on Neural Network**

(1) **Constructing deep convolution neural network**

The constructed deep convolution neural network mainly includes convolution layer, activation function layer, pooling layer, full connection layer and so on. Among them, the activation function layer converts simple linear input into complex nonlinear output to obtain better classification effect.

(2) **Training process**

In the training process, the real-time data of the image loaded with face are improved, and new training data are created from the training data by using methods such as rotation and turnover, so as to improve the scale of training data and improve the stability and training effect of the model.

Figure 4.32 shows the sample set provided by the author of this book and his students during the face recognition experiment.

(3) **Deterministic convolutional neural network model**

The input data are divided into static and dynamic categories: the static input data are the face image, and the dynamic input data are the face image in the video stream.

Face detection is to detect the input face image to avoid inputting useless information.

Preprocessing is the preliminary processing of the input data, so that the convolutional neural network can extract more useful information, including face alignment, face clipping and other operations.

Fig. 4.32 Examples of face sample set

In feature extraction, the forward propagation of convolutional neural network is used to extract the face features in the image and obtain the face feature information.

(4) **Realize face recognition**

The so-called obtaining face recognition is to apply the trained optimal convolution neural network model to video face detection and recognition, match the extracted feature information with the face information in the face data management module through feature comparison, obtain the identity information of the face in the input image and finally get the correct face detection and recognition results.

By judging the recognition accuracy and adjusting each step in time to achieve the best recognition effect, the convolution neural network model obtained at this time is the best convolution neural network model trained.

In practical technology, it is naturally necessary to establish the registration management of face database. The management system is shown in Fig. 4.33.

For the positional memory of any face training database, a matrix $\mathbf{F}_{m\times n}$ can be established. For example, after registering the face images of the author of this book and the students participating in the training in advance and storing them in the face training sample database, a positional memory matrix $\mathbf{F}_{4\times 10}$ can be established by using the face database in Fig. 4.32. Obviously, the registered sample of the author's

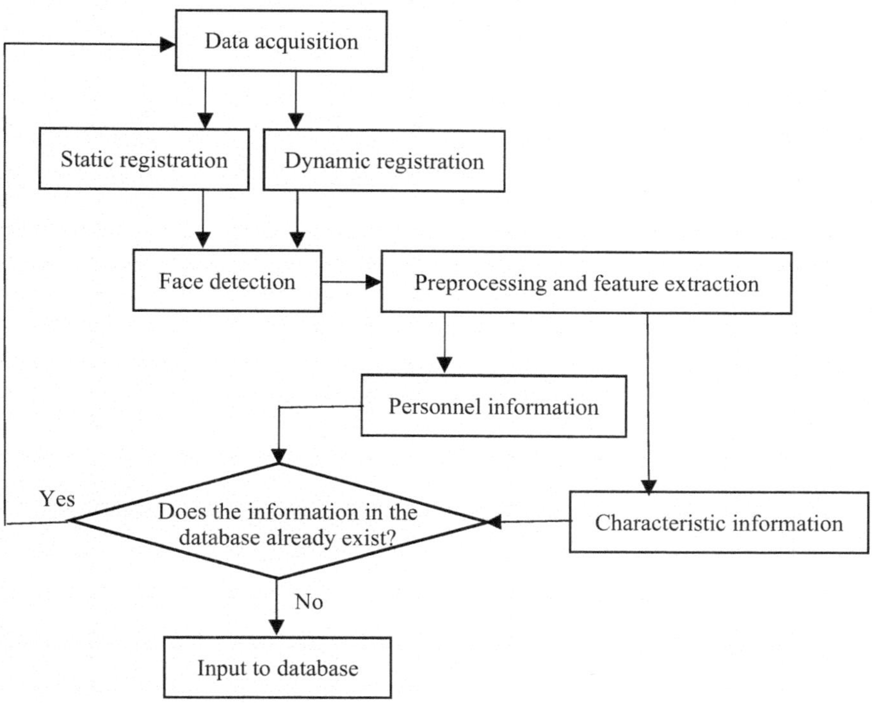

Fig. 4.33 Flowchart of face registration

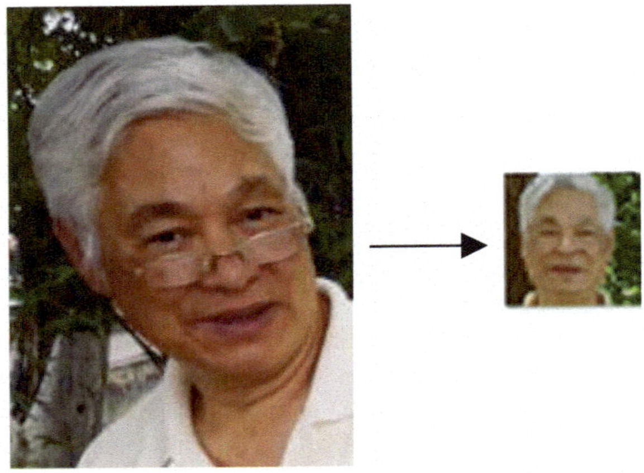

Fig. 4.34 Legend of face recognition

face image of this book is in the position of (u_2, v_5). When the measured image enters the recognition process, the recognition system will automatically perform matching operation with the training sample at (u_2, v_5) position (see Fig. 4.34).

4.3.5 Face Recognition Algorithm Based on Support Vector Machine

Support vector machine (SVM) is based on statistical learning theory. Because of its advantages, it has attracted attention and research in various fields since it was proposed. At present, the research on SVM mainly focuses on the theoretical research of its own nature, the effective algorithm of large-scale problems and the promotion of application fields. Among them, a large number of research achievements have been made in application research. In addition to the above "two-classification support vector machine" applied to the state recognition of train power system and its transmission device, the "multi-classification support vector machine" is also applied to the technical fields of face detection and recognition, network intrusion detection, handwritten character recognition and so on.

In the application of face recognition algorithm, firstly, principal component analysis (PCA) is used to reduce the dimension of histogram, and then support vector machine (SVM) is used for classification, which can improve the recognition rate. The basic flow of the recognition system is shown in Fig. 4.35.

1. **Multi-classification support vector machine**

Nonlinear support vector machine maps the sample from the input space to a feature space by establishing a nonlinear transformation, so that the original hypersurface model corresponds to the hyperplane model of the feature space and realizes the

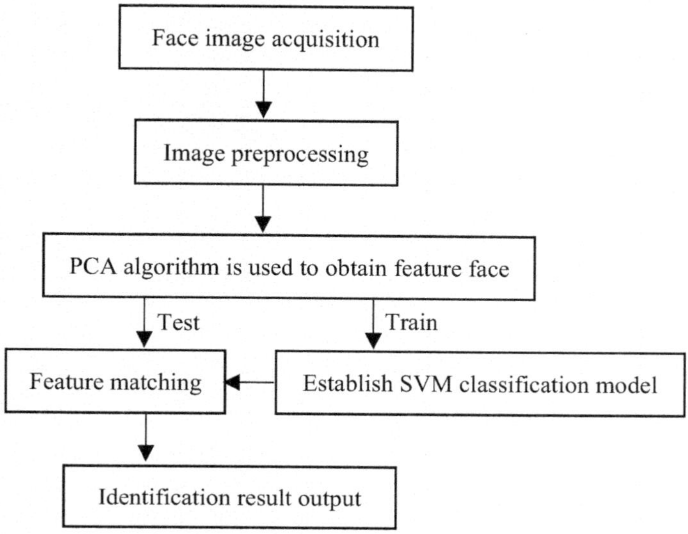

Fig. 4.35 Flowchart of SVM face recognition

linear separability of the sample in the high-dimensional feature space. Its topology is shown in Fig. 4.36.

The given training set is as follows

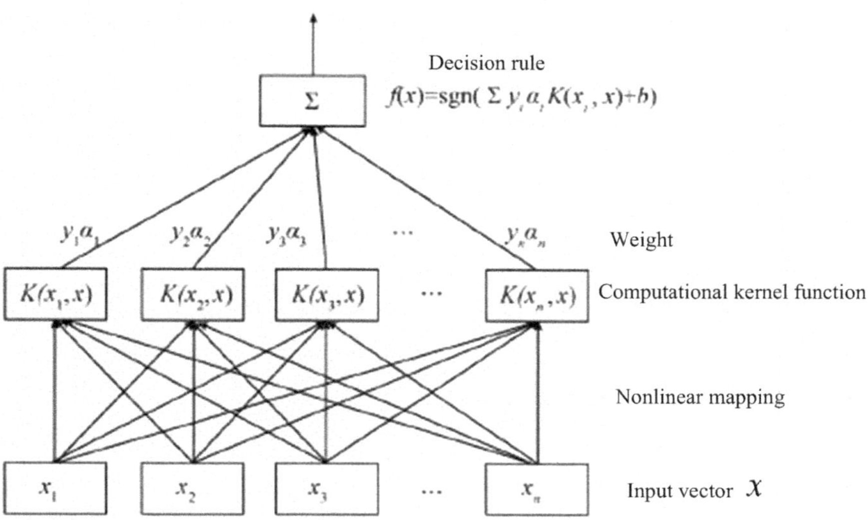

Fig. 4.36 Topology of nonlinear support vector machine

$$T = \{(x_1, y_1), (x_2, y_2), \ldots, (x_l, y_l)\} \tag{4.214}$$

Among them, $x_l \in R^n$, $y_l \in Y = \{1, 2, \ldots, M\}$、$i = 1, 2, \ldots, l$

Accordingly, a decision function $f(x) : R^n \to Y$ on space R^n is found to infer the output value y corresponding to any input x.

The above multi-classification problem is M-classification problem. The multi-classification problem can be solved by transforming it into a series of two-classification problems. The two types of support vector machines for this series of two-classification problems can comprehensively judge the category of x according to the judgment results of these two types of support vector machines for input x. Different two-classification structures correspond to different classification methods [11].

(1) **"One-to-many" classification**

The "one-to-many" algorithm, that is, the solution of the M classifier, is obtained by two kinds of problems composed of the M sample set and the rest $M - 1$ sample set. In this way, we need to construct M classifiers, and then classify and recognize a test sample.

The mathematical model of the algorithm is

$$\min_{\omega_i, b_i, \xi^i} \frac{1}{2}\omega_i^T \omega_i + C\sum_j \xi_j^i \omega_i^T \tag{4.215}$$

$$\text{s.t.} \begin{cases} \omega_i^T \varphi(x_j) + b_i \geq 1 - \xi_j^i \omega_i^T, \text{ if} : j = i \\ \omega_i^T \varphi(x_j) + b_i \geq -1 - \xi_j^i \omega_i^T, \text{ if} : j \neq j \\ \xi_j^i \geq 0, j = 1, 2, \ldots, M \end{cases}$$

The resulting decision function is

$$F_i(x) = [\omega_i^T \varphi(x_j) + b_i], i = 1, 2, \ldots, M \tag{4.216}$$

In the formula, ω represents the normal vector of the hyperplane, b represents the offset value of the hyperplane, ξ represents the relaxation variable, which can reduce the requirements of the optimal classification hyperplane, C represents the penalty factor, and when $C > 0$, it represents the penalty force for misclassification.

When solving nonlinear problems, for the constructed support vector machine, it is to construct a nonlinear mapping from low-dimensional space to high-dimensional space, and $\varphi(x_j)$ is a mapping from two-dimensional space to high-dimensional space, $\varphi(x_j) : R^2 \to R^n$.

(2) **"One-to-one" classification**

"One-to-one" classification, that is, each category and each category in the other $M - 1$ categories form two categories, so that $\frac{M(M-1)}{2}$ classification planes can be obtained. In other words, in order to distinguish the class currently waiting for judgment from another class, $\frac{M(M-1)}{2}$ two-class classifiers need to be constructed. Then, $M-1$ two-class classifiers established by this class and the remaining classes are used to comprehensively judge the category to which an input x belongs.

The mathematical description of the algorithm is

$$\min_{\omega_{ij}, b_{ij}, \xi_l^{ij}} \frac{1}{2}\omega_{ij}^T \omega_{ij} + C\sum_l \xi_l^{ij}\omega_{ij}^T \tag{4.217}$$

$$\text{s.t.} \begin{cases} \omega_{ij}^T\varphi(x_j) + b_{ij} \geq 1 - \xi_l^{ij}\omega_{ij}^T, \text{ if} : l = i = j \\ \omega_{ij}^T\varphi(x_j) + b_{ij} \geq -1 - \xi_l^{ij}\omega_{ij}^T, \text{ if} : l \neq i;\ or,\ l \neq j \\ \xi_l^{ij} \geq 0, l = 1, 2, \ldots, M \end{cases} \tag{4.218}$$

The resulting decision function is

$$F_{ij}(x) = \left[\omega_{ij}^T\varphi(x_j) + b_{ij}\right], i = 1, 2, \ldots, M;\ j = 1, 2, \ldots, M \tag{4.219}$$

When determining the category of an input sample, the above $\frac{M(M-1)}{2}$ classifiers need to be used to classify the current sample. If a classifier determines that the sample belongs to a certain category, the sample will get one vote in this category. Finally, the category with the largest number of votes is the category of the input sample.

2. **Kernel function and feature selection**

(1) **Selection of kernel function**

When $f(x) \neq 0$, the following decision function can be established for function $\int f^2(x)\mathrm{d}x < 0$.

$$f(x) = \mathrm{agn}\left(\sum_{i=1}^{l} a_i^* y_i K(x, x_i) + b^*\right) \tag{4.220}$$

Then $K(x, x_i)$ is the kernel function that can be described as an inner product operation in a feature space, $K(x, x_i) = \varphi^T(x)\varphi(x_i)$.

Where $\mathrm{agn}(\cdot)$ is the symbolic function, a^* is the optimal Lagrange coefficient, b^* is the threshold and i is the number of support vectors.

Common kernel functions have the following forms.

1) Linear kernel function

$$K(x, x_i) = x \cdot x_i \tag{4.221}$$

2) Polynomial kernel function

$$K(x, x_i) = \left(\mathbf{x}^T\mathbf{x}_i\right)^d \tag{4.222}$$

where vector $\mathbf{x} = [x]$, $\mathbf{x}_i = [x_i]$, d are positive integers given by homogeneous polynomial kernel function.

3) Radial basis kernel function

$$k(x, x_i) = \exp\left(-\frac{|x - x_i|^2}{\sigma^2}\right) \tag{4.223}$$

4) Sigmoid kernel function

$$K(x, x_i) = \tanh\left[v(x \cdot x_i) + \gamma\right] \tag{4.224}$$

Among them, when the undetermined parameter $v \to 0$ and γ take a small value, the kernel function obtains the optimal selection. Generally, $v = 0.0001$, $\gamma = 0$

When solving practical problems, the principle of selecting kernel function is to take the kernel function with the smallest error as the optimal kernel function.

(2) **Feature selection method**

Assuming that the training sample set can be correctly classified for the selected features, that is, for the training sample set, the empirical risk of the optimal hyperplane obtained each time is the same. On this basis, compare the empirical risk of the optimal hyperplane obtained this time with that of the optimal hyperplane obtained in all previous experiments on the test sample set, and then judge whether to eliminate a feature.

This assumption satisfies the basic principle of the sequence minimization method: that is, to minimize the number of features without increasing the empirical risk. In practical application, the sample set is divided into two parts: one part is used for feature selection; the other part is used to detect the effect of feature selection. They are called test sample sets. The former is called feature selection training sample set, and the latter is called feature selection test sample set.

The feature selection minimization sequence method is as follows.

1) For the original training sample set, the Gaussian radial basis kernel function is selected, the appropriate parameter σ is determined and the optimization problem is constructed and solved.
2) Remove the i-th feature of the original training sample set, repeat step 1), construct the decision function $f(x)$ and calculate the corresponding empirical risk $R_{emp}\left(\hat{f}\right)$ for the test sample set.
3) If $R_{\text{emp}}\left(\hat{f}\right) \le R_{\text{emp}}(f)$, eliminate the i-th feature and make $f(x) = \hat{f}(x)$, $R_{\text{emp}}\left(\hat{f}\right) = R_{\text{emp}}(f)$; otherwise, the i-th feature is marked.
4) Select a new feature and repeat steps 2) and 3). If all features have been marked, the calculation can be stopped [12].

3. Establish classification and matching

In other words, either the "one-to-many" algorithm is adopted, and the solution of the M classifier is obtained by two kinds of problems composed of the M sample set and the rest $M - 1$ sample set. Then, a test sample is classified and identified.

Or use the "one-to-one" classification method to form two categories of each category and the rest of $M-1$ category, so that $\frac{M(M-1)}{2}$ classification surfaces can be obtained. Then, according to the $M-1$ two-class classifiers established respectively for this class and the remaining classes, the category to which an input x belongs is comprehensively judged [13].

The face database data in Fig. 4.32 is used as the learning sample of the example.

The data of a collected dynamic face are standardized, that is, the range of the input face image reaches the predetermined error standard. Therefore, the proportional dependence of the initial weight can be eliminated, the training speed can be improved, and the possibility of falling into the local optimal state can be reduced. At the same time, weight attenuation and Bayesian estimation can be more convenient through standardized input.

The range of data standardization should be selected according to the experimental data. Generally, the default range of data standardization is $[-1, 1]$.

Recognizing and matching a face image to the face database is a typical "one-to-many" classification operation.

In order to test the practical application effect of SVM classifier on face classification and recognition, the classification model is used to classify face images. The whole matching process can be seen in Fig. 4.37.

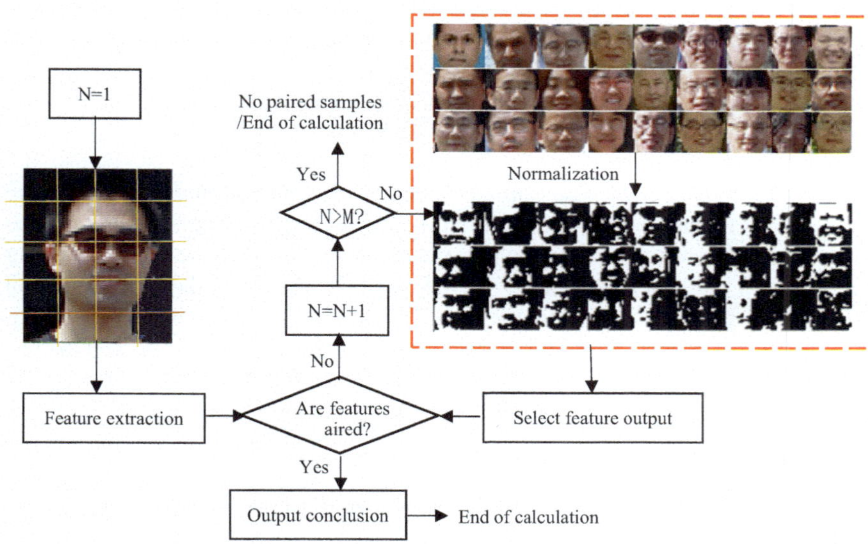

Fig. 4.37 Example of classification and matching results of "one-to-many" face database

4.3.6 *Face Recognition Algorithm Based on Hidden Markov Model*

Hidden Markov model (HMM) is used to describe a Markov process with hidden unknown parameters (States). The difficulty is to determine the hidden parameters of the process from the observed data and then use these parameters for further analysis and mining. For any face image, its pixel value is in the observable state, and its implicit state cannot be observed according to the pixel value. Therefore, a random process is needed to describe the implicit state, that is, "hidden Markov model" [14].

In recent years, the method based on Markov model has become the main research content in the field of face recognition. The face recognition method based on hidden Markov model can improve the accuracy of face recognition to a certain extent.

1. **Hidden Markov model**

Hidden Markov model is composed of two interrelated stochastic processes. One is to describe the possible changes of states by the internal Markov chain of finite states, and the other is to describe the statistical relationship between the observed values and the corresponding states. In the process of state transition, the observer only sees the output value of the random function associated with each state and cannot observe the state of the specific Markov chain, so it is called hidden Markov model.

(1) **Basic parameters of Hidden Markov model**

The essence of hidden Markov model is a probability model expressed by parameters and used to describe the statistical characteristics of stochastic processes. Hidden Markov process consists of observable observation sequence and unobservable state process. Hidden Markov model can be expressed as

$$\lambda = (N, M, A, B, \Pi) \tag{4.225}$$

where N is the number of states of the hidden Markov model in the state set $S = \{S_1, S_2, \ldots, S_N\}$; suppose that the state of time t is q_t, $q_t \in S$, $1 < t < T$, and T is the length of the observation sequence; M is the number of observations corresponding to the state in the observation set $V = \{V_1, V_2, \ldots, V_N\}$; A is the state transition probability distribution or transition matrix, $A = \{a_{ij}\}, a_{ij} = p(q_t = S_j|q_{t-1} = S_i)$, $1 < i < N, 1 < j < N$; B is the observation probability matrix or emission matrix, $B = \{b_{ij}\}, b_{ij} = p(O_t = V_k|S_t = q_j)$; Π is the initial state probability distribution vector, $\Pi = \{\pi_i\}, \pi_i = p(q_i = S_i), 1 < i < N$.

(2) **Parameter calculation and model establishment**

After the HMM model is established, there are three basic problems to be solved.

1) Given the observation sequence $O = o_1, o_2, \ldots, o_r$ and the model parameter $\lambda = (N, M, A, B, \Pi)$, evaluate the probability $p(O|\lambda)$ of the observation sequence generated by the given model;

2) Given the observation sequence $O = o_1, o_2, \ldots, o_r$ and the model parameter $\lambda = (N, M, A, B, \Pi)$, how to select an optimal sequence $Q = q_1, q_2, \ldots, q_T$ in order to best explain the observation sequence $O = o_1, o_2, \ldots, o_r$;
3) How to adjust the model parameter $\lambda = (N, M, A, B, \Pi)$ to maximize $p(O|\lambda)$.

After the above three problems are solved, the model with optimal parameters can be obtained for practical face recognition. Among them, the first problem is to estimate the probability that a given model generates an observation sequence, that is, to evaluate the matching degree between the model and the observation sequence, that is, given an unknown face, calculate which face model has the greatest probability of generating the face model through state transition, that is, which kind of face the unknown face is divided into; The second is to calculate which state transition sequence is most likely to be generated after given model parameters and observation sequence. Usually, an optimization criterion needs to be formulated. When the conditions of the optimization criterion are met, it can be determined that the state transition sequence is most likely; the third is the optimization of model parameters. After a set of observation value sequence is given, the optimization parameters need to be continuously adjusted to maximize the probability of observation value.

Face recognition based on HMM is to establish a corresponding model for each person's face after learning according to different face images of the same person. This is actually a parameter estimation problem.

The evaluation problem of problem 1) is solved by forward / backward algorithm. The forward variable $\alpha_t(i)$ is defined as follows

$$\alpha_t(i) = p(o_1, o_2, \ldots, o_t, q_t = S_i|\lambda) \tag{4.226}$$

The backward variable $\beta_t(i)$ is defined as follows

$$\beta_t(i) = p(o_{t+1}, o_{t+2}, \ldots, o_T, q_t = S_i|\lambda) \tag{4.227}$$

The steps of the forward algorithm are as follows

(I) Initialize $\alpha_t(i)$,that is

$$\alpha_t(i) = \pi_1 b_1(o_1), 1 \le i \le N \tag{4.228}$$

(II) Eecursion, that is

$$\alpha_{t+1}(j) = \left[\sum_{i=1}^{T} \alpha_t(i)a_{ij}\right] b_j(o_{t+1}), 1 \le t \le T, 1 \le j \le N \tag{4.229}$$

(III) End operation

$$p(O|\lambda) = \sum_{i=1}^{N} \alpha_T(i) \tag{4.230}$$

The steps of backward algorithm are as follows

(I) Initialize $\beta_t(i)$

$$\beta_t(i) = 1, 1 \leq i \leq N \tag{4.231}$$

(II) Recursion, that is

$$\beta_t(i) = \sum_{j=1}^{N} a_{ij} b_j(o_{t+1}) \beta_{t+1}(j) \tag{4.232}$$

Define a $\delta_t(i)$

$$\delta_t(i) = \max p(q_1, q_2, \ldots, q_t, o_1, o_2, \ldots, o_t|\lambda) \tag{4.233}$$

This represents the highest value of the total i observed values before calculation along a single path when the time t ends in the S_i state. Then we can deduce the following relationship

$$\delta_{t+1}(j) = \left[\max \delta_t(i) a_{ij}\right] b_j(o_{t+1}) \tag{4.234}$$

A two-dimensional array $\{\psi_t(j), t = 1, 2, \ldots, T, j = 1, 2, \ldots, N\}$ is used to track and record the best path of $\delta_t(j)$ in the derivation process.

The algorithm steps to solve problem 2) are as follows.

(I) Initialize $\delta_1(i)$

$$\delta_1(i) = \pi_1 b_1(o_1), 1 \leq i \leq N \tag{4.235}$$

$$\psi_1(i) = 0 \tag{4.236}$$

(II) Recursion, that is

$$\delta_t(i) = \max\left\lfloor \delta_{t-1}(i) a_{ij} \right\rfloor b_j(o_t), 2 \leq t \leq T, 1 \leq i \leq N \tag{4.237}$$

$$\psi_t(i) = \arg\max\left\lfloor \delta_{t-1}(i) a_{ij} \right\rfloor, 2 \leq t \leq T, 1 \leq i \leq N \tag{4.238}$$

(III) End operation

$$p_T^* = \max[\delta_T(i)], 1 \leq i \leq N \tag{4.239}$$

$$q_T^* = \arg\max[\delta_T(i)], 1 \leq i \leq N \tag{4.240}$$

(IV) Path backtracking (determination of optimal state chain)

$$q_T^* = \psi_{t+1}\left(q_{t+1}^*\right) \tag{4.241}$$

The result of the algorithm not only determines an optimal state chain for the observation sequence O but also can calculate the probability $p(O|\lambda)$ of the observation sequence O generated by the model λ at the same time. The resulting optimal state chain also determines the most possible correspondence between the observation sequence and each state. This correspondence is also called segmentation of the observation sequence.

How to adjust the model parameter $\lambda = (N, M, A, B, \Pi)$ to maximize $p(O|\lambda)$ is the problem of optimizing the model parameters. Due to the limitation of the number of training samples and training times of the model, the length of the observation sequence is often limited. Therefore, it is impossible to obtain the optimal parameter estimation but only one estimation maximization algorithm can be used to obtain the local optimal solution.

Definition 1 $\gamma_t(i)$ is the probability that the given HMM model and observation sequence are in state S_i at time t, then

$$\gamma_t(i) = p(q_t = s_i|O, \lambda) = \frac{\alpha_t(i)\beta_t(i)}{\sum_{i=1}^{N} \alpha_t(i)\beta_t(i)} \tag{4.242}$$

and it meets the following conditions

$$\sum_{i=1}^{N} \gamma_t(i) = 1 \tag{4.243}$$

Definition 2 The probability of being in state S_j at time $t + 1$ is

$$\xi_t(i, j) = p\left(q_t = s_i, q_{t+1} = s_j|O, \lambda\right) \tag{4.244}$$

Expressed by forward / backward probability, there is

$$\xi_t(i, j) = \frac{\alpha_t(i)a_{ij}b_j(o_{t+1})\beta_{t+1}(j)}{p(O|\lambda)} = \frac{\alpha_t(i)a_{ij}b_j(o_{t+1})\beta_{t+1}(j)}{\sum_{i=1}^{N}\sum_{j=1}^{N} \alpha_t(i)a_{ij}b_j(o_{t+1})\beta_{t+1}(j)} \tag{4.245}$$

Obviously, $\gamma_t(i)$ and $\xi_t(i, j)$ meet the following requirements

$$\gamma_t(i) = \sum_{j=1}^{N} \xi_t(i, j) \tag{4.246}$$

If $\gamma_t(i)$ is summed to a certain time period t, the number of times to access state S_i in time period t or the expected number of times to transfer out of state S_i can be obtained; similarly, by summing $\xi_t(i, j)$ to a certain time period t, the expected number of transitions from state S_i to state S_j can be obtained, that is

$$\sum_{t=1}^{T-1} \gamma_t(i) = N^{(1)} \tag{4.247}$$

where $N^{(1)}$ is the desired number of transitions from state S_i.

$$\sum_{t=1}^{T-1} \xi_t(i) = N^{(2)} \tag{4.248}$$

where $N^{(2)}$ is the expected number of transitions from state S_i to S_j.

According to the above formula, a set of estimation formulas for HMM model parameters can be obtained as follows:

$$\overline{\pi}_i = p(q_i = S_i) = \gamma_1(i),\ when,\ t = 1 \tag{4.249}$$

$$\overline{a}_{ij} = \frac{\sum_{t=1}^{T-1} \xi_t(i)}{\sum\limits_{t=1}^{T-1} \gamma_t(i)} \tag{4.250}$$

$$\overline{b}_j(o_i) = \frac{\sum_{t=1,o_i}^{T} \gamma_t(i)}{\sum\limits_{t=1}^{T} \gamma_t(i)} \tag{4.251}$$

Here, $\sum_{t=1,o_i}^{T} \gamma_t(i)$ represents the average number of observations o_i in state S_j; $\sum_{t=1}^{T} \gamma_t(i)$ represents the average number of times in state S_j.

The revaluation formula of the current model $\lambda = (N, M, A, B, \Pi)$ is defined as $\overline{\lambda} = \left(N, M, \overline{A}, \overline{B}, \overline{\Pi}\right)$. If the initial parameter of the current model is an extreme point, the model $\overline{\lambda}$ is better than the model λ in the sense of $p\left(O|\overline{\lambda}\right) > p(O|\lambda)$. That is, if a new model can be found, the probability of the new model producing the observation sequence O is greater.

Through theoretical analysis, it can be seen that if the above steps are repeated and the model parameters are continuously updated and optimized, the probability of the model generating observation sequence can be improved, which is equivalent to the continuous learning and upgrading of the model using training samples. When a certain limit condition is satisfied, the revaluation process will stop, and the learning and upgrading of the model will come to an end. At this time, the obtained model parameters are called the maximum likelihood estimation of the model.

The following auxiliary functions are defined

$$Q(\lambda, \overline{\lambda}) = \sum_{Q} p(Q|O, \lambda) \log\lfloor p(Q|O, \overline{\lambda})\rfloor \tag{4.252}$$

The revaluation formula can be obtained by maximizing $\overline{\lambda}$ from the above formula, and it can be proved that maximizing $Q(\lambda, \overline{\lambda})$ will increase the likelihood probability, that is

$$\max[Q(\lambda, \overline{\lambda})] \Rightarrow p(O|\overline{\lambda}) \geq p(O|\lambda) \tag{4.253}$$

Finally, the likelihood function will converge to the extreme point of a model function.

(3) HMM model training

In fact, hidden Markov model is a left–right face model with five states. In this model, the frontal face image is divided into five significant feature regions: hair, forehead, eyes, nose and mouth. No matter what kind of face and interference, the order of these five regions remains unchanged from top to bottom. It can be considered that each significant feature region contains the state of a human face. Therefore, it can be said that these five hidden states can produce an observation sequence, and these five states can be estimated through the observation sequence.

The purpose of hidden Markov model training is to establish a corresponding model for everyone, and through continuous training and learning, the model finally obtains the optimal model parameters, so that the model can accurately represent the significant features of the corresponding face compared with other faces. The process of HMM model training is shown in Fig. 4.38.

The calculation steps are as follows.

1) The face image is sampled by the sampling window, and the singular value decomposition is performed for each sampling window matrix, and its singular value vector is taken as the observation sequence.
2) A general HMM model $\lambda = (N, M, A, B, \Pi)$ is established to determine the number of states of the model, the allowable state transition and the size of the observation sequence vector.

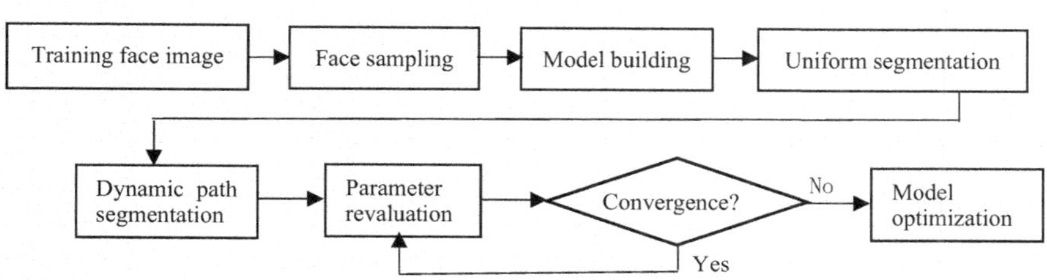

Fig. 4.38 Training flowchart of HMM model

3) The face image used for training is evenly divided into N feature regions, which correspond to N states of the model, respectively, and the initial parameters of the model are calculated.
4) Dynamic path segmentation is used to reestimate the initial value of model parameters.
5) The parameters are re estimated and iterated repeatedly until the optimal model parameters are obtained.

In the experiment, the number of iterations and iteration error are usually used to control the convergence speed at the same time, that is, when the number of iterations is greater than 15 or the iteration error of the first and second iterations is less than e^{-4}, it is considered that convergence has been achieved.

Once the maximum observation probability is obtained, the training process ends. At this time, the model parameters are the optimal model parameters, which are stored in the database as the standard model of this kind of face.

2. **Face recognition process based on Hidden Markov model**

The process of face recognition is the same as that of model training.

The image matrix in each sampling window is decomposed by singular value decomposition to obtain the observation sequence. In fact, face recognition based on hidden Markov model is to estimate the parameters of the hidden state through the pixel value (observed value) of the face image, which is the operational characteristics of the face model based on hidden Markov model. In other words, the process of estimating its implicit parameters through observations is the process of establishing face model.

The maximum likelihood probability of the sequence is generated by calculating each observation model, respectively. The model with the maximum probability is the class of the face to be recognized. It can be expressed as follows:

$$p\big(O^{(k)}|\lambda_m\big) = \max p\big(O^{(k)}|\lambda_n\big), n = 1, 2, \ldots, N, m \subseteq n \tag{4.254}$$

If the maximum likelihood probability of the sequence $O^{(k)}$ generated by the m-th model λ_m is the maximum of all probabilities, it indicates that the image k belongs to the m-th type model.

The specific identification process is shown in Fig. 4.39.

(1) **Face collection and processing**

Use the sampling window to sample the recognized image (see Fig. 4.40).

(2) **Decomposition of singular value vector**

Feature extraction is a key technology in pattern recognition. The purpose of feature extraction is to obtain the feature vector with small number of features or low feature dimension and low probability of classification error.

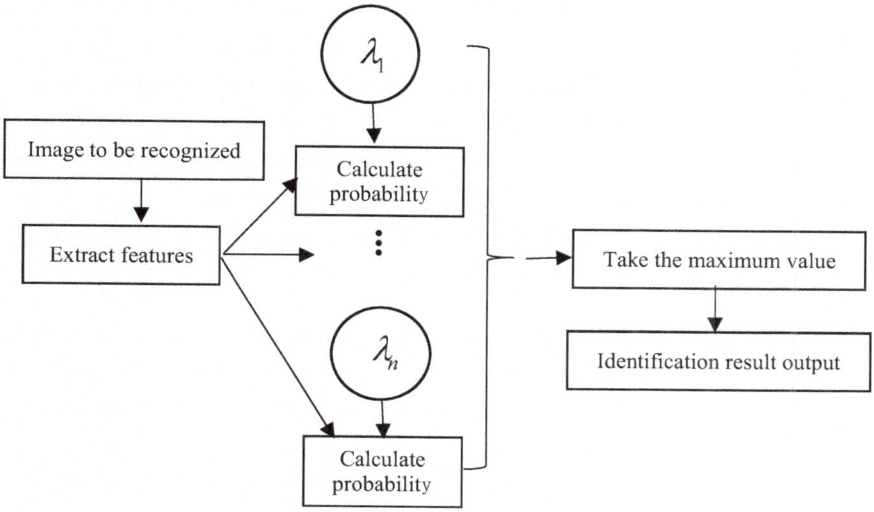

Fig. 4.39 Flow chart of face recognition based on HMM model

Fig. 4.40 Front face image

Because the most important features in practical problems are often difficult to find or cannot be measured, how to extract simple and effective features has become a very important problem in pattern recognition system.

The features used for image recognition can be divided into the following four types.

1) Intuitive features

   Intuitive features, that is, their physical meaning is clear, such as image edge, contour, texture, region and so on. Corresponding extraction algorithms can be designed for specific problems.

2) Gray statistical features

   The image is regarded as a two-dimensional random process, and the image features are described by each order moment in the statistical sense, such as seven order moment invariants.

3) Transformation coefficient characteristics

   This type of feature is to carry out various mathematical transformations on the image, and use the coefficients obtained by the transformation to represent the characteristics of the image, such as Fourier transform, Haff transform, Hadamard transform and so on.

4) Algebraic features

   Algebraic features reflect the internal attributes of the image and are invariant. When the image is treated as a matrix, various matrix decomposition or algebraic transformation can be carried out, such as principal component analysis and so on.

At present, the commonly used feature extraction method is mainly algebraic feature extraction, because algebraic features reflect the statistical law between image pixels. The gray image is regarded as a matrix composed of gray values on each pixel, which can be decomposed and algebraic transformed by various matrices.

Because the eigenvector of the matrix reflects the algebraic properties of the matrix and has stability and invariance, it can be introduced into the recognition system as a feature of the image.

Singular value decomposition of matrix is an effective algebraic feature extraction method. The advantage of this method is that the singular value has good stability, the algorithm is relatively stable and simple, the singular value has algebraic and geometric invariance, and is robust to the changes of illumination, expression and posture. Therefore, it is widely used in the field of image recognition.

Singular value decomposition (SVD) is an effective tool for Solving Least Squares Problems in the field of signal processing and pattern recognition.

Any real symmetric square matrix can be transformed into diagonal matrix by orthogonal transformation. If an image is represented by real matrix, it can be transformed into the following formula by singular value decomposition method

$$A = U \Sigma V^T \tag{4.255}$$

where U and V are two unit orthogonal matrices, and Σ is a diagonal matrix.

$$\Sigma = \begin{bmatrix} \lambda_1 & 0 & 0 & 0 \\ 0 & \lambda_2 & 0 & 0 \\ 0 & 0 & \ddots & \\ 0 & 0 & 0 & \lambda_N \end{bmatrix} \tag{4.256}$$

If A has only k nonzero singular values, then

$$A = \sum_{i=1}^{k} \lambda_i u_i v_i^T \tag{4.257}$$

where u_i and v_i are the column elements of U and V, respectively.

Take $S_{n\times 1} = \left[\lambda_1\ \lambda_2\ \ldots\ \lambda_N\ 0\ 0\ 0\right]^{\mathrm{T}}$ as the singular value vector of matrix A.

According to the definition of singular value decomposition, for any real matrix A, its singular value decomposition is unique. When λ is sequentially arranged into $\lambda_1^* \geq \lambda_2^* \geq \ldots \lambda_k^*$, the eigenvector obtained by singular value decomposition of image A should also be unique. Therefore, singular value vector can be used as the numerical feature of gray value matrix.

The singular value vector has the characteristics of stability, transpose invariance, displacement invariance, rotation invariance and mirror transformation invariance. It also has the characteristics that the singular value vector changes in proportion to the brightness of the corresponding image.

(3) **Feature extraction**

In the process of feature extraction of face recognition, the most commonly used is the left–right hidden Markov face model with five states.

The model divides the frontal face image into five significant feature regions: hair, forehead, eyes, nose and mouth. No matter what kind of face and interference, the order of these five regions remains unchanged from top to bottom. It can be considered that each significant feature region contains the state of a person's face. Therefore, it can be said that the five hidden states can generate an observation sequence, and the five states can be estimated through the observation sequence. The state structure is shown in Fig. 4.41.

In Fig. 4.41, a_{11} and a_{12} are the state transition probability distribution values of hair in the transition matrix $A = \left\{a_{ij}\right\}$, that is, the observable value matrix elements and unobservable state probability distribution value elements about hair; a_{22} and a_{23} are the state transition probability distribution values of the forehead in the transition matrix $A = \left\{a_{ij}\right\}$; a_{33} and a_{34} are the state transition probability distribution values of the eyes in the transition matrix $A = \left\{a_{ij}\right\}$; a_{44} and a_{45} are the state transition probability distribution values of the nose in the transition matrix $A = \left\{a_{ij}\right\}$; a_{55}

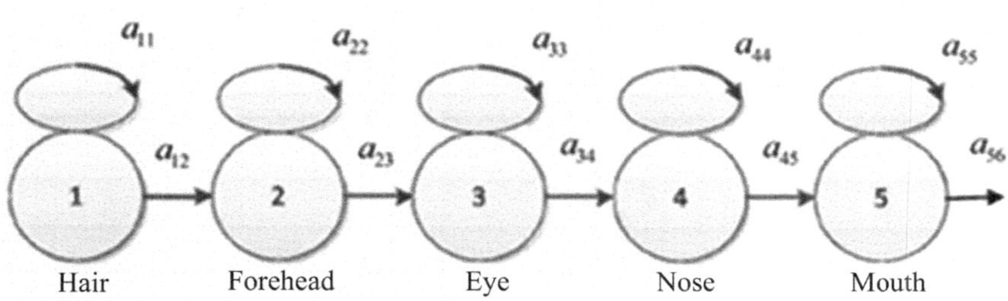

Fig. 4.41 Left–right hidden Markov face model with five states

and a_{56} are the state transition probability distribution values of the mouth in the transition moment $A = \{a_{ij}\}$.

Let the face image size be $M \times H$ and the sampling window size be $M \times L$. Sample the image from top to bottom. If the height of the overlapping part between two adjacent sampling windows is P, the number of samples is

$$T = \frac{H - L}{L - P} + 1 \tag{4.258}$$

In practice, the values of parameters L and P affect the recognition rate to a great extent. In general, the larger P, the more samples, the longer observation sequence, the more training times of the model, the more accurate parameters and the higher recognition rate. When P is large enough ($P \le L - 1$), the influence of L on the system recognition rate can be basically ignored.

Due to the stability and transpose invariance of singular value vector, compared with directly using gray value or two-dimensional discrete cosine transform coefficient, singular value vector as observation vector can obtain higher recognition rate. Because a difficult problem in face recognition is that different photos of the same person are often somewhat different. Changes in light intensity, angle and even facial expression may cause drastic changes in recognition rate. The disadvantage of gray value as observation vector is that gray value does not represent robust image features and is very sensitive to noise, illumination and rotation. In addition, large-scale observation vector will also increase the complexity of calculation and the time of system training and response.

(4) **Match result output**

Figure 4.42 shows the face images taken by the same person at different angles.

In order to reduce the change of hidden Markov model parameters caused by the change of shooting conditions and eliminate the change of state transition parameters caused by the sudden change of pixel values of pixels in the hidden state of human face (i.e., facial features), the preprocessing operation of human face image can be introduced before establishing the hidden Markov model of human face image. That

Fig. 4.42 Face images taken from different perspectives

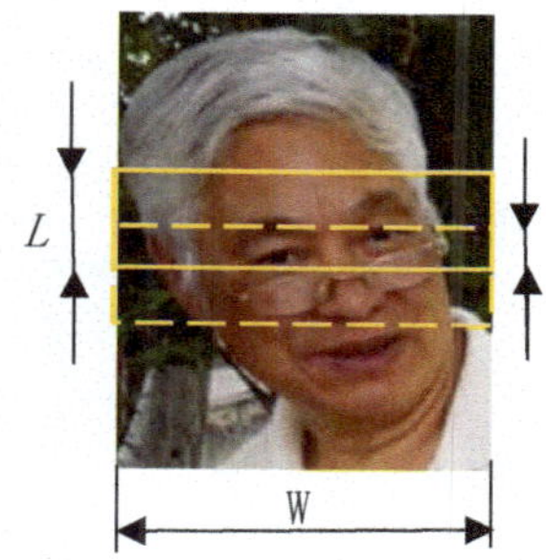

Fig. 4.43 Schematic diagram of sliding window

is, after preprocessing the face image with median filter, the preprocessing results are used as observations to estimate the parameters of hidden Markov model.

After the median filter, the pixel values of the internal pixels in the hidden state of the face will tend to be consistent. At the same time, the change of face angle, the shadow and facial features deformation in the face image are also offset by the mean filter to a certain extent. Therefore, before the establishment of hidden Markov model, the mean filtering of face image can improve the accuracy of face model.

After face image preprocessing, it is also necessary to obtain the observation sequence of face image, so as to realize the estimation of hidden Markov model parameters. In order to reduce the dimension of the observation sequence and obtain higher recognition accuracy, the singular value decomposition method can be used to obtain the observation sequence. For example, a sliding window with a width of *W* pixels and a height of *L* pixels is used to traverse the whole face image from top to bottom, and the overlapping height between two adjacent sliding windows is *P* pixels, as shown in Fig. 4.43.

Singular value decomposition is performed on the image of each sliding window area, and a plurality of matrices are obtained after decomposition as the observation values of the image of the sliding window area. The observation values of the image part passed by all sliding windows are the observation value sequence of the face image.

After the above calculation process is completed, the hidden Markov model of the face can be established, which is the process of obtaining the hidden Markov model parameters of the face through the face image in the database (Fig. 4.44).

The recognition process of face Hidden Markov model is to calculate the probability that it belongs to each face hidden Markov model. The Hidden Markov model with the largest probability can be output as the final recognition result. The specific process is as follows:

1) The filtered image is segmented by moving the sliding window, and the observed value of the image in each window is calculated;
2) Using the observed value to calculate the matching degree with each hidden Markov model;
3) The hidden Markov model with the largest probability is selected as the output of the recognition result.

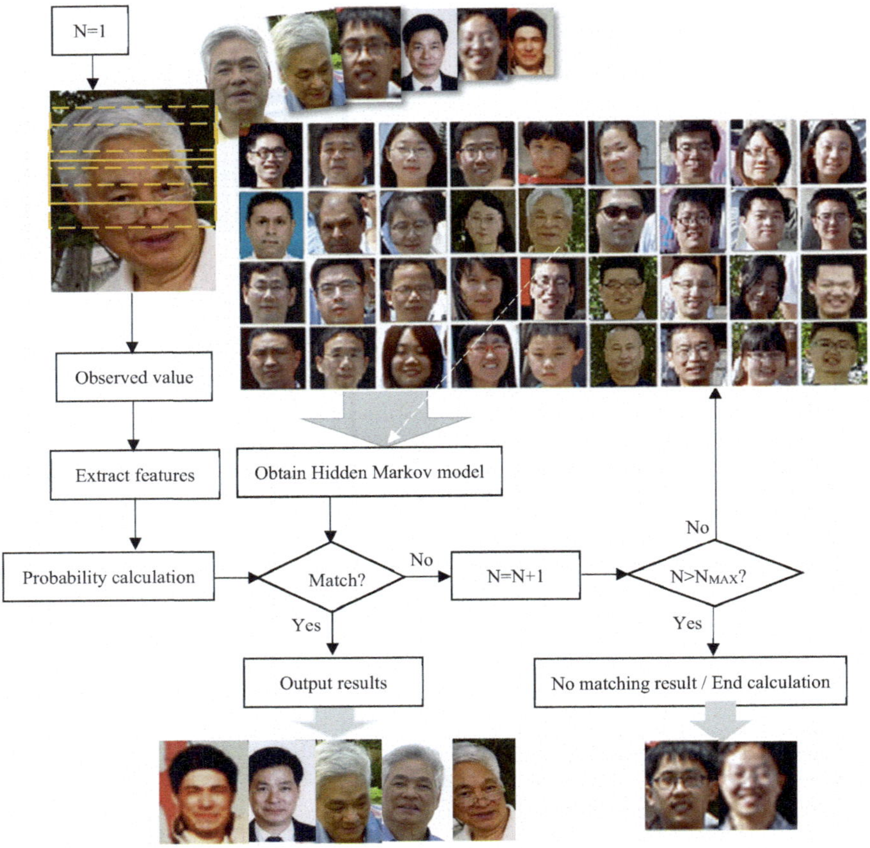

Fig. 4.44 Recognition flowchart based on hidden Markov model

As shown in Fig. 4.44, it is the process of using HMM model to recognize the accurate recognition samples obtained from the face images taken by the same person at different angles and at different periods of time.

References

1. Yan X (2019) Application of face recognition system in public security communication system of rail transit. Inf Commun 8:34–39 (In Chinese)
2. Xu H (2019) Application of wireless communication technology in rail transit communication and signal service. China New Telecommun 8:1–2 (In Chinese)
3. Zhao D (2015) A research of face recognition method through geometric feature. Dissertation, Hebei University of Technology

4. Bai D (2013) Research and improvement of clustering routing algorithms for wireless sensor networks. Dissertation, Chongqing University
5. Shen Y (2019) Face recognition based on joint sparse representation of multiple features. Inf Technol 9:154–162 (In Chinese)
6. Zhang X, Ying J (2011) Principles of automobiles intelligent technology. Shanghai Jiao Tong University Press, Shanghai
7. Gong H, Wu X, Jin Z (2019) Improved method of face recognition with the basis of illumination invariant features. Comput Digit Eng 47(9):2300–2305 (In Chinese)
8. Han J (2012) The research on face recognition algorithms based on elastic graphic matching and hidden Markov model. Dissertation, Lanzhou University of Technology
9. Peng X (2019) Research on face recognition in specific scenes based on convolutional neural network. Dissertation, Xi'an University of Technology
10. Liu N, Liu R, Yin N et al (2019) Face recognition based on deep convolution neural network. Comput Eng Softw 40(8):6–8 (In Chinese)
11. Hong C, Fu Y, Guo M et al (2019) Identification method of distribution network faults based on improved multi-classification support vector machine. J Electronic Measure Instrum 33(1):7–15 (In Chinese)
12. Xu J (2018) Research on face recognition algorithm based on sparse representation and support vector machine. Dissertation, Nanjing University of Posts and Telecommunications
13. Su X (2015) Pattern recognition based on multi class support vector machine. Comput Digital Eng 7:1202–1206 (In Chinese)
14. Meng H, Gao D, Li Y et al (2019) Face recognition based on Hidden Markov model. J Criminal Invest Police Univ China 4:124–128 (In Chinese)

Chapter 5
Application of AI in Rail Transit Operation and Maintenance

Compared with traditional methods, artificial intelligence has produced unprecedented good results in improving driving efficiency and safety.

At present, the wide promotion and application of big data technology and artificial intelligence technology make people's life more convenient and fast. On this basis, the creation of intelligent traffic management mode can obviously greatly improve the management efficiency of China's traffic field [1].

5.1 Typical Application Fields

At present, cloud computing, big data, deep learning and other artificial intelligence technologies and algorithms have been increasingly popular in rail transit.

5.1.1 Preparation of Train Operation Diagram

The compilation of train diagram is one of the core issues of rail transit system organization and the basis of rail transit system operation.

Taking high-speed railway as an example, the operation characteristics of high-speed railway are that the operation speed, operation density, punctuality rate and safety are in a "high point" state.

At the same time, the technical operation of the high-speed railway station is relatively simple. Compared with the traditional existing lines, the EMU (electric multiple units) running on the high-speed railway has two-way operation capacity. After the train arrives at the station, there is no need to carry out locomotive pickup, coupling, steering and other operations, and there is no need to enter the warehouse for maintenance every time. Therefore, the dwell time of the high-speed railway

Y. Khan, *Principles of Intelligent Rail Transit*, Advances in High-speed Rail Technology,
https://doi.org/10.1007/978-981-19-6072-7_5

train is relatively short. In addition, the high-speed train operation has a "rectangular skylight", and the time reserved for construction and maintenance on the train operation diagram is called "skylight". That is, China's high-speed railway generally operates only during the day, and the line, power supply equipment, signal equipment and EMU are constructed or overhauled at night. The high-speed rail skylight is generally overhauled from 0 a.m. to 6 a.m., which is called "rectangular skylight".

For urban rail transit, the operation control center (OCC) is responsible for the train operation command of Metro. OCC is the dispatching and command department of subway operation and production, commanding and managing the traffic, power, fire control, environmental control and other work of each subway line, as well as the emergency treatment of operation emergencies. With the increasing density of local railway network and the increasing passenger flow year by year, the efficiency of traffic dispatching and command becomes more and more important.

The train diagram widely used in China's subway signal system is an example of using the coordinate principle to illustrate the relationship between time and space of train operation. It specifies the sequence of each train occupying the section, the arrival and departure time of the train at a station, the operation time of the train section, the station dwell time, the turn back operation time, etc., which is the basis of the train operation organization.

In short, when compiling the train diagram, it is necessary to comprehensively consider the train operation parameters of subway line, line parameters, traffic capacity of signal system, passenger flow distribution and other factors. Among them, the first three factors are fixed factors, which will not change once the subway line is completed; The last passenger flow distribution factor will change with the operation, which belongs to the changing factor. The change of passenger flow distribution can be either the change of spatial distribution or the change of time distribution. After a line is opened, the passenger flow will be adjusted with the improvement of the network line or the development and changes around the line. For example, when the operation of a single line is developed to network operation, the passenger flow at the transfer station of the road network will increase significantly. For another example, at the initial stage of opening, a station on the line can only open some entrances and exits due to its relationship with surrounding commercial centers and other buildings. When all entrances and exits of the station are opened in the long term, driven by these commercial centers, the passenger flow of the station will also be increased. According to years of operation experience, the centralized travel time of passengers presents different distribution characteristics between working days and weekends in a week, and there will also be morning peak, evening peak and flat peak in a day.

When formulating the train diagram, under certain fixed factors, it is necessary to fully consider the passenger flow demand, strive to match the transport capacity provided by the train diagram with the actual demand and maximize the social benefits of the subway.

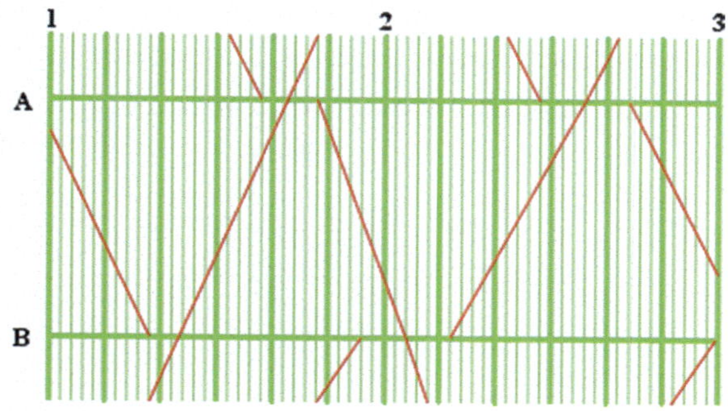

Fig. 5.1 Two-minute grid operation diagram of train

1. **Representation of train diagram**

Train diagram is a graphical representation of the train operation process in the coordinate axis. It specifies the sequence of the train occupying the section and the departure, arrival or passing time at each station.

There are two specific graphic forms: One is that abscissa represents distance and ordinate represents time, which is mainly used in European countries. The other is to use abscissa to represent time and ordinate to represent distance. This representation method is adopted in China.

In China's train diagram, it is stipulated that different types of trains are represented by different symbols and colors, and passenger trains are represented by red single line. The time division format of train diagram is also different according to the specific use conditions, mainly including the following three formats.

(1) **2-min grid operation diagram**

The interval between adjacent (thin) vertical lines is 2 min, and the vertical lines of hour grid (such as 1, 2, 3…) are thicker, which are mainly used for the preparation of new operation diagram, as shown in Fig. 5.1.

(2) **10-min grid operation diagram**

In the 10-min grid operation diagram, the interval between adjacent vertical lines is 10 min, with half hour lattices as dotted lines and hour grids as thicker vertical lines, which are mainly used for the compilation of dispatching adjustment plans and the drawing of actual performance diagrams in daily scheduling work, as shown in Fig. 5.2.

(3) **Hourly grid operation diagram**

The interval between adjacent vertical lines is 1 h, which is mainly used to prepare locomotive turnover diagram and passenger train operation scheme, as shown in Fig. 5.3.

In Figs. 5.1, 5.2 and 5.3, 1, 2, 3… represent hours, A, B… represent the center line of the station or line through which the train passes, and the slash represents the train operation line.

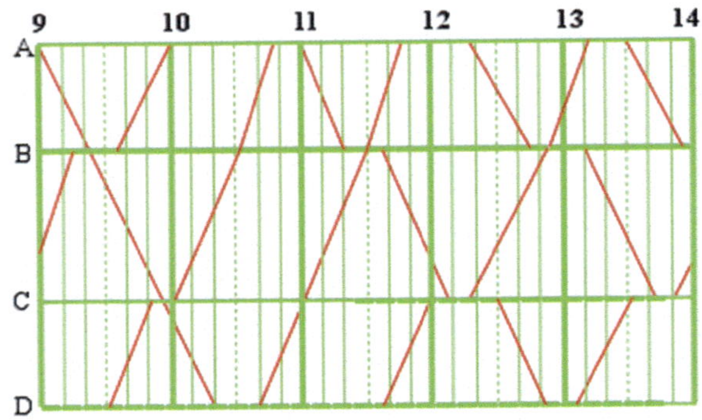

Fig. 5.2 Ten-minute grid operation diagram of train

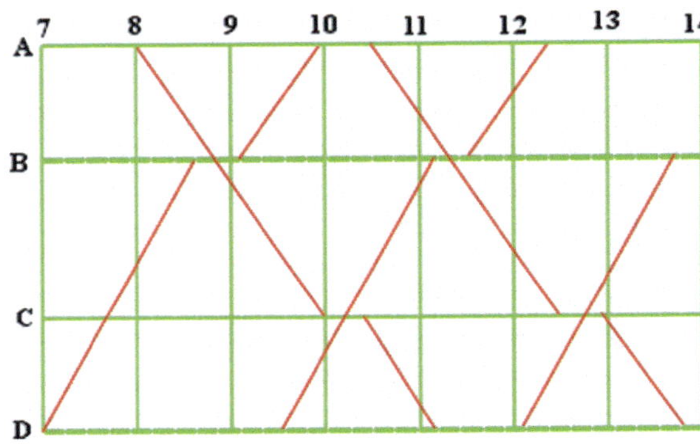

Fig. 5.3 Hourly grid operation diagram of train

2. **Basic elements of train operation**

The basic elements of train diagram include train interval operation time, train dwell time at intermediate station, dwell time at turn back station, additional starting and stopping time, station interval time, tracking train interval time and maintenance skylight opening time [2].

(1) **Train interval operation time**

The operation time of train section refers to the operation time standard of train between two adjacent stations (line posts). Generally, the locomotive department shall check and determine by combining traction calculation and actual test and calculate according to the distance between the signal passing through the station center line or line. The running time of the train shall be determined according to the up-down direction of the train. At the same time, it shall also distinguish between non-stop passing and stop passing at two stations in each section. The interval operation time required for the train to pass through two adjacent stations without stopping is called pure operation time.

(2) **Dwell time of train at intermediate station**

The dwell time of the train at the intermediate station refers to the minimum time for the train to handle necessary operations at the intermediate station, mainly including the time for handling relevant technical operations at the intermediate station, the time for handling passenger transport operations, the waiting time for train avoidance, etc.

(3) **Dwell time at turn back station**

The dwell time at the turn back station refers to the minimum time required for the locomotive to handle necessary operations at the turn back station. The dwell time standard of turn back station needs to be determined by the locomotive department according to the operation content and process of turn back station and the combination of operation process analysis and actual standard inspection.

(4) **Additional time for starting and stopping**

The additional starting and stopping time refers to the time when the interval operation time is longer than the pure interval operation time due to the start or stop of the train. When compiling the train diagram, the additional starting and stopping time of the train is generally a given constant, and its value is determined according to the traction process calculation of the actual situation.

(5) **Station interval operation time**

Station interval operation time refers to the minimum interval time required for the arrival, departure or passing operation of two adjacent trains at the station. When determining the interval operation time of the station, it shall comply with the provisions of relevant regulations and the technical operation time standard of the station, so as to ensure the traffic safety and make the most effective use of the section carrying capacity.

(6) **Interval time of tracking train**

The interval time of tracking trains refers to the minimum interval time between two trains in tracking operation. The determination of the tracking train interval depends on the starting and braking performance of the train, the technology of the train control system, the station signal interlocking equipment, the slope of the line and other factors.

(7) **Opening time of maintenance skylight**

The maintenance time of skylight refers to the time used for repairing and maintaining railway technical equipment such as vehicles, locomotives and lines. The skylight time of high-speed railway is generally arranged between 0:00 and 6:00 at night, and fixed maintenance skylights are generally not set during the day.

3. **Basic requirements for preparation of train operation diagram**

(1) **Ensure driving safety**

The preparation of train diagram must comply with relevant railway regulations, strictly implement various technical operation procedures and strictly abide by various operation time standards.

(2) **Adapt to market demand**

In order to meet the market demand, we first need to transport passengers efficiently and conveniently. The train diagram proposed by the passenger transport department must predict the passenger flow density during operation, so as to establish the train operation scheme, EMU routing plan and the time standard for stopping at the station to complete passenger boarding and landing or other technical operations.

(3) **Make full use of railway capacity**

This requires the rational use of EMUs. The laying of train operation line shall minimize unnecessary parking time and strive to improve the travel speed of the train; EMUs should be used economically and reasonably to maximize their benefits.

(4) **Ensure that the train diagram can be adjusted flexibly**

When compiling the train diagram, it is necessary to reasonably control the utilization rate of section capacity to ensure that the train diagram has adjustment flexibility and meet the adjustment requirements when the train operation order changes.

(5) **Ensure that the train distribution meets the passenger flow demand**

The number and distribution of train operation lines shall be based on the passenger demand to meet the passenger demand to the greatest extent. At the same time, the preparation of train operation diagram shall also have a certain predictability to predict the development and change of market demand, so as to improve transportation efficiency.

(6) **It is necessary to realize the balanced operation of station and section trains**

The preparation of train diagram shall consider the balance of train operation and try to achieve the balanced arrival and departure of station trains in each time period, so as to make full use of the receiving and departure line and section capacity of the station, so as to improve the railway transportation capacity.

(7) **Reasonably arrange the work and rest time of staff**

Of course, it is necessary to reasonably arrange the work and rest time of train staff in order to ensure that the staff have a good mental state, which is an important link to ensure the safe operation of the train.

4. **Mathematical model of train operation**

For the mathematical model of train operation, we must first make assumptions about the model, then clarify its constraints and finally determine the objective function.

The following describes the relevant technical contents by taking the operation of high-speed railway as an example.

(1) **Model assumptions**

(I) It is applicable to a single double track high-speed railway, and trains of different grades are operated on the line.
(II) Assuming that the trains in the up direction and down direction of high-speed railway are completely independent of each other and will not affect each other, only a single direction needs to be considered in modeling.
(III) It is assumed that all stations on the line in a single direction can receive and send trains at the same time.
(IV) Only consider the path of the train on the line in a single direction.
(V) When the train in the same direction runs in the section, assuming that the train of the same grade does not overtake, the high-grade train can overtake the low-grade train, and the low-grade train can be overtaken twice at most in one station.
(VI) The data of high-speed railway train operation scheme (including operation section, train grade, number of trains, stop scheme, etc.) are determined and known, and the number of arrival and departure trains at the station is also known.
(VII) Various safety interval standards for train operation, additional starting and stopping time at the station, stopping time, pure operation time between districts, etc., are known, and the standards of the same level are consistent.
(VIII) Ignore some factors considered in the layout of train operation diagram, such as the number of operating vehicles of EMU, passenger flow change and transfer demand.

(2) **Constraints**

(I) Constraint of train passing section sequence

It must run in the specified interval sequence, that is

$$a_s^l - d_s^l > 0 \tag{5.1}$$

where a_s^l represents the arrival time of the train at the station and d_s^l represents the departure time of the train at the station;

$l \in L$ represents the mathematical set element of the train passing through the station, and L represents the mathematical set of all trains in the line, $L = \{l_1, l_2, l_3, \ldots, l_n\}$;

$s \in r_l$; $r_l = \left\{s_i^l \middle| 1 \le i \le m_l, l \in L, s_i^l \in S\right\}$;

S represents the mathematical set of stations on the line, $S = \{s_1, s_2, s_3, \ldots, s_m\}$.

(II) Train departure interval constraint

When the train is running, at most one train is allowed to enter a block section at the same time, that is, when the train is running in the same direction, the subsequent train l_j needs to start after the previous train l_i starts for a period of time, that is, the front and rear trains should meet a certain departure interval.

(III) Train arrival interval constraint

When the train in the same direction runs, the front train l_i and the rear train l_j should meet a certain arrival interval.

(IV) Train interval operation time constraint

It is specified that the actual operation time of the train in the section shall be greater than or equal to the standard operation time of the train in the section. The train interval operation time includes pure operation time and additional starting and stopping time.

(V) Station dwell time constraint

The specified train dwell time at the station shall not be less than the lower limit of the minimum parking time required to meet the passenger boarding and landing operations and corresponding technical operations. On the other hand, in order not to affect the subsequent trains and improve the efficiency of the train diagram, it shall not exceed the upper limit of the parking time.

(VI) Skylight time constraint

As the high-speed railway adopts the rectangular skylight for night maintenance, it is not allowed to drive within the skylight range.

(VII) Departure time range constraint

The departure time of the train shall be restricted by the effective time band. The reasonable departure range of trains with different operation schemes will also be different; especially for cross-line trains, the reasonable departure time range needs to be considered.

(VIII) Station arrival departure track constraints

During train operation, the number of arrival departure tracks occupied by the train shall not exceed the number of arrival departure tracks used for receiving and sending such trains at the station.

(IX) Overrun constraint

When the speed of the subsequent high-grade train l_j is greater than that of the previous low-grade train l_i, the high-grade train can (need to) overtake the low-grade train.

(3) **Objective function**

Railway travel demand is considered from the perspective of improving the transportation capacity of the train diagram, and the optimization goal is to minimize the time occupied by the train line. The time occupied by the train operation line in the train diagram refers to the maximum value of the sum of the time occupied by the train in each station.

$a_s^{\min}$, $a_s^{\max}$, $d_s^{\min}$ and $d_s^{\max}$ are used to represent the earliest arrival time, the latest arrival time, the earliest departure time and the latest departure time of station S, respectively.

Then, the earliest and the latest time to occupy the station are.

$$t_s^{\min} = \min\{a_s^{\min}, \quad d_s^{\min}\} \tag{5.2}$$

$$t_s^{\max} = \max\{a_s^{\max}, \quad d_s^{\max}\} \tag{5.3}$$

Further, the minimum value $\min Z_1$ of the time occupied by the target train line in the train diagram can be expressed as

$$\min Z_1 = \max\{t_s^{\max} - t_s^{\min} | s \in S\} \tag{5.4}$$

In order to improve the quality of train operation mapping, the total travel time of the train should be reduced as much as possible. Therefore, a function $\min Z_2$ that minimizes the total travel time of the train can be expressed as follows

$$\min Z_2 = \sum_{l \in L} \left(a_{s_{ml}^l}^l - d_{s_1^l}^l\right) \tag{5.5}$$

where the subscript m_l represents the number of stations the train passes through, s_1^l represents the departure station of the train, and s_{ml}^l represents the final arrival station of the train.

For the multi-objective optimization problem, the objective weighting method can be used to give different weight coefficients ω_1 and ω_2 to different optimization objective functions, and the multi-objective optimization problem can be transformed into a single objective optimization problem. The overall objective function of the model is

$$\min Z = \omega_1 Z_1 + \omega_2 Z_2 \tag{5.6}$$

$$\omega_1 = \frac{1}{Z_1^{\max} - Z_1^{\min}} \tag{5.7}$$

$$\omega_2 = \frac{1}{Z_2^{\max} - Z_2^{\min}} \tag{5.8}$$

5. **Basic algorithm of model solving**

Algorithms can be divided into two categories: deterministic optimization methods and heuristic methods.

Deterministic optimization method refers to the method based on the traditional mathematical programming method, using the theory of operations research to establish a mathematical model and accurately solve the model to obtain the optimal solution. Such algorithms are usually based on the idea of enumeration and obtain the optimal solution through global search. They can obtain satisfactory results for small-scale scheduling problems, but for large-scale complex scheduling problems, the amount of calculation of global search will be very huge and it is difficult to obtain satisfactory results. The commonly used methods are usually branch-and-bound method and dynamic programming method.

Heuristic method is based on heuristic reasoning and obtains the optimal solution by local search or neighborhood search. The commonly used methods are genetic algorithm, tabu search algorithm, simulated annealing algorithm, neural network algorithm and so on.

(1) **Characteristics of genetic algorithm**

Genetic algorithm is a highly parallel, random and adaptive search algorithm developed according to the biological evolution process of natural genetic selection and natural elimination [3].

The basic idea of genetic algorithm is "survival of the fittest", which represents the solution of the problem as "chromosome". Through a series of genetic operations on the chromosome group, it makes continuous evolution from generation to generation and finally obtains an individual "most suitable for the environment"; that is, the optimal solution of the problem is obtained at this time.

Because of its simple idea, easy implementation and good robustness, as an important intelligent computing technology, it is widely used in many fields such as scientific research, production and manufacturing.

As a general search algorithm, genetic algorithm has the basic characteristics of general search algorithm: First, generate the initial candidate solution; calculate the fitness value of the candidate solution according to the constraints; and select and leave some candidate solutions according to the fitness value and discard other candidate solutions. By operating the remaining candidate solutions, a new candidate solution is generated.

Genetic algorithm combines the basic characteristics of general search algorithm in a special way, which distinguishes it from general search algorithm and forms its unique characteristics.

(I) Genetic algorithm operates on the coding of problem parameters. Traditional optimization algorithms generally calculate the actual value of problem parameters, while genetic algorithm is to calculate and process some form of coding of problem parameters. The algorithm itself has high flexibility and can play a role in a wide range of problem solving.
(II) Genetic algorithm starts from the problem solution set, not from a single solution. It has the characteristics of parallel search and adaptability. The traditional algorithm usually starts from a single initial solution, but it is easy to enter the local extreme point and the efficiency is not high. Genetic algorithm starts

from the initial solution group composed of multiple individuals. The search range is wide, which can effectively avoid entering the local extreme points and better carry out the global optimization.

(III) Genetic algorithm directly takes fitness as search information without other auxiliary information. In order to determine the search direction, the traditional search algorithm needs not only the value of the objective function, but also some auxiliary information (such as the derivative value of the objective function). Genetic algorithm mainly determines the search direction according to fitness, and the calculation of individual fitness cannot rely on the accurate estimation of objective function, so it has a wide range of applications.

(IV) Genetic algorithm uses probability search technology. Many traditional search algorithms often use deterministic search methods, that is, transfer the search points through the determined transfer relationship and way. Its limitation is that it is easy to fall into local extreme points. Genetic algorithm uses probability search mechanism and has uncertainty; that is, it carries out selection, crossover, mutation and other operations with a certain probability, so as to expand the search scope and have global optimality [4].

(2) **Basic principle and flow of genetic algorithm**

When genetic algorithm solves the optimization problem, first encode the feasible solution according to certain rules to form chromosomes and then generate multiple initial chromosomes to form an initial population. On this basis, follow the principle of "survival of the fittest" for subsequent operations.

The genetic algorithm screens the population according to the individual fitness value, and the individuals with high fitness value have a higher probability of being selected. At the same time, the next generation population is formed through selection, crossover, mutation and other operations. The optimal solution is obtained by iterating continuously until the termination criterion is met.

A series of operations of genetic algorithm better maintain the population diversity; the solution process is not easy to fall into local extreme points and has strong global search ability. During this period, excellent individuals can be retained and continuously evolve.

Genetic algorithm is to find the optimal solution through cyclic iteration. Its basic process is shown in Fig. 5.4 [5].

5.1.2 Preparation of Train Diagram from the Perspective of Energy Saving

With the rapid advancement of China's high-speed railway construction process and the continuous expansion of the scale of road network, China's energy is relatively scarce, and energy conservation and emission reduction have been paid more and

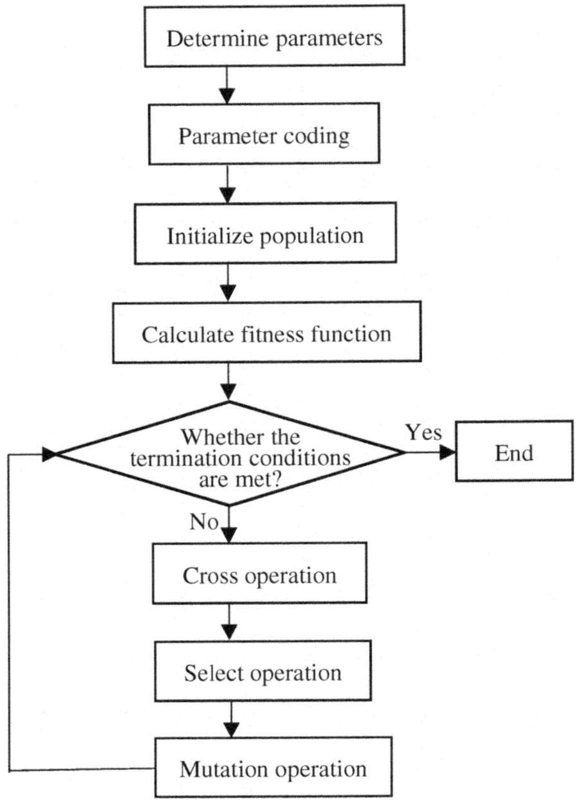

Fig. 5.4 Basic flowchart of genetic algorithm

more attention. The rapid development of railway has not only driven economic growth, but also caused many energy consumption problems.

The artificial intelligence system can obtain the spatial and temporal distribution of all passenger flow in the line network in a certain period through the ticket sorting system of the subway. Then, according to the network structure of the line network, the actual train supply and the actual passenger flow, the supply and demand of transport capacity are compared, and the time period of concentrated passenger flow (which can be divided into peak and flat peak), station and detention time are analyzed to further obtain the matching degree of transport capacity. According to the conclusions drawn by the system, the personnel preparing the train diagram can optimize the train diagram according to the passenger flow distribution, adjust the stop time of the train at some stations, maximize the traffic density during the peak period of passenger flow, appropriately increase the traffic density during the peak period and reduce the train density during the peak period. In this way, it can not only meet the actual transport capacity demand, but also reduce the waste of transport capacity in flat peak and maximize the economic benefits of operation.

1. **Basic theory and method of train energy-saving diagram adjustment**

(1) **Relationship between train operation curve and energy consumption**

Based on the latest wireless train-ground communication technology, the main purpose of train on-board ATO (automatic train operation) system is to control the train to run from one station to another according to the timetable. This requires the cooperation of ATO and ATR (automatic train regulation). In order to effectively save energy and operate according to the schedule, ATO must calculate the required speed from a station to the next station when starting from a station.

In order to accurately calculate this track, the following basic train data shall be considered: various information about the line; distribution information of track profile (including slope) and curve; detailed information required for parking at the next station (track circuit number, operation time, etc.); and vehicle weight information.

In ATO system, cruise/coasting is an economic operation mode designed for energy-saving operation of train; that is, under the condition of known train operation track, the energy-saving mode of automatic operation is implemented according to vehicle parameters, planned route to the next parking point and other information.

(2) **Energy consumption characteristics of train operation curve**

The railway automatic control system needs to be able to save the cost of electric energy in many ways, including the power consumption of vehicles, the power demand during starting and the power loss during braking. In practice, the energy-saving method used by trains is not only adjusting the train timetable while drawing up the middle note, but also adjusting the running grade curve in the middle note of train operation to achieve the purpose of energy saving. The operation level curve aims to reduce the energy consumption of traction by reducing the requirements of train traction system.

The time division curves of four basic speeds of trains running in straight and infinite speed track section are shown in Fig. 5.5 [6].

The maximum allowable speed of the train in the section is related to the type of train, track construction conditions and section length. Generally, $V_{\max}$ of each section will be given by engineers before the preparation of operation diagram. Whether the train can reach the maximum speed $V_{\max}$ during section operation is related to section length L and section travel time T_s. Specifically, $L_{\text{ab}}(V_{\max})$ is used to represent the travel distance of the train that accelerates from 0 to $V_{\max}$ and then decelerates to 0 immediately (see curve form a_1) or decelerates to 0 slightly (see curve form a). For any section whose length meets $L \leq L_{\text{ab}}(V_{\max})$, since the train speed cannot reach $V_{\max}$, the control strategy of the train can only accelerate first and then decelerate. For the interval whose length meets $L > L_{\text{ab}}(V_{\max})$, there may be three control strategies: no inertia condition (see curve form b); no cruise condition (see curve form c); and including coasting and cruise conditions (see curve form d).

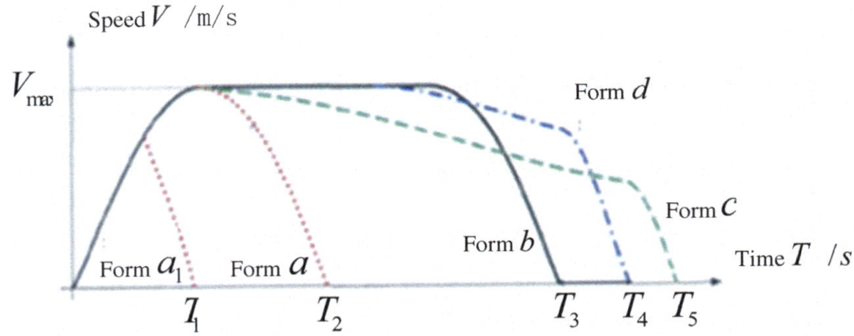

Fig. 5.5 Time division curve of train speed in the section

The use of train operation grade curve can reduce the power consumption of a single train, and the technologies adopted include the following three items.

(I) **Reduce acceleration time to reduce peak speed**

The train reduces the total time of its acceleration, which reduces the traction request and reduces the energy demand of the train.

(II) **Reduce the acceleration to reduce the line running speed**

The train adopts a small acceleration, which reduces the power required for traction, and finally reduces the energy demand of the train.

(III) **Coasting**

When the train leaves the station, it accelerates first, then coasts, then slows down and finally stops at the next station.

The train coasts after reaching the maximum line speed, which will increase the running time according to a specific proportion. Compared with the normal specified speed, the train reduces the power required for acceleration request and traction, thus reducing energy consumption.

Figure 5.6 shows the comparison of the simplified curve of the running distance characteristics of a single train. The area formed by the curve (realization or dotted line) and the abscissa distance line in the figure is equivalent to the value of energy consumption.

Obviously, the single vehicle energy-saving operation is relatively simple. Under the condition of meeting the running time, the coasting curve should be used as much as possible instead of "pursuing" the minimum running time curve, otherwise the residence time at the station will be increased.

(3) **Energy-saving operation under tracking conditions**

In practice, most trains are tracking operation, and their operation control is affected by the front signal or train.

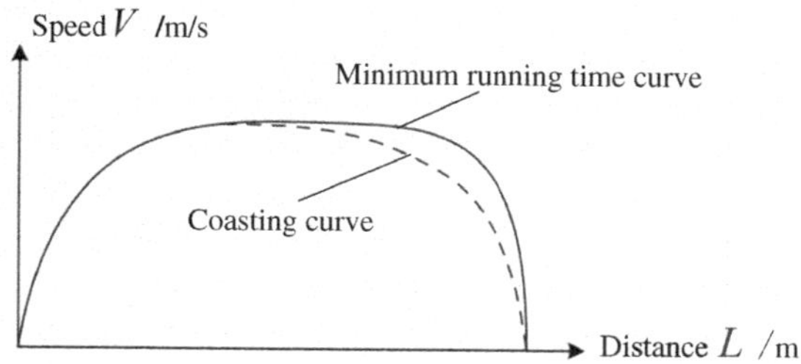

Fig. 5.6 Running distance curve of single train

During the train tracking operation, the front signal display is a command indicating the train operation. The front and rear trains are located on different plane and vertical sections of the line, and their space–time interval is in dynamic change. The change of the position of the front train will bring the change of signal display and further affect the operation and manipulation of the tracking train.

In the fixed block system, the partition interval of the train is long, and one partition can only be occupied by one train. Although it is not conducive to shorten the operation interval of the train, it can reduce the energy consumption. The characteristic of flexible train operation interval of moving block system makes its energy-saving effect remarkable in peak hours. Generally, when operating in peak hours, the departure density is high and the passenger mobility is fast. The delay of the train ahead will cause the possibility of frequent braking of subsequent trains. The frequency and duration of this braking are greatly reduced in the moving block system due to the shorter and uniform minimum train interval. The uniform minimum train interval makes the rear train closer to the front train and improves the operation efficiency.

2. **Necessary conditions for energy-saving operation of high-speed train**

After introducing Pontryagin's extreme value principle, the Hamiltonian function corresponding to the minimum energy consumption model can be obtained, that is, the following formula

$$
\begin{aligned}
H = {} & [u_t F_t(v) - \alpha u_b F_b(v)] - \frac{\lambda_1}{v} \\
& - \lambda_2 \left[\frac{u_t F_t(v) - u_b F_b(v) - w(v) - w(s)}{M \cdot v} \right]
\end{aligned}
\tag{5.9}
$$

where u_t indicates the level information of traction handle of high-speed train. For example, the traction handle is divided into 10 levels, $u_t = 0.2$, indicating 2 levels of traction; u_b indicates the level information of the brake handle of high-speed train, for example, the brake handle is divided into five levels, $u_b = 0.2$, indicating level 1 braking; $F_t(v)$ is the maximum traction force that high-speed train can exert at speed v; $F_b(v)$ is the maximum braking force that the high-speed train can exert at speed v; $w_t(v)$ is the rolling resistance of high-speed train operation at speed v; $w(s)$ is the line additional resistance of the high-speed train at the distance S; and α is the ratio of regenerative braking force to total braking force during braking of high-speed train.

If the regenerative braking efficiency of high-speed train is assumed to be 1, α can be defined as braking utilization rate; t is the running time of high-speed train, M is the total mass of high-speed train, and λ_1 and λ_2 are Lagrange operators.

Equation (5.9) contains inequality constraints of state variable v. In order to remove this constraint, the relaxation operator P is introduced to obtain the adjoint state equation and complementary relaxation equation corresponding to the Hamiltonian function as follows

$$\begin{cases} \frac{d\lambda_1}{ds} = \frac{\partial H}{\partial t} + \frac{dP}{ds} \\ \frac{d\lambda_2}{ds} = \frac{\partial H}{\partial v} + \frac{dP}{ds} = \mu_t F'_{t(v)} - \alpha\mu_b F'_{b(v)} + \frac{\lambda_1}{v^2} - \lambda_2\left[\frac{\left[\mu_t F'_{t(v)} - \mu_b F'_{b(v)} - w'(v)\right]\cdot M\cdot v}{M^2 v^2}\right] - \\ \lambda_2\left[\frac{-\left[\mu_t F_{t(v)} - \mu_b F_{b(v)} - w(v) - w(s)\right]\cdot M}{M^2 v^2}\right] + \frac{dP}{ds} \\ (v - \overline{v})\frac{dP}{ds} = 0 \\ \frac{dP}{ds} \geq 0 \end{cases} \tag{5.10}$$

where $\overline{v}$ is the limit value of the line to the running speed of high-speed train; μ_t and μ_b are the energy utilization rate of high-speed train transmission system during traction and braking, respectively. For simplified analysis, they can be assumed to be constant 1.

The following formula can be obtained by collecting the subitems of μ_t and μ_b contained in H function

$$H = \left(1 - \frac{\lambda_2}{M\cdot v}\right)\mu_t F_{t(v)} - \left(\alpha - \frac{\lambda_2}{M\cdot v}\right)\mu_b F_{b(v)} - \frac{\lambda_1}{v} - \lambda_2\left[\frac{-w(v) - w(s)}{M\cdot v}\right] \tag{5.11}$$

According to Pontryagin's extreme value principle, the necessary condition for taking the minimum value of energy consumption calculation model is that H takes the minimum value. According to formula (5.11), it can be deduced that the necessary condition for H to be minimized is

$$\begin{cases} \frac{\lambda_2}{M\cdot v} > 1; u_t = 1, u_b = 0 \\ \frac{\lambda_2}{M\cdot v} = 1; 0 \leq u_t \leq 1, u_b = 0 \\ \alpha < \frac{\lambda_2}{M\cdot v}; u_t = 0, u_b = 0 \\ \frac{\lambda_2}{M\cdot v} = \alpha; u_t = 0, u_b = 0 \\ \frac{\lambda_2}{M\cdot v} = \alpha; u_t = 0, u_b = 1 \end{cases} \tag{5.12}$$

Formula (5.12) expresses five working conditions, which shows that when the high-speed train operates in energy-saving mode, at least one of the brake handle and traction handle remains at level 0, which is consistent with the requirements of train control. The first, second and fifth equations correspond to the maximum traction, coasting and maximum braking conditions of high-speed train, respectively. The second and fourth equations are the singular points of energy-saving operation of high-speed train, which correspond to partial traction and partial braking conditions, respectively. However, these two conditions cannot determine the detailed operation state of high-speed train, which can be determined only after further discussion as follows.

For some traction conditions, the following formula can be derived from formulas (5.10) and (5.12)

$$\frac{\lambda_1}{v^2} + w'(v) + \frac{\mathrm{d}P}{\mathrm{d}s} = 0 \tag{5.13}$$

When the high-speed train operates in an energy-saving manner under some traction conditions, its operating speed is either equal to the line speed limit $\overline{v}$ or less than $\overline{v}$. When $v \leq \overline{v}$, since $\frac{\mathrm{d}P}{\mathrm{d}s} = 0$, the expression (5.13) can be simplified to

$$\frac{\lambda_1}{v^2} + w'(v) = 0 \tag{5.14}$$

For some braking conditions, it can be deduced that when the high-speed train operates in energy saving under some braking conditions, its running speed is either equal to the line speed limit $\overline{v}$ or less than $\overline{v}$ and equal to v^*. v^* is determined by λ_1 and α, and its calculation formula is

$$\frac{\lambda_1}{v_b^{*2}} + (1-\alpha)u_b F_b'(v_b^*) + w'(v_b^*) = 0 \tag{5.15}$$

In a word, from the above derivation and calculation, it can be concluded that the necessary condition for energy-saving operation of high-speed train is to reasonably use the adjustment strategies under the four working conditions of maximum traction, maximum braking, coasting and constant speed. When the high-speed train is running at a constant speed, its running speed should be determined by formulas (5.14) and (5.15) [7].

3. **Utilization methods of renewable energy**

In fact, there is a higher-level train diagram. That is, there is a train operation timetable with optimized energy distribution. This timetable has begun to enter the practical stage.

The so-called train operation timetable with optimized energy distribution means that when the regenerative electric energy generated by train coasting and braking deceleration is fed back to the power grid, the regenerative electric energy can be absorbed and utilized by other online trains in real time, so it has the effect of high efficiency and energy saving of the whole train operation network.

"Train operation timetable with optimized energy distribution" includes the following contents.

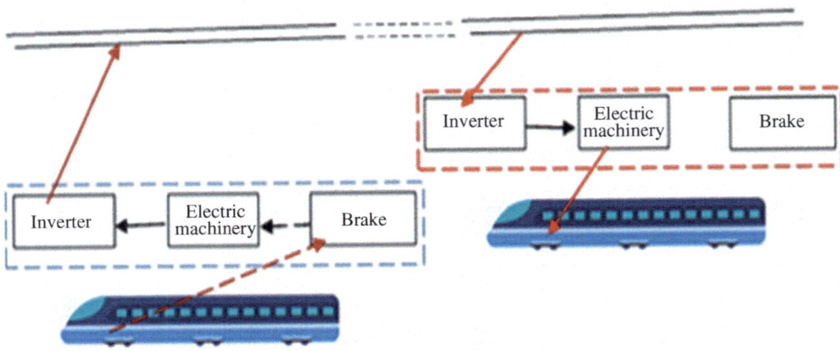

Fig. 5.7 Schematic diagram of regenerative electric energy transmission of train

(1) **Optimized operation timetable of renewable energy**

This is a "steady-state operation timetable" based on the train diagram formulated by comprehensively considering the train operation parameters of subway line, line parameters and the driving capacity of signal system, and considering the optimal utilization of renewable energy.

(2) **Stochastic intelligent adjustment of timetable**

This is an intelligent adjustment technology for dynamic operation schedule according to random disturbances (changes in passenger flow distribution, temporary changes in operation schedule, local equipment faults, etc.).

Of course, this requires the technical transformation of the train power supply system; that is, the locomotive power transmission system must have a two-way inverter to ensure the contactless switching function of grid power supply and regenerative inverter feedback and send the regenerative electric energy generated by train coasting and braking deceleration to the grid in real time.

Figure 5.7 shows the principle and process of regenerative electric energy transmission formed by train braking.

After the train renewable energy recovery technology is adopted, the recovery and reuse of renewable electric energy can be obtained, which makes the energy consumption of the whole train operation system play an excellent intelligent regulation role and can receive a good system energy-saving effect (see Fig. 5.8) [8].

5.1.3 Telephone Blocking Method

At present, the rail transit system, especially the subway operation in the city, adopts ATC (automatic train control) to organize the train operation. ATC realizes the automatic monitoring, protection and operation of the train. When ATC system fails,

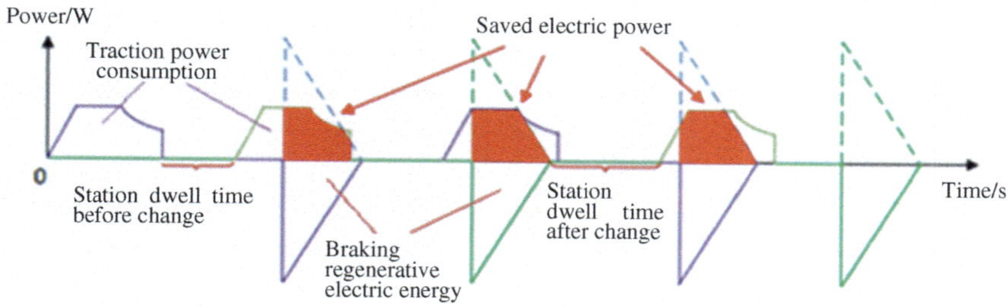

Fig. 5.8 Energy-saving effect interval diagram after intelligent adjustment of train operation timetable

telephone blocking method is often used to organize train operation in the fault section. The telephone blocking method is that the subway staff organize the train operation by means of telephone and passing the road ticket according to the driving principle of the telephone blocking method. At this time, the train position information completely depends on the naked eye (including video monitoring), and the information dissemination depends on telephone communication and road ticket transmission, which makes the accuracy of information acquisition and the real-time performance of information dissemination relatively low. Combined with the possible manual misjudgment, it directly affects the safety and efficiency of train operation. Therefore, based on computer network and sensor equipment, the information-based driving assistance system following the principle of telephone blocking method came into being.

Telephone blocking method is a manual traffic organization method in which two adjacent subway stations confirm that the blocking section is free through telephone contact and handle the blocking by sending out telephone record numbers.

1. **Basic principle of telephone blocking method**

Telephone block method is a kind of fixed block, and its block section can be divided into "one station and one section", "two stations and two sections" or "multiple stations and multiple sections". In contrast, the section divided by "one station and one section" has high train passing capacity, but its fault tolerance is low, while the section divided by "two stations and two sections" is the form adopted by many subway operation companies. The blocking handling process of telephone blocking method is that the rear station communicates with the front station by telephone. After confirming that the front "two stations and two sections" of the rear station are free, the front station grants the departure permit of the rear station. Then, the personnel of the rear station fill in the road ticket and give the road ticket to the driver. The driver holds the road ticket and commands to drive into the blocked section to the next station according to the gesture of the departure hand. The new system obtains the train position information through the sensing equipment and marks the position of each train on the station diagram, thus eliminating the process of telephone confirmation that the section is free. In addition, the station personnel control the

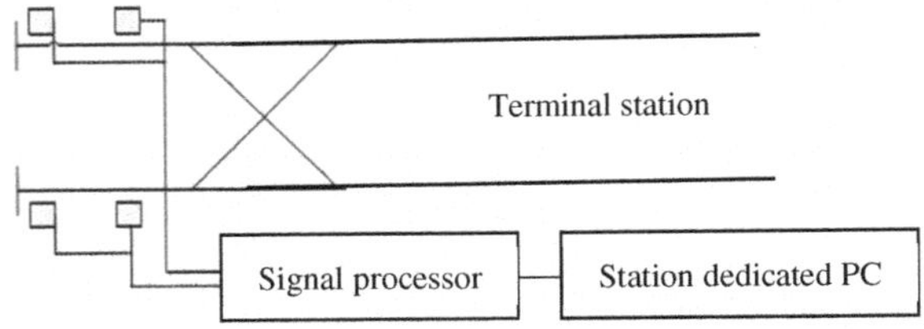

Fig. 5.9 Layout of axle counter sensor

departure indicator of the station through this system and use the departure indicator signal as the section occupation certificate, eliminating the filling and transmission of road tickets.

2. **Key components of the system**

For telephone blocking method, axle counter sensor, photoelectric sensor and departure indicator are indispensable components in telephone blocking method.

(1) **Axle counter sensor**

Axle counter sensor is a train position sensing device commonly used in the railway industry. It is generally electromagnetic. The sensor is triggered when the train axle passes by [9].

The advantage of axle counter sensor is that it can be used in harsh environment, such as poor track bed, rusty rail and humid environment.

The axle counter adopts the detection method at both ends; that is, axle counter sensors are installed at both ends of the section to detect the occupation of the section. The installation layout of axle counter sensor is shown in Fig. 5.9.

The axle counter sensor communicates with the host computer through serial port and adopts specific coded message protocol.

(2) **Photoelectric sensor**

Photoelectric sensors can be divided into opposite type and reflection type according to their working form. The opposite type refers to the working form of light entering the receiver from the transmitter. When an object blocks the light path, the sensor triggers. Reflective type refers to the working form in which light is emitted from the transmitter and received by the receiver after being reflected by the object.

In the platform track area and interstation track area, diffuse reflection photoelectric sensor is generally used to collect train position information.

In the tunnel, the sensor is installed on the tunnel wall. When a train passes the sensor, it will cause light reflection and trigger the sensor. For a station, in one driving direction, it is necessary to install a photoelectric sensor in the center of the platform area and the interstation area about 400 m away from the center of the platform; that is, there are two photoelectric sensors. Because a station is divided into two driving directions: up and down, a total of 4 photoelectric sensors are installed.

When the train is about to enter the station, it will trigger the sensor in the interstation area. When the train arrives at the parking space, it will trigger the central

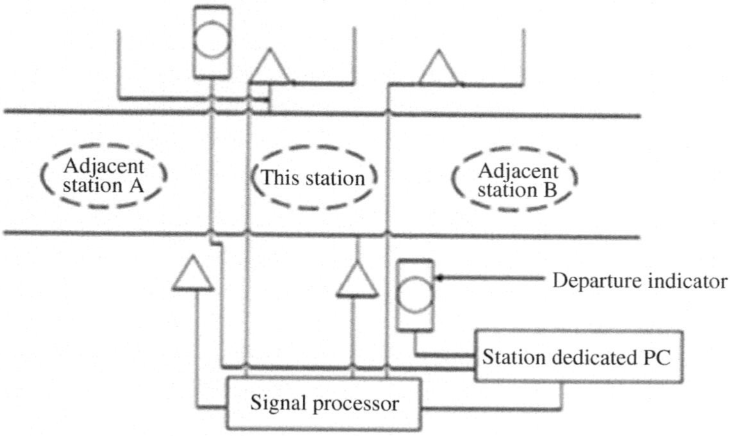

Fig. 5.10 Layout of photoelectric sensor

position sensor in the platform area. The sensor trigger has an anti-shake processing; that is, if the sensor is in the trigger state for 80% of the time within a period of time, such as 4 s, then the sensor is considered to be triggered by the train. The specific deployment of sensors and other hardware is shown in Fig. 5.10.

In the communication protocol between photoelectric sensor, signal processor and PC, three bytes can be used as a communication frame. The first two bytes are the data area, and the third byte is the check value.

Since the trigger state of each sensor is different in the corresponding binary bits in the data frame, there will be no conflict when each sensor is triggered at the same time.

(3) **Departure indicator**

Departure indicator is the direct controller of section block, which controls the stopping of train.

Generally, the departure indicator can use LED display screen to represent the departure signal by drawing various graphics on the LED screen. There are two circular drawing areas on the LED display screen, which are left and right drawing areas, respectively. Each drawing area can display three colors: red, yellow and green. The last 4 bits of the data frame of one byte of the departure indicator are the repetition of the first 4 bits, which is used as a check to increase the reliability redundancy [10].

3. **Generation of system instructions**

According to the existing operation process, at the issue link of telephone blocking dispatching order, the traffic dispatcher will issue corresponding dispatching orders to all station attendants in the telephone blocking area through the special dispatching telephone. The attendants need to make corresponding records after listening, and the traffic dispatcher will select a attendant to repeat them. The dispatching order will be issued successfully after both parties check it. According to practical experience, when dealing with emergencies, both traffic controllers and station attendants will be in an extremely busy state, and the number of parallel services will surge. At this time, the operator on duty will record the dispatching order manually, especially the dispatching order with more contents, which often leads to low efficiency. From the perspective of correspondence, the dispatcher and the attendant have a one-to-many relationship. When repeating, only a certain station is selected, and the consistency of the dispatching command at the receiving end has potential security risks. The successful issuance of dispatching order can only be established on the premise that all stations receive dispatching orders normally. No matter which station has a problem, it will affect the final traffic safety.

At present, artificial intelligence can provide real-time speech recognition function. When the traffic dispatcher dictates the dispatching command to the attendant, the intelligent terminal can collect the dispatcher's audio signal, directly convert the audio of the dispatching command into corresponding text through voice recognition technology and display it to the traffic dispatcher and all stations receiving the dispatching command at the same time through the display of the intelligent terminal and keep the record in the database. After listening to the dispatching order, the station attendant does not need to write a record and directly repeats it to the dispatcher according to the display of the intelligent terminal to check the correctness of the dispatching order. The dispatcher can also check the dispatching contents displayed on the intelligent terminal while the attendant repeats and check the dispatching repeated by the attendant from two aspects of voice and text. This can ensure the consistency of dispatching orders received by all stations and avoid the wrong non-safety factors of inconsistent dispatching orders and dispatching orders. At the same time, it also eliminates the time for the attendant to copy manually, reduces the interference of other parallel businesses, greatly improves the efficiency of the attendant to accept the dispatching order, shortens the time of actually issuing the dispatching order and ensures the driving safety. The original voice and text data in the database can also provide support for traffic analysis in the future.

Similarly, the traffic dispatcher can also use the voice recognition function to send dispatching commands to the train driver.

4. **Real-time generation of train diagram**

Of course, before starting the intelligent telephone blocking method, the traffic dispatcher needs to issue the dispatching command of enabling the telephone blocking method to the train driver and the station at the same time through the wireless voice dispatching console and the special dispatching telephone. In the process of organizing the train operation according to the telephone blocking method, the station attendant shall confirm that the departure route is ready after receiving the point announcement sent by the train from the adjacent departure station, that is, send the block request instruction to the adjacent receiving station. After checking that "one station and one section" are free (the receiving station platform, the section between the departure station and the receiving station are free), specifically, the receiving station attendant can agree to the blocking request of the departure station only after confirming that the three conditions of this station platform are free, the receiving route is ready, and the ticket of the previous train has been canceled.

After the train is sent from the station, the duty officer of the departure station shall immediately report the departure time to the duty officer of the receiving station and the arrival and departure time of the train to the traffic controller. For the intelligent system, the traffic controller inputs relevant information and data into the system software according to the train arrival and departure time reported by the attendant. The whole process can realize the adjusted real-time train diagram under the support of intelligent auxiliary technology.

When implementing the traditional telephone blocking method to organize traffic, because the automatic train monitoring system cannot automatically generate the actual train diagram, the traffic dispatcher cannot monitor the operation of online trains in real time. In view of this, the station attendant needs to report the arrival and departure time of the train to the traffic dispatcher. The dispatcher can manually draw the train diagram based on this, so as to realize the real-time monitoring of online train operation safety. Under the premise of increasing traffic density, this painting method has several disadvantages.

(1) When the train operation interval is very small and the density is very large, the frequency of the attendant reporting the arrival and departure time of the train will be very high. At this time, the special dispatching telephone will usually be occupied all the time, affecting the reporting of other services.
(2) When there are two or more electric buses leaving the station at the same time in the telephone blocking area, the attendants at two or more stations will report the arrival and departure time at the same time. The traffic dispatcher can only make all stations queue up to report the point according to the order of telephone access and draw the train diagram in turn. At this time, if a station handles the blocking wrongly and is in the queue state, the traffic dispatcher cannot supervise the train operation in time. Once it is found that the wrong time for blocking is delayed, it is very prone to traffic accidents.
(3) A traffic controller is required to draw the train diagram from beginning to end, which makes it difficult to handle other traffic services and greatly reduces the efficiency of traffic command.

In recent years, with the continuous development of computer software algorithms, artificial intelligence can assist people to complete the production of plane data such as graphs and tables according to the manually input data.

At this time, after the telephone block is started, the station attendant only needs to input the station name of the station, the train number of the departing train and the arrival and departure time in the intelligent terminal of the station, and the intelligent system will convert the input data into the train operation diagram with abscissa as time and ordinate as station according to the drawing program, which will be displayed in the display terminal of the traffic dispatcher. This intelligent terminal does not depend on the interface of the signal system. It only needs to manually input the station name and the arrival and departure time of the train and automatically generate the actual train diagram in real time. This kind of intelligent system allows all stations to input the data required by the train diagram at the same time without queuing. The system can display the actual operation track of all trains on the train diagram according to the input data. The traffic controller can check the actual running track of all trains, eliminate the time delay, help to find potential driving hazards at the first time and greatly improve the driving safety. The station attendant does not need to report the arrival and departure time of the train to the traffic dispatcher through the special dispatching telephone, which greatly reduces the occupancy rate of the special dispatching telephone. The traffic dispatcher can carry out other traffic services at the same time, which improves the command efficiency.

5. **Improve the operation safety of telephone blocking method**

As an emergency degraded operation mode, the traditional telephone blocking method has great man-made potential safety hazards in practical application. There have been tragic cases of serious traffic accidents after adopting the traditional telephone blocking method at home and abroad. The introduction of artificial intelligence into the traditional telephone blocking method process can eliminate the potential safety hazards caused by human factors, such as assisting the traffic dispatcher to send dispatching orders accurately and quickly, ensuring the consistency of receiving dispatching orders at the station and preventing traffic accidents caused by wrong receiving dispatching orders. At the same time, it can also reduce the delay of manual paving, monitor the train in time and prevent traffic accidents caused by wrong block handling at the station [11].

5.2 Popularization of Audio and Video Comprehensive Intelligent Recognition Technology

The rapid development of artificial intelligence and its harmonious combination with all walks of life make people realize that artificial intelligence will be a profound driving force to change social life and even the world, both now and in the future. The change of artificial intelligence to society will not only combine with the economy and bring economic benefits, but also promote the construction of a harmonious society and the field of social public security and virtually provide everyone with a safe and reliable intelligent protection network.

5.2.1 Basic Model of AI Security System

Under the background of the rapid development of cloud computing, big data, artificial intelligence and other technologies, promoting the construction of informatization and intelligence has become the only way for the reform of modern police mode.

As rail transit is an important "channel" for social flow of people, there are bound to be many uncertain factors. With the increasing efforts of the country's "opening up" to the outside world, various complex factors are implied in the entry and exit of overseas personnel. Therefore, the country needs to strengthen security technologies and measures in all aspects of rail transit.

At present, in the specific deployment of artificial intelligence in the field of security, any country also clearly emphasizes the need to "focus on the urgent needs of comprehensive social governance, new crime investigation and counter-terrorism, develop intelligent security and police products integrating a variety of detection and sensing technologies, video image information analysis and recognition technologies and biometric recognition technologies and establish an intelligent monitoring platform".

Some scholars have created an open vision technology that can combine face recognition, pedestrian recognition, vehicle recognition and intelligent video analysis. This technology can monitor and control key places (especially the entrance/exit of rail transit), so as to provide basic video data for urban security work (see Fig. 5.11).

Security work is different from the processing, analysis, recognition and decision-making of simple "single signal" (image, video, voice, sound, action behavior, etc.). The processing of information in security work is a comprehensive process, and finally, the target data need to be extracted through intelligent algorithm. Among them, the attention perspective and analysis/processing method of information are realized through structured technology.

1. **Information structure technology**

Taking image and video information as an example, video structuring technology is an information technology derived from massive data information.

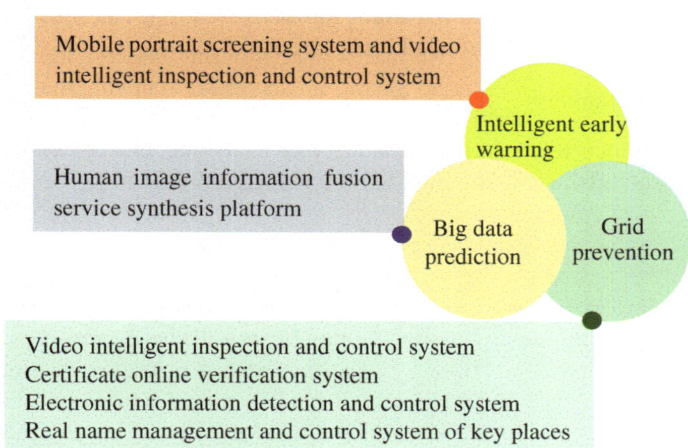

Fig. 5.11 Basic model of AI security system

Due to the unstructured characteristics of video, the original information is not enough to form a data chain. If this problem cannot be solved, the intelligent system cannot solve this kind of problem. Whether it is video or image, it needs to be extracted and recognized in time. This is the need for structured processing to convert the video information into structural information and effectively apply it to the computer system. This involves the second problem, that is, the technology of implementing structured description problem.

The traditional technology is generally to convert unstructured content into structured language, but it cannot meet the requirements of intelligent analysis system. Therefore, it is necessary to introduce the deep learning algorithm into the structured technology and carry out intelligent mining and linkage according to the semantic relationship, so as to form the content data with intelligent characteristics such as automatic retrieval and automatic extraction (see Fig. 5.12).

Intelligent video technology mining determines the depth of intelligent analysis of unstructured data.

2. **Deep learning technology**

The concept of deep learning originates from the research of artificial neural network. It is an algorithm of computer simulating human learning mechanism. The structure of deep learning is more complex than that of shallow learning, which includes a multi-layer perceptron with multiple hidden layers.

Deep learning is different from the previous shallow learning. Deep learning adopts artificial intelligence, introduces neural analysis and uses hierarchical structure to go deep into intelligent learning. The whole deep learning system breaks through the traditional structure and introduces a multi-layer combination network including input and output, which can not only connect the node layers, but also promote the links between different nodes. It is more important to find the key to imitating the human brain and turn learning into active learning and active extension, which is closer to the role of artificial brain (see Fig. 5.13).

3. **Big data technology**

Big data technology refers to the technology of analyzing and processing a large number of structured, semi-structured and unstructured data to obtain new value,

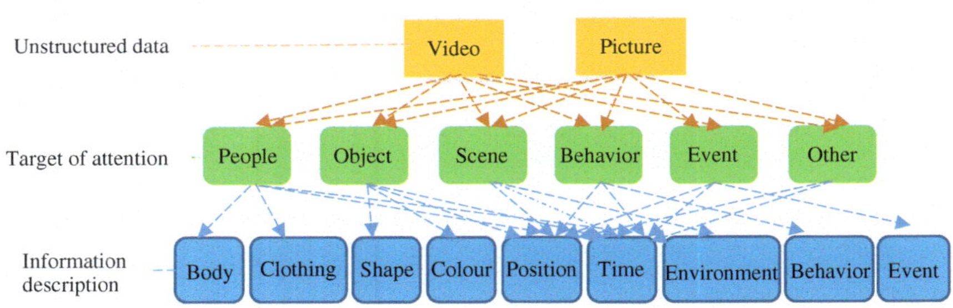

Fig. 5.12 Information description of unstructured data

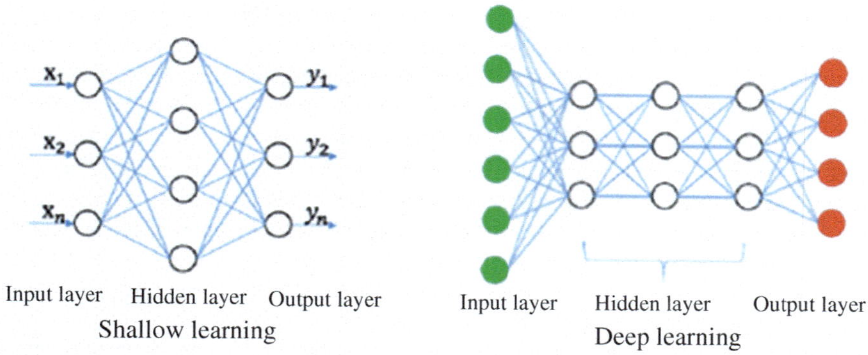

Fig. 5.13 Structural differences between shallow learning and deep learning

which requires a large number of storage devices and computing resources. The big data technology framework has the characteristics of distribution, clustering, unified storage, unified access and dynamic capacity expansion. Its technical framework is shown in Fig. 5.14.

4. **Video cloud storage**

Video cloud storage technology is an evolutionary form of general cloud storage system, but it is different from general cloud storage. It adopts the design idea of business application system on-demand allocation and integrates the technologies of computing resource cluster application, load balancing scheduling, computing/memory/network resource virtualization, cloud service, distributed storage and so on. It can provide unified external storage for different types of storage resources in the data center to the cluster through distributed storage software, so as to realize high-performance, high reliability, high capacity and high availability business access services [12].

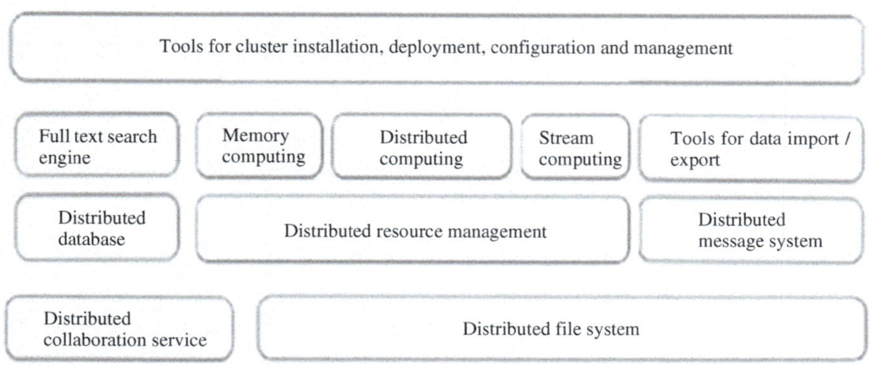

Fig. 5.14 Big data technology framework

Intelligent security is the general trend of building a new smart city and a safe city. In the future, artificial intelligence will not only be a key weapon to maintain social stability, but also rise to a new level of national defense application. Intelligent cloud storage technology can widely cover all kinds of places such as society and government with advanced technologies such as intelligent identification, big data, cloud storage and cloud computing and build an interconnected coupled system architecture, so as to fully tap and utilize the data value of the four elements of "people, cars, things and cases". It can be said that the establishment of an integrated business operation process before, during and after the event is the core for the national public security organ to realize the transformation and upgrading of public security prevention and control.

Such innovative security thinking is more suitable for every traffic link in rail transit.

5.2.2 Role of Voiceprint Recognition in Transportation System

Voiceprint, like fingerprint, palmprint, DNA, retina, iris and vascular pattern, is the inherent biological feature of everyone, with uniqueness and stability. In recent years, voiceprint recognition technology, which extracts everyone's unique voiceprint information from voice for identity identification, has attracted the attention and research of many scholars at home and abroad.

Voiceprint recognition also involves technical processes such as signal preprocessing, voiceprint feature extraction and voiceprint feature matching.

1. **Voiceprint signal preprocessing**

The quality of speech signal affects the accuracy of voiceprint recognition system to a certain extent.

In the real environment, hardware equipment, transmission media, environmental noise and other speakers will affect the quality of voice signals. In addition, the moving and changing sound source signal will also have a negative impact on the acquisition work. Before transmitting the speech signal, it is necessary to denoise the speech signal to extract the pure speech sample signal. Therefore, the application of noise reduction technology in the preprocessing stage is very important. Preprocessing the original speech signal is the initial step and a crucial process in the voiceprint recognition system, which directly affects the accuracy of the voiceprint recognition system.

Speech signal belongs to one-dimensional signal, which changes with time. As a non-stationary time-varying signal, it contains rich content information. Preprocessing is usually to perform denoising, frame windowing, endpoint detection and other operations on the original speech signal. Researchers often use "short-time stationary technology", which is because the one-dimensional speech signal originally belongs to non-stationary time-varying signal. It is obviously impractical to use the method of processing stationary signal to process non-stationary signal. By studying the formation process of speech signals, researchers found that the formation of human voice is closely related to the movement of oral muscles, but the movement speed of oral muscles is relatively slow, which is far less than the speed of sound vibration. Therefore, in a very short time range, the speech signal can be regarded as having the property of short-term stability. In this preprocessing process, the main technical methods include microphone array denoising, pre-emphasis, framing and windowing, endpoint detection and other processing technologies, which are very practical.

(1) **Microphone array denoising**

In the complex and changeable practical background, the speech signal collected by the traditional single microphone is often a mixed speech signal composed of environmental noise and multiple sound sources. After the 1990s, in order to weaken the influence of noise on speech signal in real environment, researchers developed speech signal acquisition technology based on microphone array. Microphone array refers to a group of microphones placed according to different positions and certain space spacing. Its principle is to locate the direction of the sound source through the relative time delay between the sound source signal and each microphone.

Judging the direction of this sound source involves beamforming technology (BF). Beamforming refers to the delay compensation and amplitude weighting of each microphone in the output phase and time and finally produces a beam pointing to the direction of the sound source. Microphone array technology has strong selectivity for sound source and can judge the propagation direction and source direction of speech signal relatively accurately. Microphone array can effectively avoid interference and noise outside the signal direction. Because of its obvious anti-noise effect, it is now widely used in headphones, speaker manufacturing, voice communication technology and various electronic products.

(2) **Preaggravation**

According to the structure of human vocal organs and the principle of sound signal generation, the speech process is affected by mouth and nose radiation and glottic excitation, and the high-frequency signal in speech will be weakened. To sum up, we need to pre-emphasize the high-frequency region of the original sound signal before extracting the characteristic parameters of the sound signal. Its working principle is to input the original sound signal into a first-order high-pass filter, so that the signal amplitude in the high-frequency region of the speech signal will increase and the amplitude in the low-frequency region will decrease. The main function of pre-emphasis processing on the original speech signal is to improve the resolution of high-frequency region, which is conducive to subsequent feature extraction and analysis.

(3) **Framing plus window**

As a typical non-stationary time-varying signal, speech signal usually keeps a syllable unchanged in 10 to 30 ms. In this instantaneous range, the movement mode of the vocal organ is relatively constant, and the various characteristic parameters of the speech signal are basically stable. Therefore, the short-time stationary technology is used to analyze the sound signal. In the continuous speech signal, the signal with the length of 10 to 30 ms is intercepted as a frame. In order to prevent the change difference between two adjacent frames from being too large, the cross-overlap processing is generally carried out during frame division, and the area of the overlap area is usually one-third to one-half of the whole frame area. In order to prevent missing information between frames, the audio signal after framing needs to be windowed. Using reasonable window function can better reflect the characteristic changes of audio signal when analyzing the characteristic parameters of short-term speech frame. Rectangular window and Chinese name window are often used window functions, but the disadvantage of using rectangular window is that its spectrum is easy to be omitted. Because the main discrimination width of the Chinese name window is wider, the low-pass performance is better, and the characteristic parameters of the voice signal can be better retained, so the selection of the Chinese name window is more reasonable.

(4) **Endpoint detection**

The sound signal in time domain includes not only pure and clean voice signal, but also mute and noise part. When collecting voice signal, it is not all effective information. In order to remove the mute part, the starting point of speech signal is usually recognized by endpoint detection. This can not only extract the effective voice signal and delete the mute part, but also weaken the noise and reduce the amount of data calculation to a certain extent. At present, endpoint detection has achieved some research results. The commonly used research methods include zero-crossing rate method, short-term energy method and double threshold method based on zero-crossing rate and short-term energy.

Zero-crossing rate is a relatively simple method. This method generally uses the level signal for positive and negative alternating display. Its judgment basis is the sound with low zero-crossing rate is voiced sound, and the sound with high zero-crossing rate is unvoiced sound. The short-term energy method distinguishes voiced sound and unvoiced sound according to the energy function. The sound with larger frame energy is voiced sound, and the sound with smaller frame energy is unvoiced sound. However, in practical application, it is not very accurate to determine the starting position of voice signal only by relying on zero-crossing rate or short-term energy. These two characteristics are often considered to determine the starting position of voice signal.

2. **Feature extraction in voiceprint recognition**

The purpose of feature extraction is to describe the speech signal with less information. In other words, extracting features from the original speech signal is to ensure the robustness of describing the main features of speech. At the same time, the extracted feature parameters are required to be reliable, stable and easy to extract. After years of research on voiceprint recognition technology by experts, some reliable feature parameters have been found to describe speech signals.

Almost all voiceprint recognition models use the characteristic parameters at the acoustic level as the recognition standard, but there are many personality factors to judge the speaker's identity, including the acoustic characteristics related to the structure of human vocal organs, such as timbre, amplitude, frequency, formant, reflection coefficient, spectrum and cepstrum; personal speaking habits, such as speaking speed, volume, high intonation and so on; and personal education level, such as rhythm, modification, semantics and words influenced by teachers or parents.

(1) **Criteria for selecting characteristic parameters**

Generally speaking, the most original speech signal collected by the speech acquisition device cannot be directly used as the input of voiceprint recognition model. There are three main reasons. The first is that the original speech signal has not been processed, including many unclear and unstable factors. These unknown factors may have a great impact on the training and accuracy of voiceprint recognition system model. Second, because the most original voice signal contains a large amount of data, the calculation amount and running time of the system model will be greatly increased, and the storage burden of data will also be increased. Third, it is affected by the input requirements of the system model. For example, the input of voiceprint recognition based on convolutional neural network model is generally a two-dimensional spectrogram rather than the most original speech signal.

In order to get the input characteristic parameters of voiceprint recognition model, we generally need to perform framing operation on the original speech signal. Since the voiceprint feature is formed by the feature parameters in each frame of the speech signal, the selection of the type of feature parameters is particularly important to the accuracy of the system.

(2) **Classification of characteristic parameters**

In voiceprint recognition technology, due to the instability of speech signal, easy loss of speech signal characteristic parameters, complex actual environment, easy imitation of speech, insufficient sample parameters and other problems, the extraction method and method of characteristic parameters of voiceprint recognition technology need to be further optimized. Choosing different characteristic parameters will lead to different recognition rate. Choosing reasonable characteristic parameters can not only improve the stability and robustness of voiceprint recognition system, but also improve the recognition rate. Since the development of voiceprint recognition technology, the characteristic parameters used can be roughly divided into three categories.

The first is the data directly derived from the speech spectrum, including formant, perceptual linear prediction coefficient (PLP) and Mel frequency cepstrum coefficient (MFCC).

The second is the linear prediction coefficient and its derived coefficients, such as linear prediction coefficient (LPC), linear prediction cepstrum coefficient (LPCC) and its combination parameters.

The third is the mixed parameter, which is the feature vector composed of the above different feature parameters.

3. **Voiceprint pattern matching**

The selection of recognition model is the focus of voiceprint recognition technology. Different models will lead to different effects of voiceprint recognition. Voiceprint recognition is a typical pattern recognition. Pattern matching in voiceprint recognition system is the classification of algorithms in pattern recognition. In the voiceprint recognition system, a model base is established for the speaker, the model parameters of the speaker are obtained in the training stage, and the final score is obtained through pattern matching in the test stage.

The commonly used models of voiceprint recognition include probability and statistics method, artificial neural network (ANN), vector quantization method (VQ), dynamic time warping method (DTW), Gaussian mixture model (GMM), GMM-UBM (Universal Background Model), decision rules and so on [13].

Face recognition + voiceprint recognition will undoubtedly further increase the safety redundancy in the process of passenger flow in rail transit and then realize the full scientific and technological coverage of AI technology in rail transit system.

5.2.3 Basic Principles of Ensuring Audio and Video Information Security

Although the algorithms of "audio and video" information collection, feature extraction and analysis have become increasingly mature and perfect, and have gradually entered all aspects of national economy, industrial and agricultural production, scientific and technological research and development, national defense, military and social life, and achieved remarkable beneficial results. However, people's "worries" and "concerns" also follow: Will "personal privacy" be leaked because human biological information is "widely" collected and transmitted?

This is a common social dispute at home and abroad.

Of course, it should be admitted that everything in the world has its positive/negative characteristics. This is an indisputable fact, but with the emergence of a new problem (contradiction), people have the wisdom to limit the "problem (contradiction)" to a minimum and reduce the negative impact to a minimum. Therefore, while the country improves the legal system and the rule of law, there are more effective technical means to solve it.

Among them, the most basic scientific method is to implement scientific "encryption" in the whole process of human biological information collection and processing, because any collected source information can be expressed and stored in mathematical way, and the "encryption" of mathematical formula is an "easy thing".

For example, in face recognition technology, the digital image obtained by collecting the original face image is expressed (described) by two-dimensional matrix (gray image) or high-dimensional matrix (color image), which can use the method of mathematical encryption to encrypt the original data.

Take a two-dimensional gray digital image as an example.

If the two-dimensional matrix of the gray-scale digital image of the original image is **A**, the information security of all subsequent steps can be realized through the encryption matrix **E**, that is

$$\mathbf{A} \cdot \mathbf{E} = \mathbf{B} \tag{5.16}$$

Once the collected image is stolen "halfway", it is also "irrelevant", because what is stolen is **B** rather than real **A**. When "opening" **B**, what you see is only a "messy" gray image (or a pile of random codes). Only by "decrypting" (or deciphering) can these "random codes" be restored to the original figure **A**, that is

$$\mathbf{A} = \mathbf{B} \cdot \mathbf{E}^{-1} \tag{5.17}$$

By analogy, there are other "endless" encryption/decryption algorithms, as well as "multi-level" and "multi-dimensional" encryption methods and algorithms, which makes it more difficult for the "decoder" to "decrypt", even to the extent of "impossible decryption".

References

1. Shu T (2018) Application of artificial intelligence in Urban rail transit. J Liaoning Provincial Coll Commun 20(6):25–27 (In Chinese)
2. Yang H (2015) Railway transportation organization. China Railway Publishing House, Beijing
3. Zhou M, Sun S (1999) Genetic algorithms theory and applications. National Defense Industry Press, Beijing
4. Yun Q (2000) Evolutionary algorithms. China Machine Press, Beijing
5. Ren X (2018) Research on the method of train working diagram of high speed railway. Dissertation, Shijiazhuang Tiedao University
6. Vuchic V (2007) Urban transit systems and technology. Wiley, New York
7. Wang T (2014) Research on high speed railway train dispatch command startegy and methods from the perspective of energy saving. Dissertation, China Academy of Railway Sciences, Beijing
8. Yang G (2019) Research on energy optimization and real-time reconstruction algorithm of stochastic disturbance timetable in Metro System. Dissertation, Sahng Jiao Tong University, Sahnghai
9. Gong J (2011) Analysis on engineering application of various electronic axle counter equipment. Railway Signal Commun 47(6):34–37 (In Chinese)
10. Qiu Y (2015) A train supervision system base on telephone block system and unstable network. Dissertation, South China University of Technology, Guangzhou
11. Tian S, Wei W (2017) Using artificial intelligence to build a public intelligent monitoring, early warning and control system. China Safety Prev 8:102–106 (In Chinese)
12. Geng Z (2018) Intelligent analysis system of police video image features. Dissertation, Heilongjiang University, Harbin
13. Ding D (2019) Research on intelligent voice print recognition method. Dissertation, Yangtze University, Wuhan

Chapter 6
Intelligent Maintenance of Internal and External Environment of Train

The intelligent maintenance of the train carriage environment includes the intelligent adjustment of the carriage temperature, the intelligent maintenance of harmonious and comfortable environment and the intelligent supervision of passenger boarding and alighting safety.

6.1 Intelligent Adjustment of Compartment Temperature

At present, the general indoor air conditioning temperature regulation basically adopts the single closed-loop feedback regulation of room temperature sensing under the set temperature, and the train carriage is no exception. This "traditional" temperature regulation method obviously does not achieve the degree of energy-saving optimization, and it will also cause discomfort (high or low temperature) for indoor personnel due to improper room temperature setting. Therefore, the intelligent regulation of compartment room temperature has begun to appear in the real rail transit operation.

6.1.1 Double Closed-Loop Intelligent Regulation Principle of Carriage Room Temperature

The so-called double closed-loop regulation of room temperature is an automatic temperature regulation system composed of feedback inner loop and feedback outer loop. The former is the traditional negative feedback, while the latter is the intelligent feedback formed by the algorithm according to the local temperature and other environmental conditions. Its regulation system is shown in Fig. 6.1. Among them, the temperature controller mainly controls the temperature output of the central air conditioner and assists in adjusting the air volume at the air outlet of the carriage.

Y. Khan, *Principles of Intelligent Rail Transit*, Advances in High-speed Rail Technology,
https://doi.org/10.1007/978-981-19-6072-7_6

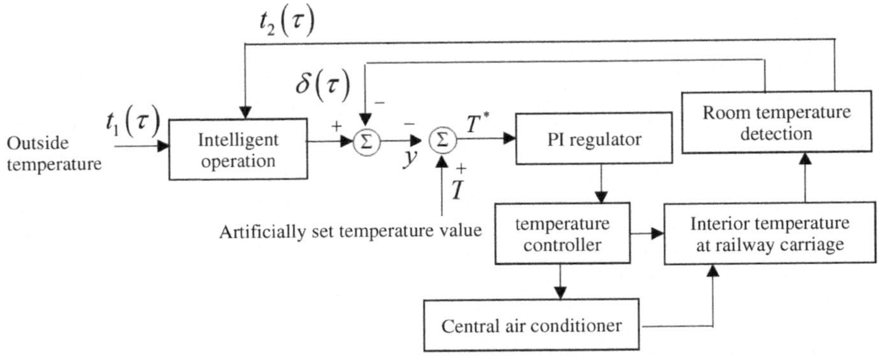

Fig. 6.1 Block diagram of double closed-loop regulation system for interior temperature at railway carriage

Among them, the so-called intelligent calculation, that is, the temperature outside the railway carriage, the average room temperature inside the railway carriage and the heat value emitted by the personnel inside the railway carriage are considered as a whole to form a set temperature intelligent correction value. Then, it is compared with the "artificially set temperature value" to obtain the intelligent reference value of temperature regulation. The method of adjusting the room temperature can make the comfort of passengers in the compartment reach the "suboptimal" level.

If the temperature change curve function outside the railway carriage is $t_1(\tau)$, the average room temperature change curve inside the railway carriage is $t_2(\tau)$, and the intelligent correction value of temperature is y (i.e., intelligent operation output), and the basic operation equation is as follows:

$$y = \frac{1}{2}[t_1(\tau) + t_2(\tau)] - \delta(\tau) \tag{6.1}$$

where $\delta(\tau)$ is the dynamic value function of heat temperature emitted by the passenger group in the railway carriage.

When the intelligent output y is subtracted from the external artificially set temperature T, a real-time intelligent reference value T^* for temperature regulation is obtained, that is

$$T^* = T - y \tag{6.2}$$

The function curve of the whole intelligent regulation is shown in Fig. 6.2.

$\delta(\tau)$ is the measured value of human body radiation temperature from the built-in thermal imager or infrared sensor thermometer in the subway carriage. In two different space conditions, the impact of heat emitted by human body on the subway carriage environment is obviously different. At the same time, summer is different from winter.

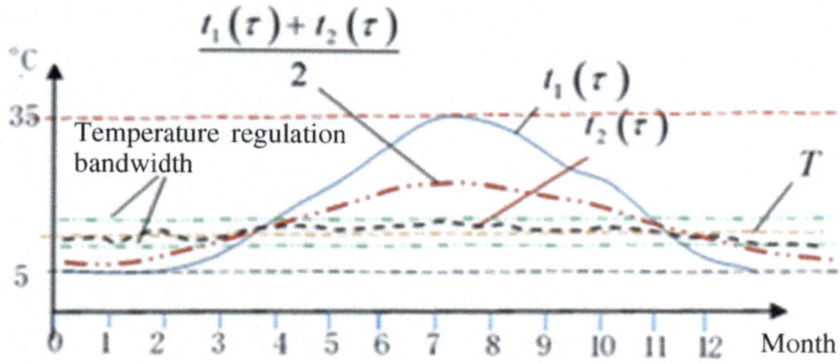

Fig. 6.2 Temperature regulation curve inside a subway carriage

In the adjustment of the ambient temperature of the carriage, the physical significance of adopting intelligent control technology lies in the progress of thinking mode. That is to "evolve" the mechanical inertial thinking mode of simple negative feedback $T - t_2(\tau)$ into a new form of intelligent thinking.

6.1.2 Advantages of Intelligent Temperature Regulation in the Carriage

After adopting intelligent regulation technology, the internal temperature of the train carriage has the following obvious advantages.

(1) Overcome the inertia thinking defect of "artificially setting temperature value".
(2) It can change the "reference value" of compartment temperature regulation in real time according to the actual changes of external temperature throughout the year and even at any time, so as to make it intelligent.
(3) It can automatically change the temperature "reference value" completely "humanized" according to the characteristics of people wearing different clothes throughout the year (such as thin in summer and thick in winter).
(4) It can sense different heat dissipation effects of human body according to the "density" of different passengers in the carriage.

Intelligent temperature control can make the temperature inside the train carriage achieve the best comfort effect [1].

6.2 Intelligent Technology of Ride Safety Guarantee

In order to improve the safety environment, nowadays, many subway platforms in cities are basically equipped with screen doors.

Although the installation of PSD (platform screen door) on subway or light rail platform can meet the basic conditions of providing passengers with a safe riding environment, PSD directly faces passengers and is the only way for passengers to get on and off. Its operation mode and reliability are directly related to the quality of service provided by the station to passengers, as well as the convenience and personal safety of passengers. At the same time, its working performance will also directly affect whether the train and its whole line can run normally and on time. However, the operation of subway and light rail stations that have installed platform screen door in cities such as Shanghai and Shenzhen has not reached the level of intelligence. Due to the design defects of PSD control system, there have been many incidents of passengers being caught between train doors and PSD and even casualties of passengers [2].

For example, the Oriental Morning Post reported on July 20, 2007 that "the trial installation detector of Shanghai Metro screen door prevents passengers from being caught in the door crack". In order to avoid passenger casualties, Shanghai Metro Operation Company has begun to install detectors on the platform screen doors of stations to detect whether passengers or objects are trapped by the platform screen doors. After the detector is installed and the train door and platform screen door are completely closed, the infrared light curtain detectors set on both sides will automatically emit infrared rays to form a light curtain barrier. If a foreign object is sandwiched between the two doors to block the light curtain, the detector will terminate the departure signal at the first time. According to the working principle of the design, the infrared detector does not emit light continuously for 24 h. At ordinary times, it is in a dormant state. Before the PSD is closed and ready for departure, the transmitter will automatically emit 7 beams of infrared light for 2-s scanning. This is a sensing device to detect whether there is "foreign matter blocking the light curtain" between the train door and the platform screen door. However, the device "forms a light curtain barrier" by a pair of infrared transmitting/receiving devices "set on both sides", which makes the on-site installation more complex. On the other hand, the biggest disadvantage of using infrared detector is that it is unable to identify and distinguish the shape and type of foreign objects.

Of course, there are "5-level automatic control and 3-level manual door opening modes". The so-called 5-level automatic control includes the following: The first level is the driver's control in the cab; the second level is the control of the front end of the platform for the train driver to manually operate and control the PSD; the third level is the local control of each PSD, which is mainly used for maintenance; the fourth level is the main control computer in the equipment room to control the PSD; and the fifth level is the operation indicator panel in the station integrated control room. The so-called 3-level manual door opening modes include that the platform attendant can open the door in the station with the key; passengers can open the door on one side of the train by holding the handle; and passengers can open the emergency door on one side of the train by pressing the crossbar on the emergency door.

It should be said that the above operation mode and control method of platform screen door have comprehensively considered the service environment and various possible conditions. This technical scheme can be applied to the general normal operation environment. However, there are still the following three technical defects.

(1) In addition to sensing when the "screen door meets an obstacle", there is a lack of automatic detection means for the operation state of the screen door. The whole plant system is only controlled by program and monitored manually. Therefore, the overall degree of automation is not high.
(2) It is precisely because the system needs to intervene and control through manual monitoring, which has the problem of artificial response lag when accidents occur. Therefore, the tragedy is irreparable when unexpected events occur.
(3) In case of emergency, passengers can open the door by "holding the handle" or "pressing the crossbar on the emergency door". This auxiliary means is undoubtedly necessary, but it is unreliable, because most passengers often do not master these "emergency common sense" or even know nothing.

As a result, intelligent PSD control came into being.

6.2.1 Platform Screen Door Intelligent Control System

Platform screen door intelligent control system includes CMOS camera, signal processor, controller and driver. The working process of the system is that, firstly, the CMOS camera outputs the collected image to the signal processor, and then, the signal processor performs A/D conversion, enhancement processing and information recognition on the collected image signal and converts the recognition results into control instructions and outputs them to the controller. The controller converts the digital signals of control command and emergency control command into analog control signal and sends the analog control signal to the driver. The driver amplifies the analog control signal and drives the mechanical servomechanism to operate the opening or closing of PSD (see Fig. 6.3).

The intelligent control method of platform screen door consists of the following steps. Firstly, the images of parked trains and non-parked trains on the platform track are collected by CMOS camera, and the images are output to the signal processor. The signal processor extracts the image feature information and makes it into two basic matching templates. The threshold gray error and correlation between the field image and the basic matching template are established, and then, the criterion formula and

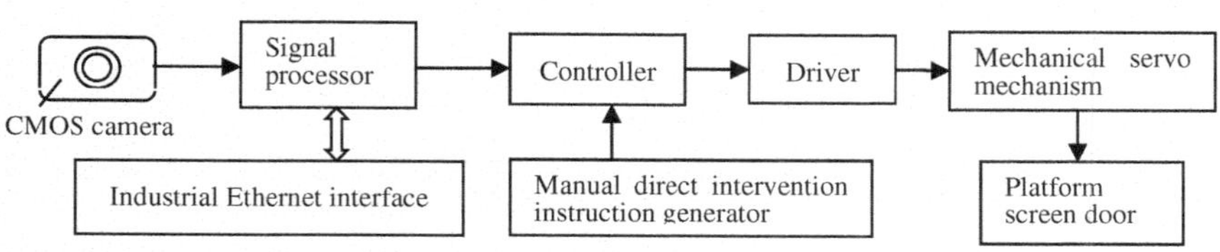

Fig. 6.3 Schematic block diagram of platform screen door intelligent control system

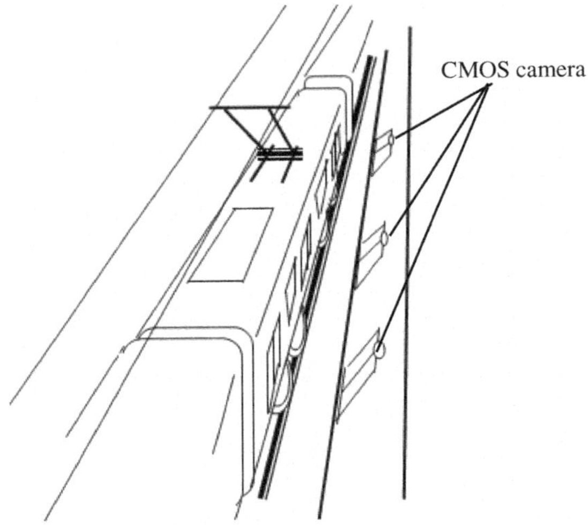

Fig. 6.4 Actual installation position of CMOS camera

its judgment index are generated. The field image is processed and recognized by the signal processor, and then, accurate control instructions can be generated and sent to the controller. Finally, the controller converts the received digital control command into analog control command and outputs it to the driver. The driver control command is amplified to drive the servomechanism of PSD.

CMOS camera is installed on the door frame of each PSD and faces the train side. Adjust the top view angle to ensure that the monitoring area is within the field of view of the camera (see Fig. 6.4).

Generally speaking, the technical characteristics and basic performance parameters of CMOS camera adopted include the following contents. The lens diameter is 1 / 3 inch, and the photosensitive device is black-and-white CMOS, 510 × 492 pixels, definition of 380TV line, minimum illumination of 2.0 lx, focal length of 2.9 mm, field angle of view of 140°, PAL / NTSC, output voltage of $V_{pp} = 10$ V, characteristic impedance 75 Ω and power supply 200 mA / 12VDC.

The signal processor includes an A/D conversion module, an image enhancement module, an image recognition module and a storage unit. Among them, the input of A / D conversion module receives the analog image collected by CMOS camera, converts the analog image signal into digital image signal and outputs the digital image signal to the image enhancement module. The image enhancement module is responsible for enhancing the digital image signal and outputting the enhanced digital image to the image recognition module. The image recognition module is responsible for extracting the characteristic information of the digital image, collecting the characteristic information of the image under the two conditions of parked train and non-parked train on the platform track as the basic matching template and storing it in the storage unit. The image recognition module will compare and recognize the field image with the two basic templates in the storage unit, respectively, according to the threshold gray level of the field image and the template image and their mutual error, correlation criterion and index and finally convert the recognition results into control instructions and output them to the controller. The storage unit is used to store digital images and their characteristic information and realize the reading/writing of data in the storage unit by the image recognition module. The signal processor has an industrial Ethernet interface to exchange information with the host computer server through industrial Ethernet. The controller has two input ports. The first input port is connected with the signal processor and is responsible for receiving control instructions, and the second input port is responsible for receiving emergency control instructions.

In case of maintenance shutdown, fire fighting, combat readiness and other special conditions during abnormal operation, the controller directly executes the emergency control command without receiving the control command output by the signal processor. The emergency control command directly operates the opening or closing action of PSD. At the same time, the CMOS camera and signal processor enter the sleep state and do not stop the sleep and resume the active state until the controller receives the command to cancel the emergency control again.

Emergency control command refers to the manual direct intervention command from outside the system. The station manager inputs the emergency control command to the second input port of the controller through the manual direct intervention command generator. The controller converts the digital signal of the control command into an analog control signal and sends the analog control signal to the driver.

The so-called manual direct intervention command generator refers to the device that generates wired switch control commands or wireless switch control commands under maintenance shutdown, fire fighting, combat readiness and other abnormal operating conditions. The second input port of the controller is matched with a corresponding wired or wireless instruction receiver according to the communication protocol of the switching instruction output type of the manual direct intervention instruction generator.

During the working process of the system, the system must be initialized first. The specific steps are as follows.

(1) The CMOS camera collects the images of parked trains and non-parked trains on the platform track and outputs them to the signal processor.
(2) The signal processor performs analog-to-digital conversion, digital image enhancement processing and image feature information extraction on the above two status images through its internal A/D conversion module, image enhancement module and image recognition module and finally stores them in the storage unit in the signal processor as two basic templates. The first basic template is the digital image and characteristic information of no train parking on the station track. The second basic template is the digital image and characteristic information of train parking on the station track. The feature information includes object edge line, image gray histogram and template average gray.
(3) Establish the threshold gray level of field image and template image, as well as the error and correlation index between them.

After initialization, the system enters the running state. At this time, the signal processor receives the field image collected by the CMOS camera in real time and converts it into a digital image signal through the A/D conversion module and sends it to the image enhancement module. The digital image signal enhanced by the image enhancement module is transmitted to the image recognition module. The image recognition module extracts the image information and compares the collected field image with the two basic templates in the storage unit according to the threshold gray level of the field image and the template image and their mutual error, correlation criteria and indicators. Finally, the recognition results are converted into control instructions and output to the controller. The controller converts digital control commands into analog control commands and outputs them to the driver. The analog control command is amplified by the driver to drive the opening/closing of PSD [3].

According to the two basic image templates and image recognition correlation criteria and indicators established by the system initialization, the on-site image is collected in real time through the CMOS camera, and the actual situation between the train and the PSD is determined through the information processing, judgment and recognition of the signal processor. Finally, send control commands to the controller and driver to control the opening/closing of platform screen door.

6.2.2 Algorithm Steps

Step 1: The CMOS camera is installed on the upper door frame of the PSD, facing the train side, and the top view angle is adjusted to 60° to ensure that the monitoring area is fully within the field of view of the camera.

Step 2: Collect the images of parked trains and non-parked trains on the platform track through the CMOS camera and output the images to the signal processor.

Step 3: The signal processor converts the received analog image signals of parked trains and non-parked trains on the platform track into two digital image signals through the A/D conversion module and sends them to the image enhancement module. Then, the image enhancement module carries out automatic gain reinforcement of image light source for the two digital image signals delivered and carries out technical processing for brightness, white balance, color saturation and contrast adjustment, edge enhancement, etc., and then sends them to the image recognition module for image feature information extraction. Finally, they are stored in the storage unit of the signal processor as two basic templates. The two basic templates are as follows: The first basic template is the digital image and characteristic information of no train parking on the station track, and the second basic template is the digital image and characteristic information of train parking on the station track. The feature information includes object edge line, image gray histogram and its average gray.

Step 4: Establish the image threshold gray level of the field image and the basic template, as well as the error and correlation criterion formula and index between them.

(1) Image threshold gray level

In order to segment the brightness value of the image, the pixels brighter than the threshold gray level and the pixels darker than the threshold gray level are divided into black-and-white groups. The result of segmenting the luminance value should make the black-and-white components in the image basically equal, the boundary of the image clear and the main body basically distinguishable. On the gray level histogram, the threshold is represented as a vertical separation line. All gray levels on the left side of the separation line will become black and the right side will become white. On the one hand, the separation line should make the area of the left and right sides equal to ensure that the histogram has the same number of black-and-white pixels. On the other hand, it is assumed that the gray level probability distribution can be approximated by two Gaussian distributions, one representing the subject foreground and the other representing the unwanted background object. The threshold should be selected appropriately to ensure that the binary boundary is clear and distinguishable, and the foreground region and background region can be segmented correctly.

In the specific recognition process, the embodiment provided here first searches the valley point between the two peaks of the gray curve and determines it as the gray level threshold of the corresponding image and then uses the threshold as the basis of image segmentation to segment the foreground and background areas of the image.

(2) Allowable error

When judging whether the images match or the degree of similarity, it is necessary to compare the image gray histogram and its average gray feature value between the field image and the basic template. If the difference between the two is too large, it is considered that they are not the same scene. If the error is within the allowable range, it is considered that they are very likely to be related and can be recognized as the same scene. The embodiment provided here defines the error as the ratio of the absolute value of the difference between the field image and the template feature information value to the template feature information value, which is expressed as a percentage. The error value < 2.5% is determined as the allowable error.

(3) Correlation criterion formula and index

Based on the threshold gray level, the first basic template matrix $\mathbf{A}1$, the second basic template matrix $\mathbf{A}2$ and the field image matrix $\mathbf{B}$ are established, as well as the cross-correlation function under discrete conditions, including their respective variance, normalized correlation coefficient and the definition of inner product operation in the linear space composed of all two-dimensional matrices. Starting from the linear space, according to the norm and included angle concept of Euclidean space, the correlation $sim1$ between the first basic template matrix $\mathbf{A}1$ and the field image matrix $\mathbf{B}$ and the correlation $sim2$ between the second basic template matrix $\mathbf{A}2$ and the field image matrix $\mathbf{B}$ are defined as

$$sim1 = \cos\angle(\mathbf{A}1, \mathbf{B}) = \frac{\mathbf{A}1 \cdot \mathbf{B}}{\|\mathbf{A}1\|\|\mathbf{B}\|} = \frac{\mathbf{A}1 \cdot \mathbf{B}}{\sqrt{\mathbf{A}1 \cdot \mathbf{A}1}\sqrt{\mathbf{B} \cdot \mathbf{B}}} \tag{6.3}$$

$$and\ sim2 = \cos\angle(\mathbf{A2}, \mathbf{B}) = \frac{\mathbf{A2} \cdot \mathbf{B}}{\|\mathbf{A2}\|\,\|\mathbf{B}\|} = \frac{\mathbf{A2} \cdot \mathbf{B}}{\sqrt{\mathbf{A2} \cdot \mathbf{A2}}\sqrt{\mathbf{B} \cdot \mathbf{B}}} \tag{6.4}$$

where $\boldsymbol{A}1 = \left[a1_{x,y}\right]_{M \times N}$, $\mathbf{A2} = \left[a2_{x,y}\right]_{M \times N}$, $\boldsymbol{B} = \left[b_{x,y}\right]_{M \times N}$; M and N represent the number of rows and columns of each frame image represented by matrices **A**1, **A2** and **B**, respectively. The elements in the matrix, respectively, represent the gray value of the pixel points corresponding to their row and column coordinates, and their footmark x, y represents the row and column coordinates (x, y) of the pixel points.

The practical criterion formula obtained by using the definition of inner product is as follows

$$sim1 = \frac{\sum_{x=0}^{M-1}\sum_{y=0}^{N-1} a1_{x,y} \cdot b_{x,y}}{\sqrt{\sum_{x=0}^{M-1}\sum_{y=0}^{N-1} a1_{x,y}^2}\sqrt{\sum_{x=0}^{M-1}\sum_{y=0}^{N-1} b_{x,y}^2}} \tag{6.5}$$

$$and\ sim2 = \frac{\sum_{x=0}^{M-1}\sum_{y=0}^{N-1} a2_{x,y} \cdot b_{x,y}}{\sqrt{\sum_{x=0}^{M-1}\sum_{y=0}^{N-1} a2_{x,y}^2}\sqrt{\sum_{x=0}^{M-1}\sum_{y=0}^{N-1} b_{x,y}^2}} \tag{6.6}$$

In linear space, the smaller the included angle is, the greater the similarity is, the more obvious the image correlation is, and the more similar the image is. If the correlation reaches 100%, the first basic template matrix **A**1 and the field image matrix **B**, the second basic template matrix **A2** and the field image matrix **B** can be linearly expressed with each other.

For example, taking the similarity between the template image matrix and the field image matrix $\geq 95\%$ as the correlation index, the similarity between the field image and the template image can be confirmed by judging the correlation index, so as to judge whether there is foreign matter on the site.

Step 5: After completing steps 1 to 4, the system enters the running state, and the field image collected in real time by CMOS camera is sent to the signal processor.

Step 6: The signal processor converts the received analog image signal into digital image signal through the A/D conversion module and sends it to the image enhancement module. The image enhancement module performs enhancement technology processing on the delivered digital image signal and sends the processed digital image signal to the image recognition module through the output channel.

Step 7: The image recognition module establishes the field image matrix **B** for the received enhanced digital image signal, searches the valley point between the two peaks of the gray curve for the template image and the field image, respectively, and determines it as the gray level threshold of the corresponding image and then uses the threshold as the basis of image segmentation to segment the foreground and background areas of the template image and the field image. Judge the working condition between the PSD and the train side according to the allowable error, correlation criterion and index, and judge whether the current train is entering or leaving the station according to the movement of the object edge line before and after the train stops at the station, so as to judge the specific condition of the PSD train side, which has the following four conditions.

In the first condition, $sim2 \geq 95\%$, when the train stops at the platform and there is no foreign matter between the PSD and the train, the train can start and leave the station.

In the second condition, $sim2 < 95\%$, if the train stops at the platform and there is foreign matter between the PSD and the train, the PSD shall not be closed and the train shall not be started at the same time.

In the third condition, $sim1 \geq 95\%$, there is no train stopping at the platform and there is no foreign matter between the PSD and the track. The PSD must be closed and the train behind must be notified that it can enter the station normally at this moment.

In the fourth condition, $sim1 < 95\%$, if there is no train stopping at the platform and there is foreign matter between the platform screen door and the track, the platform screen door must be opened, and manual personnel must be assigned to deal with the situation on the site urgently and inform the train behind that it is prohibited to enter the station at the moment.

The movement determination of the object edge line refers to that the image recognition module determines the specific operation state of the current train in or out of the station according to the movement of the object edge line (specifically the edge line of the train carriage) before and after the train stops at the station, including the following three states.

The first state, from no train image in the station to the appearance of train image until the appearance of still train image, indicates that the train stops at the station.

In the second state, the moving train image appears from the still train image until there is no train image at the station again, indicating that the train starts to leave the station.

The third state, from no train image in the station to the moving train image, and continuously until there is no train image in the station again, indicates that the train does not stop and directly passes through this platform.

Step 8: The image recognition module converts the judgment result into control instruction and sends it to the controller through the output port of the signal processor. The controller converts the received digital control instruction into analog control instruction and outputs it to the driver. The driver amplifies the analog control instruction and drives the mechanical servomechanism to operate the opening or closing of the PSD.

The control instructions include the following five items.

The first instruction is to open the train door first and then the platform screen door when the train enters the station and stops at the platform.

The second instruction is when passengers get on and off the train, confirm that they are in "the first condition, the train stops at the platform and there is no foreign matter between the PSD and the train", and then close the PSD.

The third instruction is when passengers get on and off the train, confirm that they are in the "second condition, the train stops at the platform and there is foreign matter between the PSD and the train", do not close the PSD, send out an alarm sound and prohibit the train from starting.

The fourth instruction is to notify the train behind to enter the station only after confirming "the third condition, no train stops at the platform and no foreign matter between the PSD and the track".

The fifth instruction is, in the case of confirming "the fourth condition, no train stops at the platform and there is foreign matter between the PSD and the track", the train behind must be notified not to enter the station.

In case of maintenance shutdown, fire fighting, combat readiness and other special conditions during abnormal operation, the controller directly executes the emergency control command instead of receiving the control command output by the signal processor. The emergency control command directly operates the opening or closing action of PSD and inputs the emergency control command into the controller. At the same time, the CMOS camera and signal processor enter the sleep state. The sleep is not stopped until the controller receives the instruction to cancel the emergency control again, and the activation state is restored.

The emergency control instruction refers to the manual direct intervention instruction from outside the system. That is, the station manager inputs the emergency control command to the controller through the manual direct intervention command generator. The controller converts the digital signal of the control command into an analog control signal and sends the analog control signal to the driver.

The manual direct intervention command generator refers to a device that generates wired switch control commands or wireless switch control commands in maintenance shutdown, fire fighting, combat readiness and other abnormal operation states. The controller matches the corresponding wired or wireless command receiver according to the communication protocol of the switching command output type of the manual direct intervention command generator [4].

Compared with the existing similar technologies, the image recognition module establishes the field image matrix **B** for the received enhanced digital image signal, searches the valley point between the two peaks of the gray curve for the template image and the field image, respectively, and determines it as the gray level threshold of the corresponding image and then uses the threshold as the basis of image segmentation to segment the foreground and background areas of the template image and the field image. Then, the working condition between PSD and train side can be determined according to the allowable error, correlation criterion and index. According to the movement of the object edge line before and after the train stops at the station, it can be judged whether the current train is entering or leaving the station, so as to judge the specific situation of the train side of the platform screen door. Only in this way can we ensure that the accuracy of system recognition can reach 100%. Only by monitoring whether there are foreign matters between the PSD and the track after the train leaves the station, can we accurately send information and control instructions to the supervisor or the upper computer server. These are the technical guarantee to ensure the comfort and safety of passengers getting on and off the train and the safe and punctual operation of the train.

6.3 Intelligent Assistance of the Environment in the Carriage

As we all know, in order to advocate the social virtue of respecting the old and loving the young and jointly build a beautiful and harmonious living environment, we have specially set up "special seat for love" in the appropriate positions of our buses, subways, rail transit and other public transport vehicles. The purpose is to enable people who "need social care, care and help" to get appropriate care when traveling by bus. However, things are often not like the original intention of setting up the "special seat for love", but there will be some uncivilized phenomena: Often, some young people turn a blind eye to the text prompt of the "special seat for love" as soon as they enter the bus compartment and sit on it as long as they are empty. It can often be seen that some young people sit on the "special seat for love". Even if there is a "special seat for love" later, and even a pregnant woman with a "big stomach" stands next to him, he can turn a blind eye, act as if there is no one, or pretend to be sleepy, and let the pregnant women around him stand until he gets off. If someone kindly reminds or persuades those who occupy the "special seat for love" to give up their seat, they will often be maligned by those who occupy the "special seat for love" and even fight. There is no lack of effectiveness of advanced technical means to improve this bad social phenomenon.

6.3.1 Device Structure and Setting Method

The special seat intelligent prompt device system is composed of pinhole camera, signal processor and voice prompt. Among them, the pinhole camera is set at an appropriate position on the carriage wall opposite to the "special seat for love" and directly facing the passenger's face position on the seat, which is used to collect the passenger's face position image on the seat in real time (see Fig. 6.5). The signal processor undertakes the whole operation task of image information processing, recognition and decision-making. The signal processor receives the image signal of the face and body shape of the passenger from the pinhole camera and processes, recognizes and judges it. Once it is confirmed that the passenger in the current special seat belongs to healthy young and middle-aged passengers (excluding the possibility of pregnant women), the signal processor will generate control instructions in real time according to the recognition results and friendly prompt the young and middle-aged passengers through the voice prompt: "this is a 'special seat for love', would you please give it to the passenger in need of help? Thank you for your assistance!".

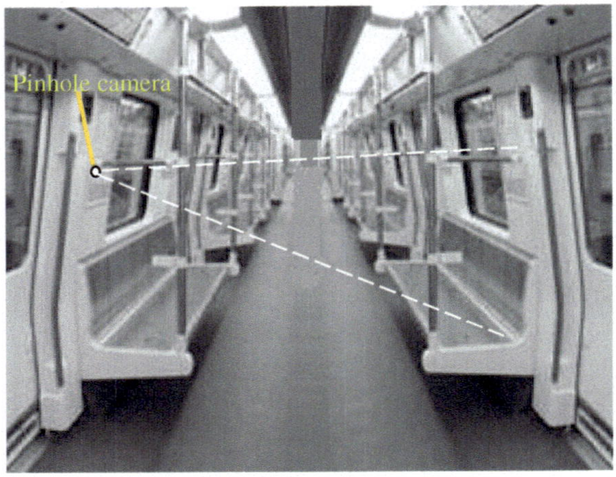

Fig. 6.5 Schematic diagram of pinhole camera setting method

In practical application, there is no need to configure a signal processor for each carriage. In practical application, there is no need to configure a front signal processor for each compartment. For each train, it only needs to configure a central signal processor to centrally process the personnel flow information collected by all carriages, and identify the type of passenger to whom the passenger belongs according to the analysis and operation of the passenger's face and body image on the special seat in real time.

At this time, the structure of the device system is shown in Fig. 6.6. In the figure, N represents the number of special seats in the train carriage. For example, if there are 6 special seats in each carriage, there are 48 special seats in 8 trains, that is, N = 48. The number of pinhole camera and voice reminder represents the corresponding seat number. The gating time and interval of each channel of "input channel gating" and "output channel gating" are determined by the "signal acquisition, scanning and operation" software of the signal processor. The gating channel of "output channel gating" corresponds to "input channel gating". The signal processor has "multi-task function". Only when "knowing" that a special seat is occupied by other passengers, the signal processor will gate the corresponding "voice prompt" and send the "voice prompt" instruction. While gating any "voice prompt", the signal processor is already processing the next "scanned" image signal.

6.3.2 Core Algorithm

The special seat intelligent prompt method includes two steps: training sample learning and online recognition.

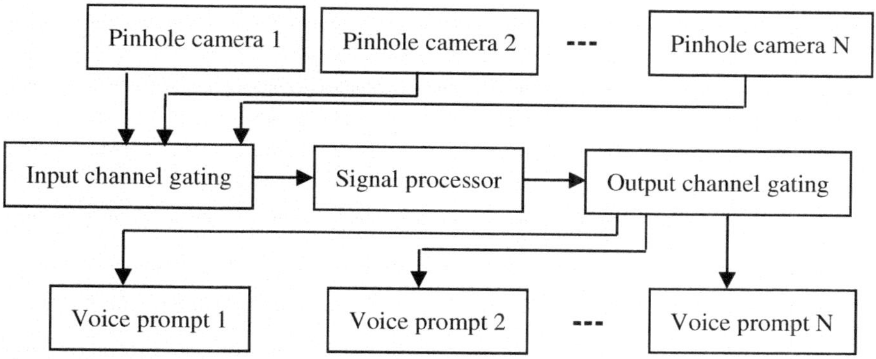

Fig. 6.6 System structure diagram of implementation device

1. **Learning of Training Samples**

The so-called learning of training samples means that the training samples must be learned in advance before the intelligent system enters online operation. When the system is in the learning state, the system must collect and process the face and body shape images of passengers aged 16 to 55.

The learning of training samples consists of 9 steps.

(1) Collect the face and body shape images of young and middle-aged passengers and convert the color space images

The face and body shape images of young and middle-aged passengers sampled by pinhole camera are enhanced and then transformed from RGB color space to HSV color space for expression. In other words, it is necessary to continuously collect multiple such images, including the images of passengers' body shape and faces with different expression characteristics at different ages from 16 to 55, and then convert each enhanced image from RGB color space to HSV color space for expression.

(2) Distinguish the face and body shape of young and middle-aged passengers from the background

In the sampled face images of young and middle-aged passengers, the scene hue of the pixels is scanned from left to right and from top to bottom according to the pixel coordinates, and the pixel points corresponding to the hue set of the hue interval [2°,47°] in the HSV color space are drawn up as the face area. Therefore, other areas and backgrounds in the young and middle-aged face image can be accurately distinguished from the face area.

This is because there are obvious tonal differences between the face skin color in the character image and the clothing and background. At the same time, the hues of face skin color, clothing and background are distributed in different HSV angle regions, and the hues of face skin color are relatively stably concentrated in a certain angle region in HSV color space. Experiments have confirmed that the angular distribution of the hue H of the skin color is basically maintained between [2°,47°] in the HSV color space, whether it is irradiated by natural light or artificial light source, and whether the performance of the camera system is different. Therefore, the difference of the skin color, clothing, background and other scenery of the human face can be distinguished by the hue value of the person image in the HSV space. It is further confirmed by experiments that the probability that the hue value of face skin color is 11° value is the highest, so the hue value of 11° is called the probability peak of face skin color. When the distribution probability of face skin tone in interval [2°,47°] is $P(H)$, the probability of $H = 11°$ reaches the highest, that is, $P(11°)=P_{\max}$. In other words, when the hue of a scene is 11°, the confidence of identifying the scene as human face skin color reaches the highest.

(3) Determine the central position and area of the face of the sampled middle-aged and young passengers

Using the HSV color space expression of the sampled face images of young and middle-aged passengers, in the face skin tone set, take the pixel coordinates corresponding to the tone value closest to 11° as the face center point. For example, through the face area search results, it is obtained that the tone set of the tone interval [2°,47°] falling into the HSV color space is {..,9.7°,10.1°,9.5°,..}, and the pixel coordinates corresponding to the set are $\{\ldots, (i_{k-1}, j_{k-1}), (i_k, j_k), (i_{k+1}, j_{k+1}), \ldots\}$, in which the tone value closest to 11° is 10.1°, and its corresponding pixel coordinate is (i_k, j_k), and then, (i_k, j_k) can be determined as the coordinate of the center position of the face. i represents the column coordinates of pixels, j represents the row coordinates of pixels, subscript represents the number of columns and rows, k of i_k represents the k th column, and k of j_k represents the k-th row. Finally, taking the face center point as the base point, expand $\frac{u}{2}$ row pixels up and down and $\frac{v}{2}$ column pixels left and right as the face area of $u \times v$ size, which is referred to as the face area of $u \times v$ for short.

(4) Import the training samples of facial region images of young and middle-aged passengers

After the $u \times v$ face region is intercepted from the face images of middle-aged and young passengers, it is imported into the sample training system database as a training sample.

The $u \times v$ face region training samples of the collected face images of young and middle-aged passengers have 256 Gy levels. Generally, the number of training samples is $n = k \times l \geq 300$. Among them, k and l are the number of sampled young and middle-aged images and the number of samples in the sampled face area, respectively. For example, $k = 30$ and $l = 10$, respectively, represent that the number of sampled young and middle-aged images is 30 and the number of samples in the sampled face area of each young and middle-aged passenger is 10.

(5) Convert the training sample image into one-dimensional vector

Each two-dimensional training sample image data of middle-aged and young passengers' face area is transformed into a one-dimensional vector, and the face features of middle-aged and young passengers are defined as class 1 face features, and the face features of others are class-1 face features. The "others" include children, the elderly, women, etc.

The one-dimensional vector $\mathbf{X}_i$ of the ith image of class 1 can be expressed as follows

$$\mathbf{x}_i = \begin{bmatrix} x_{i1} \; x_{i2} \; \ldots \; x_{im} \end{bmatrix}^T = \begin{bmatrix} x_{ij} \end{bmatrix}^T \tag{6.7}$$

where x_{ij} represents the gray value of the j th pixel of the i th sample of class 1; $i = 1, 2, \ldots, n$ is the serial number of class 1 face area sample; $j = 1, 2, \ldots, m$ is the number of pixels taken from each sample image, and $m = u \times v$, u and v are the number of column pixels and row pixels of the sample image, respectively.

(6) Calculate the eigenvalues and eigenvectors of training samples

Calculate the mean $\overline{x}$ of class 1, that is

$$\overline{x} = \frac{1}{n \times m} \sum_{i=1}^{n} \sum_{j=1}^{m} x_{ij} \tag{6.8}$$

The mean $\overline{x}$ thus obtained is called class 1 mean face.

After normalizing the above training samples, it can be expressed as

$$v_i = x_i - \overline{x}; \; i = 1, 2, \ldots, n \tag{6.9}$$

The normalized vector $\mathbf{V}$ of class 1 average face composed of training samples is expressed as follows

$$\mathbf{v} = \begin{bmatrix} v_1 \; v_2 \; \ldots \; v_n \end{bmatrix}^T \tag{6.10}$$

At this time, the covariance matrix of class 1 average face is

$$\mathbf{Q} = \begin{bmatrix} v_1 \; v_2 \; \ldots \; v_n \end{bmatrix}^T \begin{bmatrix} v_1 \; v_2 \; \ldots \; v_n \end{bmatrix}; \; \mathbf{Q} \in R^{n \times n} \tag{6.11}$$

Use formula (6.11) to calculate the eigenvalue λ_l and its eigenvector of $\mathbf{Q}$ and rearrange them from large to small to generate the eigenvector as follows

$$\mathbf{p} = \begin{bmatrix} \lambda_1 \; \lambda_2 \; \lambda_3 \; \ldots \end{bmatrix}^T \tag{6.12}$$

Among them, $\lambda_1 \geq \lambda_2 \geq \lambda_3 \geq \ldots \lambda_1 \geq \lambda_2 \geq \lambda_3 \geq \ldots$

(7) Establish face feature criteria

Because the feature vector corresponding to the larger eigenvalue contains more face feature information, the vector space composed of the feature vector corresponding to the first s larger eigenvalues can approximately represent the main information of the face image. The value of s is determined by experiment.

Select the normalized value corresponding to the first s larger eigenvalues from $\mathbf{v} = \begin{bmatrix} v_1 & v_2 & \ldots & v_n \end{bmatrix}^T$ to form a new normalized vector as

$$\hat{\mathbf{v}} = \begin{bmatrix} v_1 & v_2 & \ldots & v_s \end{bmatrix}^T \tag{6.13}$$

Therefore, $\hat{\mathbf{v}}$ can be directly used to represent class 1 facial feature vector $\mathbf{v} = \begin{bmatrix} v_1 & v_2 & \ldots & v_n \end{bmatrix}^T$, that is, the facial region feature vector of young and middle-aged passengers.

That is to say, after establishing the face feature normalization vector, it can be used as the criterion to identify whether the current passenger is a young and middle-aged passenger in turn.

The calculation process from formula (6.8) to (6.13) is also called the projection of the face image to the feature space.

(8) Establish female feature normalization vector

This is to establish the female feature normalization vector for class 1 face features. That is, after obtaining the feature vector of class 1 face at the first level, repeat the steps (1) to (7) above and then establish the distinguishing features between women and men in class 1 face features. The so-called distinction is to distinguish the female feature elements in the "class 1 face feature" vector and then establish the feature vector of subclass 2 with "female feature" in the "class 1 face feature".

In other words, the feature vector of subclass 2 can be decomposed by formula (6.13), which is expressed as follows

$$\hat{\mathbf{v}}_2^* = \begin{bmatrix} v_1^* & v_2^* & \ldots & v_k^* \end{bmatrix}^T \tag{6.14}$$

The rest is the male feature score vector in class 1, and its formula is expressed as follows

$$\hat{\mathbf{v}}_1^* = \begin{bmatrix} v_{k+1}^* & v_{k+2}^* & \ldots & v_s^* \end{bmatrix}^T \tag{6.15}$$

where k is any one of s eigenvalue sequences arranged from large to small, and it can be taken as $k = \frac{s}{2}$.

(9) Distinguish whether pregnant women are pregnant or not according to the eigenvalues of the subclass 2

The so-called body shape discrimination for the subclass 2 of eigenvalues is to obtain the feature points and feature vectors of "pregnant woman shape" from its body shape image.

The technical key point of determining the feature points of "pregnant women's form" is to measure the proportional relationship between the length of connecting lines between the main feature points of body contour.

As shown in Fig. 6.7, the body shape of pregnant women has obvious characteristic curve, and the connecting line of corresponding characteristic points constitutes a specific proportional relationship, as shown in the following formula

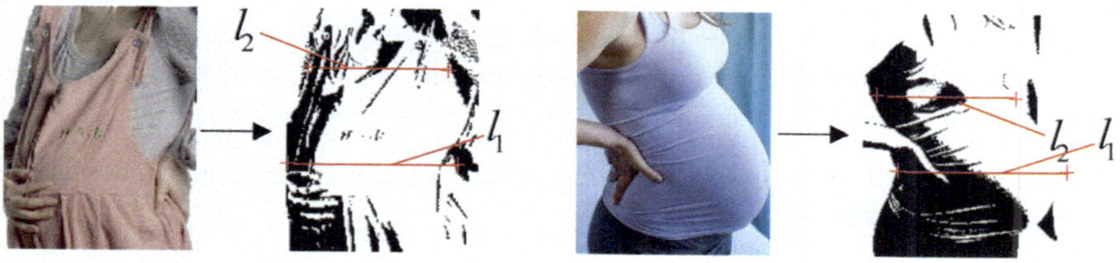

Fig. 6.7 The characteristic curve of pregnant women's body shape and the connecting line between its characteristic points

$$\frac{l_1}{l_2} > \varepsilon_2 \tag{6.16}$$

where ε_2 is the judgment threshold. As long as the characteristic scale of a female passenger is greater than the judgment threshold, it can be confirmed that the female passenger is pregnant.

2. **Online Identification**

When the above subway special seat intelligent prompt device system enters (is) in the online working state, it will monitor the passengers on the "love special seat" in real time, determine the control command output for the passengers on the special seat according to the recognition results and decide whether to prompt the passengers by the voice prompt. The whole process consists of the following steps.

(1) **Collect the passenger's face position image on the special seat online and convert the color space**

After the real-time image of the passenger's face position on the special seat collected by the pinhole camera is strengthened, it is converted from RGB color space to HSV color space for expression.

The passenger's face position image includes several situations; that is, there is a passenger sitting on the special seat, there is no passenger sitting, or the special seat is blocked by other objects.

(2) **Distinguish the passenger's face image on the special seat from the background**

In the collected passenger face position image on the special seat, scan the scene tone of the detected pixels from left to right and from top to bottom according to the pixel coordinates and draw up the pixel points corresponding to the tone set of the tone interval [2°,47°] in the HSV color space as the face area. If the pixel points corresponding to the tone interval [2°,47°] can be found, it indicates that there are passengers on the special seat. Therefore, the images of other areas of passengers in the special seat and their background are accurately separated from their face areas. Continue to the next step (3).

If no pixel corresponding to the [2°,47°] hue in the HSV hue range is found, it indicates that no one has taken the seat at present, or the field of view of the pinhole camera is temporarily blocked by the body parts of the people standing in the carriage. At this time, the calculation program returns to the above substep (1) and continues to monitor the state on the special seat.

(3) **Determine the face center position and eye area of the collected passenger face image on the special seat**

Using the HSV color space expression of the face, taking the pixel coordinates corresponding to the hue value closest to H1 as the face center point and taking the face center point as the base point, expand $\frac{u}{2}$ row pixels up and down and $\frac{v}{2}$ column pixels left and right as the face area of $u \times v$ size. It is commonly known as $u \times v$ face area; that is, the face tracking area of $u \times v$ has been obtained at this time.

(4) **Track the passenger's face area on the special seat**

The first-order prediction algorithm is used as the method to track the face area of passengers in special seats.

Set the face movement speed of passengers in the current special seat as $V(t_k) = \left[\, V_i(t_k)\ V_j(t_k) \,\right]^T$, and

$$\begin{bmatrix} V_i(t_k) \\ V_j(t_k) \end{bmatrix} = \begin{bmatrix} \frac{i_k - i_{k-1}}{t_k - t_{k-1}} \\ \frac{j_k - j_{k-1}}{t_k - t_{k-1}} \end{bmatrix} \tag{6.17}$$

That is, under the interval Δt, the central position of the face is determined through two operations before and after, $\Delta t = t_k - t_{k-1}$.

The first-order prediction estimate shall be

$$\begin{bmatrix} \tilde{V}_i(t_k) \\ \tilde{V}_j(t_k) \end{bmatrix} = \begin{bmatrix} \frac{i_{k-1} - i_{k-2}}{t_k - t_{k-1}} \\ \frac{j_{k-1} - j_{k-2}}{t_k - t_{k-1}} \end{bmatrix} \tag{6.18}$$

It is predicted that the pixel coordinates of the passenger's face target on the special seat are

$$\begin{bmatrix} \hat{i}_{k+1} \\ \hat{j}_{k+1} \end{bmatrix} = \begin{bmatrix} \tilde{V}_i(t_k)(t_k - t_{k-1}) \\ \tilde{V}_j(t_k)(t_k - t_{k-1}) \end{bmatrix} + \begin{bmatrix} i_k \\ j_k \end{bmatrix} \tag{6.19}$$

where $V_i(t_k)$ and $V_j(t_k)$ are the components of speed $V(t_k)$ on the two coordinate axes i and j in the pixel coordinate system at kth time, respectively; $\tilde{V}_i(t_k)$ and $\tilde{V}_j(t_k)$ are the estimated component values of speed $V(t_k)$ on the two coordinate axes i and j at $k-1$th time, respectively; i_k, i_{k-1} and i_{k-2} are the i coordinate values at kth, $k-1$th and $k-2$-th time, respectively; j_k, j_{k-1} and j_{k-2} are v coordinate values at k-th, $k-1$th and $k-2$-th time, respectively; and $\hat{i}_{k+1}$ and $\hat{j}_{k+1}$ are the estimated coordinate values of i and j at $k+1$th time, respectively.

Therefore, when the face center position of t_k at time k is (i_k, j_k), it can be predicted by the first-order prediction algorithm that the face center position of t_{k+1} at time $k+1$ is $\left(\hat{i}_{k+1}, \hat{j}_{k+1}\right)$.

(5) **Import test samples**

According to the face center coordinate (i_k, j_k) tracked by the system, expand $\frac{u}{2}$ row pixels up and down and $\frac{v}{2}$ column pixels left and right in turn, intercept the $u \times v$ face area as the test sample of the passenger face area of "love special seat" and import it into the recognition system.

(6) **Calculate the eigenvector of the test sample**

Repeat the operations of formulas (6.8) to (6.13) to complete the calculation of the image eigenvalues and eigenvectors of the test samples and obtain the normalized values corresponding to the first s larger eigenvalues selected from $\mathbf{v} = \begin{bmatrix} v_1 & v_2 & \ldots & v_n \end{bmatrix}^T$ to form a new normalized vector as follows

$$\tilde{\mathbf{v}} = \begin{bmatrix} v_1 & v_2 & \ldots & v_s \end{bmatrix}^T \tag{6.20}$$

Therefore, $\tilde{\mathbf{v}}$ can be directly used to represent the facial region features of the current passengers in the special seat, that is, the feature normalization vector of the test sample.

(7) **Type identification**

The following operations can be used to identify the category of passengers on special seats. The $u \times v$ face region feature $\tilde{\mathbf{v}}$ projected into the feature space is compared with the training sample feature $\hat{\mathbf{v}}$ through the distance classification function to determine the category of the sample to be recognized, that is

$$G\left(\tilde{\mathbf{v}}, \hat{\mathbf{v}}\right) = \left\|\tilde{\mathbf{v}} - \hat{\mathbf{v}}\right\| \leq \varepsilon_1 \tag{6.21}$$

This shows that the passengers in the current special seat belong to class 1. Otherwise, the passengers in the current special seat belong to class-1.

In formula (6.21), $\tilde{\mathbf{v}}$ and $\hat{\mathbf{v}}$ represent the face feature normalization vectors of test samples and training samples, respectively.

(8) **Further subdivision of types**

For passengers belonging to category 1, it is necessary to further distinguish whether they belong to category 2 and exclude male passengers in category 1.

Repeat steps (1) to (7) above to match the recognized object with the feature vector of type 2. If it does not belong to class 2, turn to step (10); otherwise, continue with step (9).

(9) **Calculation of characteristic points**

Extract the morphological feature points of the target image and calculate the proportion between the connecting lines of the key feature points to determine whether they belong to class 2. If $\frac{l_1}{l_2} > \varepsilon_2$, they are considered as pregnant women and return to step (1); otherwise, continue (10).

(10) **Control decision**

According to the identification results, the control command output is determined for the passengers in the current special seat; that is, once it is determined that the passengers in the special seat are "young men" or other "non-pregnant middle-aged and young women", the system outputs the control command in real time. Driven by the control command, the voice prompt sends a friendly prompt to the "recognized target" passenger: "this is a 'special seat for love', would you please give it to the passenger who needs help? Thank you for your assistance!". Otherwise, the system will not give a prompt, including the case of no one in the special seat.

From step-by-step (1) to (10) cyclic operation, online real-time identification of special seats can realize the whole process monitoring of special seats.

In addition, if the identified object is a "young male" or other "non-pregnant young and middle-aged female" but is a sick passenger or disabled person, it can still be "corrected" by voice recognition and other means. If someone (including himself or surrounding passengers) says, "he or she is a patient...", the system will immediately "correct" and broadcast the language of care.

Of course, people often stand in the passage of subway cars. At this time, how to collect and recognize the images of passengers on the "special seat for love"? Once the "special seat for love" is blocked for a long time, the image information on the "special seat for love" will be captured in real time as long as someone moves, even if there is only an instant gap. Therefore, this technology is also known as the "single visual sensor" of "never fatigue" and "never blink", which plays a positive role and social benefits that non-technical methods cannot play in creating a comfortable and harmonious riding environment [5].

References

1. Zhang X (1975) Thermal measurement principle and its modern technology. Sahng Jiao Tong University Press, Shanghai
2. Zhang X, Zhu X, Chen T et al (2007) An auxiliary electronic control device that assists in the management of traffic. Chinese Invention Patent CN200710037035.2. February 2007
3. Guo J, Zhang X, Ying J (2007) Intelligent control system of platform screen door. Chinese Invention Patent CN200710045171.6. August 2007
4. Zhang X, Ying J, Guo J et al (2007) Intelligent control method of platform screen door. Chinese Invention Patent CN200710045172.0. August 2007
5. Zhang X, Ying J, Zhang X et al (2010) Intelligent prompt device and method for bus special seats. Chinese Invention Patent CN201010501545.2. October 2010

Chapter 7
Future Technological Development of Rail Transit

China's rail transit has also become the forefront of the application of advanced technologies, such as informatization, intelligence, energy conservation, safety, environmental protection and so on. Compared with the mature application of intelligent system in other fields, there are still many deficiencies in the application of intelligent production, operation and maintenance system in rail transit system, which needs further technical improvement and upgrading.

The following takes high-speed railway as an example.

7.1 Technical Status and Development of High-Speed Train

Driven by the wave of rail transit development, a large number of advanced technologies and new equipment have been widely used, which has effectively promoted the high-quality development of modern rail transit system characterized by safety, efficiency, green, intelligence and sustainability.

Therefore, it is necessary to review the development and technical status of high-speed railway.

7.1.1 Review of High-Speed Rail Development

The development of high-speed train can be basically divided into four important stages.

The first stage is represented by Shinkansen power decentralized high-speed EMU in Japan. Japan is the first country in the world to open high-speed rail operation and promote its application on a large scale. In 1964, Shinkansen in Japan created a milestone in the commercial operation of high-speed rail with an operation speed of 210 km/h. Since then, they have been developing and testing new technologies

Y. Khan, *Principles of Intelligent Rail Transit*, Advances in High-speed Rail Technology,
https://doi.org/10.1007/978-981-19-6072-7_7

almost every 14 years and formed the development characteristics of small axle load and low energy consumption.

The second stage is represented by TGV power centralized high-speed EMU trains in France. France is the second country in the world to open high-speed rail operation (1981) and promote its application on a large scale. In 1989, it increased the maximum operation speed to 300 km/h, creating a new situation for the promotion of high-speed rail in Europe. Representative technologies in the same period also include hinged machine body, pendulum technology, double-layer locomotive body, etc. The typical characteristics of its high-speed railway line are large ramps (35 ‰), few tunnels and low construction cost (The construction cost of the first high-speed railway kilometer is less than 20% of that of Shinkansen in Japan.).

The third stage is represented by ICE3 power distributed high-speed EMU in Germany. Germany opened the commercial operation of high-speed railway in 1991 and increased the maximum operation speed to 330 km/h in 2002. The technologies in the same period also include linear eddy current braking, multi-standard power supply, flexible marshaling, etc. ICE3 makes the advantages of power distributed EMU return again and extends it to many countries including Russia and China, promoting the rapid development of high-speed railway in the world.

The fourth stage is represented by China Fuxing power decentralized standard EMU. China opened its first high-speed railway operation in 2008, catching up from behind and developing rapidly. At present, the total mileage of high-speed railway in China has accounted for more than 60% of the total mileage of high-speed railway in the world. The train running speed ranks first in the world and has become a "golden card" made in China to the world. Its high-speed EMU is famous all over the world for its wide body, large passenger capacity, high stability, high punctuality and other advantages.

7.1.2 Weak Links of Existing Technologies in China

1 The R&D ability of basic technology is not strong

It shows that key materials and processes still depend on the technology of other countries.

Due to the unique characteristics and design of rail transit vehicles in the complex assembly site, it is necessary to adopt a specific assembly process to realize accurate and rapid automatic identification of vehicle component information. However, the reality is that the types of real-time monitoring are incomplete during data acquisition. There is a phenomenon of failure omission or false report in the production and manufacturing process. In the system operation and maintenance, the vehicle has the problem of over repair or under repair.

2. **Lack of R&D capability in common technology**

This problem is reflected in the inadequate role of the simulation platform and the lack of unified product collaborative R&D and simulation verification. The so-called platform includes vehicle mechanical system simulation platform, noise analysis and diagnosis platform, locomotive body tolerance research platform, traction, braking, signal and power integrated simulation platform, etc. In addition, it is necessary to train the network test platform, establish the communication network and wireless communication network, carry out the research of non-destructive testing technology, form relevant technical standards and realize the perfect technical links of R&D and testing.

3 **Lack of R&D capacity in key technologies**

As for the life cycle of train, the research on independent safety protection is still insufficient. The research on Fault Analysis and prediction, safety control and guarantee system technology of rail transit system in complex environment is insufficient. The integrated technology of intelligent safety guarantee, operation and maintenance of urban rail transit system needs to be constructed, optimized and improved. In addition, the relevant research of high-speed rail intelligent sensing system is not deep enough, the greening degree of rail transit energy utilization mode is not enough, and the complete interconnection system of train control has not been realized.

4 **The construction of practical application product platform is not in place**

At present, the platform design of relevant systems has not been completed. The platform has the characteristics of standardization, modularization and whole process management. It can collect, analyze, test and share information and data of products in the whole cycle, realize the integrated chain connected with design, production and service, and promote the modern and intelligent development of urban rail transit system.

Due to disorderly competition and repeated construction, insufficient industrial supporting capacity, insufficient product quality types, insufficient construction of intelligent platform for product production, operation and maintenance, insufficient R&D capacity and other problems, China's urban rail transit vehicle industry has a bottleneck of strategic upgrading, which is an urgent problem to be solved.

7.1.3 International High-Speed Train Technology Status and Development Direction

1 **European Union (EU) "Shift2 rail" plan**

As China joins the competition in the development of international rail transit science and technology, in order to improve the global competitiveness of European railway products, the EU has put forward the "Shift2 rail" plan to consolidate its leading position in the global product and service market. There are three quantitative indicators, that is, to achieve: The life cycle cost of rail transit system is reduced by 50%, the capacity is improved by 100%, and the reliability and punctuality of service are improved by 50% [1].

In order to ensure the realization of technical objectives, seven guiding directions are put forward, including traction system based on silicon carbide and independent rotating wheels, train control and management system based on wireless transmission technology, application of lightweight locomotive body materials, running structure of new structure and lightweight materials, new braking system with higher braking rate and lower noise, door system of new lightweight structural materials and internal layout of modular carriage. In addition, Shift2 Rail also focuses on common issues such as smart materials, interoperability, energy and sustainability, human capital and so on.

2 Representative commercial locomotive form abroad in the future

(1) Research in Japan

Japan's N700S locomotive is the most high-end model in Shinkansen N700 series. It continues to uphold the concept of safety, comfort, reliability and environmental protection, and has been put into operation in 2020. This model is the first high-speed train equipped with SiC (silicon carbide) drive system in the world [2].

The locomotive model will be combined with running air cooling technology, and the main converter device will be greatly miniaturized and lightweight. It is 600 kg lighter than N700 (with blowing mechanism) and can be installed on the same vehicle as the main transformer. This simplifies the types of vehicles and enables them to realize flexible marshaling of 8, 12 and 16 vehicles according to customer needs.

In addition, due to the use of SiC devices, it can generate large current in the high-frequency domain, reduce the weight of each driving motor by 70 kg on the basis of the original power and reduce the total power consumption by 7%. While realizing the world's lightest axle weight of 11 T, it greatly reduces the occupied space and weight constraints under the floor, so as to realize the installation of unified equipment under the vehicle and form a standard vehicle, which is conducive to mass production, improve quality and reduce cost. At the same time, N700S has also carried out optimization design in the aspects of optimizing the shape of the vehicle head, shortening the braking distance in case of earthquake, adopting herringbone gear transmission, installing full active damping control for first-class vehicles, optimizing semi-active structure, adopting flexible sliding plate pantograph, strengthening vehicle condition monitoring and early warning, installing emergency battery pack and improving the comfort of the carriage, which has greatly improved the performance of the vehicle [3].

Alfa-X in Japan is the next generation high-speed train platform built by JR East Japan for Shinkansen. It is planned to carry passengers at an operating speed of 360 km/h in 2025. A typical feature of Alfa-X is locomotive head design. In order to reduce the pressure wave generated when entering the tunnel at high speed and reduce wind resistance, the front of the vehicle is designed to be 22 m long. In the experiment, through the pressure wave detection and test of locomotive head under various working conditions, it is proved that this design can significantly improve the energy absorption space of locomotive head collision. Alfa-X is also characterized by the use of active vertical and horizontal hydraulic shock absorbers to improve operation stability and comfort. Alfa-X will also adopt a new "seismic shock absorber", which can start automatically in case of strong seismic vibration, so as to stabilize the operation of the train and prevent the train from derailment. In order to deal with the risk of frequent earthquakes, Alfa-X also applies two kinds of emergency braking devices, namely the wind resistance braking deceleration plate on the roof and the linear eddy current brake on the frame, so that the high-speed train can complete deceleration or braking as soon as possible in unexpected situations such as natural disasters.

In addition, the predecessor of L0 maglev train in Japan is MLX-01 maglev test vehicle, which adopts superconducting maglev technology [4].

The faster the speed of maglev vehicle, the greater the suspension force and guidance force. At low speed, it depends on rubber wheels for support and guidance. When the vehicle speed reaches 150-180 km/h, it will enter a completely contactless suspension state with a suspension air gap of more than 100 mm [5].

Japan began to carry out research on maglev technology in 1962 and began the test of maglev train in 1972. In 1977, it completed 7 km superconducting maglev Miyazaki test line, and in 1997, it completed 18.4 km double track Yamanashi test line. Finally, in 2003, mlx01-2 high-speed train set a record of 581 km/h. In 2013, the whole line of 42.8 km Yamanashi test line was completed, and the test line will become the actual operation line of the central Shinkansen in the future. On April 22, 2015, Japan's l0 maglev train set a new world record at 603 km/h (manned operation) on the track of Yamanashi test line. In 2014, the Japanese government officially approved the construction of the central Shinkansen maglev railway from Tokyo to Osaka via Nagoya. The total length of the section from Tokyo to Nagoya is 286 km, and it is planned to operate at a speed of 500 km/h in 2027. The 152 km section from Nagoya to Osaka is planned to be completed in 2045. The total length of the project is 438 km, and the total cost is about 518.4 billion RMB yuan.

(2) **Research in France**

TGV Avelia is the next generation of high-speed train developed by French State Railway Corporation in cooperation with Alstom. It is planned to be put into commercial operation in 2022. The high-speed train adopts technologies such as centralized power EMU, permanent magnet motor, articulated bogie, double-layer locomotive body, friction stir welding, carbon fiber composite (end beam), modern passenger system, SiC-based traction system and so on. TGV Avelia does not pursue high performance or new functions as the selling point, but pays more attention to how to reduce the unit transportation cost and daily maintenance cost and improve the profit margin of high-speed train service. Through technological innovation and optimization, TGV Avelia aims to reduce the construction and operation cost by 20% compared with the previous generation of trains, the recycling utilization rate of materials and structures is higher than 90%, the energy consumption is reduced by

25%, and the interior layout is optimized to expand the available space of trains by 20%, greatly reducing the cost of the whole life cycle of trains [6].

(3) **Research in Germany**

Velaro Novo in Germany is a new high-speed train developed by Siemens after five years. It can realize flexible marshaling configuration at the operating speed of 250-360 km/h. It is expected to be officially put into operation in 2023.

Its main features are as follows.

(1) The traction and braking system adopts silicon carbide traction auxiliary converter and permanent magnet synchronous traction motor to increase the traction power by 10%, braking power by 70% and efficiency by 5%. The braking distance is reduced to 4700 m, and the per capita power is 14.81kw/person.
(2) In terms of structure, the locomotive body applies friction stir welding technology and optimizes the shape of bogie and roof, and the new electrical system design reduces the occupied space, increases the passenger compartment space and can be quickly and flexibly marshaled to meet the configuration requirements of different traffic volumes and destinations.
(3) The application of intelligent technology includes intelligent sensors, which can automatically give early warning of maintenance, railway conditions, damage and other conditions, so as to reduce maintenance costs.
(4) Modular design, following the principle of "air traffic control", internal accessories are standardized and fixed, and all seats, tables and layouts can be fully customized and replaced.
(5) MVB/PROFINET bus is adopted for network technology, and Ethernet is adopted for vehicle maintenance network and passenger information system network.

Through the above optimization design, the passenger capacity is increased by 10%, the weight is reduced by 15%, the vehicle cost is reduced by 25%, the energy consumption is reduced by 30%, the maintenance and repair cost is reduced by 30%, and the operation cost is greatly reduced.

German NGT is the next generation high-speed train project carried out by German Aerospace Center, with a designed maximum operating speed of 400 km/h. NGT adopts innovative designs such as 100% low floor structure similar to tram, independent rotating wheel, single axle structure, wheel set radial position control, double-layer locomotive body structure, locomotive body combined with plate and truss structure, glass fiber reinforced plate (GFK) sandwich filling structure, light metal structure (30% less mass than metal structure), new HVAC and so on; and effectively reduce wheel rail and pneumatic noise, improve ride comfort, improve vehicle safety and reduce wheel rail wear and life cycle cost. Modular design method and system integration design technology are adopted to speed up the product design and manufacturing process, improve the development efficiency of new products and shorten the product development cycle. The ultimate goal is to increase the reliability and punctuality rate by 50%, reduce the energy consumption by 50% and reduce the life cycle cost by 40% compared with ICE3 operating at 300 km/h.

(4) **New technology of magnetic levitation and low vacuum**

Due to the problems of air resistance, wheel rail adhesion, train Snake Movement instability, operation noise and pantograph catenary contact power collection, the operation speed of wheel rail train is difficult to be greatly improved, and the operation energy consumption of locomotive and the wear caused by mechanical friction will also increase significantly with the increase of speed. The latest test results show that the per capita energy consumption per 100 km of EMU operating at 400 km/h is about 30% higher than that at 350 km/h.

In the 1920s, the technical concept of maglev train was born in Germany, and maglev technology was used to solve the problems of wheel rail adhesion, friction, vibration and high-speed current collection of wheel rail train [7].

Using German technology, China built the world's only commercial high-speed maglev line in Shanghai in 2002, and its maximum operating speed has reached 430 km/h.

Japan started the construction of the first high-speed maglev line from Tokyo to Nagoya in 2014. Its designed maximum operating speed is 505 km/h and is planned to be opened in 2027.

At present, China is developing a high-speed maglev system technology with a speed of 600 km/h.

However, both wheel rail and maglev trains will face huge air resistance (proportional to the square of speed) and noise (proportional to the sixth to eighth power of speed) when they operate in an open space atmosphere. Especially when the speed exceeds 400 km/h, the air resistance will account for more than 80% of the total train operation resistance, which will bring great challenges to economic and environmental protection for commercial operation. Therefore, in order to obtain higher economic operation speed, on the basis of reducing wheel rail friction and vibration by using magnetic levitation technology, building a low vacuum operation environment to reduce air resistance and noise is an important direction for the development of higher speed rail transit technology in the future.

The shortcomings of low vacuum pipeline high-speed maglev system mainly include two aspects: First, as a high-speed transportation system with different technology from wheel rail high-speed railway, it is unable to use the domestic railway with a length of more than 100,000 km for joint operation; and second, restricted by high-speed maglev turnout and related technologies, the train tracking interval is long, which affects the train operation efficiency.

American scholar Robert David put forward the idea of "vacuum pipeline transportation" as early as 1904. In the 1970s, Switzerland proposed an underground low vacuum high-speed maglev Metro system with tunnel as pipeline, which is called Swissmetro System. The system evacuates the underground tunnel into a low vacuum close to a standard atmospheric pressure, uses linear propulsion technology and magnetic levitation and guidance technology, and the design speed is 500 km/h [8].

In the 1990s, American engineer Auster registered the trademark and invention patent of vacuum pipeline transportation system (ETT). The system is a small-decentralized pipeline "automobile" mode. The pipeline is pumped into low vacuum. The cylindrical maglev cabin that can carry 6 passengers is adopted, and its operation speed can reach 6500 km/h.

In 2004, Chinese scholars also put forward a technical scheme for the development of low vacuum high temperature superconducting maglev high-speed system [9].

However, the low vacuum pipeline high-speed magnetic levitation system still has many key systems, such as overall system, low vacuum and line system, vehicle system, suspension guidance system, drive and braking system, communication and operation control system, safety guarantee and disaster prevention and rescue system, which need to be deeply studied.

(1) **Overall system**

There is no successful design experience for the overall level of low vacuum pipeline high-speed magnetic levitation system. The key technologies to be overcome are integrated design technology, complex multi-physical field system coupling analysis technology, system safety and reliability technology, system simulation optimization and test verification technology, etc.

(2) **Low vacuum and line system**

When the vacuum scale of the system is large, it is difficult to dissipate heat in time through convection in the vacuum. At the same time, high-speed operation will put forward higher requirements for the smoothness, reliability, structural characteristics and other parameters of the line. The key technologies to be overcome include construction and control of large-scale pipeline in low vacuum environment, construction of high-precision track, design and manufacture of lines, bridges and tunnels, high-speed maglev turnout technology, instantaneous large-area and high-density heat dissipation technology of track, etc. [10]

(3) **Vehicle system**

Under the condition of high-speed movement, due to the coupling effect of force, heat, vibration and noise, the vehicle will inevitably put forward new requirements for the train body and running mechanism. Among them, the key technologies to be overcome include integrated design technology of low resistance/low noise/low heat body shape, design technology of lightweight and high load-bearing body structure, design technology of high natural frequency running mechanism, etc.

(4) **Suspension guidance system**

When the train is moving at high speed, it puts forward higher requirements for the stability, loss suppression and structural strength of the suspension system. The key technologies to be overcome include large and heavy load stable magnetic levitation design technology, high-speed dynamic levitation force fluctuation suppression technology, new high-performance superconductor development technology, etc.

(5) **Drive and brake system**

When the train is running at high speed, it puts forward higher requirements for the working efficiency of the driving system under the conditions of high frequency, high voltage and large current, as well as the braking force density and speed range of eddy current brake at different speeds. The key technologies to be overcome include high thrust high-speed motor development technology, high-power high-speed drive converter control technology, high-density and reliable energy storage and power supply technology, wide range and high-density eddy current braking technology, etc.

(6) **Communication and operation control system**

Due to the strong coupling between the high-speed magnetic levitation system of low vacuum pipeline and its high requirements for control complexity and reliability, the key technologies to be overcome include high-speed and high reliable operation control technology, high reliable wireless communication technology between locomotive and ground in low vacuum closed environment, etc.

(7) **Security system**

The security system mainly includes monitoring and evaluation of state and disaster, emergency disposal and safety rescue technology, etc.

7.2 Overall Development Trend of World Rail Transit

At present, the development of high-speed rail transit technology has formed two distinct roads: One is a mature technology facing the market of sustainable development and pursuing to reduce the life cycle cost, such as NGT in Germany and Avelia in France, and the other way is to face the future high-speed rail transit market, adopt disruptive technology, pursue high speed and take into account low cost, such as Japanese superconducting maglev train and American super high-speed railway.

7.2.1 Development Trend of Wheel Rail Technology

At present, the maximum line test speed of wheel rail test train is 574.8 km/h (such as French high-speed EMU test vehicle V150), but the maximum operation speed of wheel rail train under research is only 400 km/h.

For foreign wheel rail high-speed railways, due to the limitations of existing line conditions, regional distance, construction funds and aviation competition, the pursuit of the operation speed of traditional wheel rail transit is not strong, but more from the perspective of sustainable development and improving international competitiveness, emphasizing energy conservation and consumption reduction, environmental protection and noise reduction, safety, comfort, intelligence and efficiency.

There are obvious trends in structural innovation, material innovation, control technology innovation and cutting-edge technology application innovation. For example, carbon fiber structure is used to reduce the mass of bogie and vehicle body, SiC-based chip is used to reduce the volume and mass of control system, intelligent sensor is used for more effective machine and vehicle condition monitoring, fully enclosed built-in axle box bogie is used to reduce noise, and independent wheel structure is used to reduce unsprung mass, The articulated common bogie is adopted to reduce the mass and noise of the whole vehicle.

Due to the problems of wheel rail train in wheel rail adhesion, instability of locomotive snake movement, running noise and pantograph catenary contact current collection, mechanical wear and energy consumption will increase significantly with the increase of speed. Furthermore, it is generally believed in the world that the commercial operation technology risk of wheel rail high-speed trains above 400 km/h is greater.

7.2.2 Development Trend of Magnetic Levitation Technology

Maglev vehicle realizes contactless suspension and guidance between the train and the track through magnetic force, which has the advantages of fast acceleration, small vibration and good comfort, and effectively avoids the problems of wheel rail train.

1 Mechanism principle of maglev train

Fig. 7.1 shows a physical and mathematical model of EMS (electromagnetic suspension) maglev train.

The dotted line (arrow and circle) in the figure is the "thought expression line" of the abstract (virtual) process.

The (a) in Fig. 7.1 is the structural schematic diagram of EMS train. In the (b) of Fig. 7.1, the fixed electromagnet is at the top, representing the long stator core electromagnet set on the track, and the ferromagnetic ball at the bottom represents the train simulator. When the coil winding on the electromagnet is connected with current i, the generated electromagnetic field forms suction $F(i, x)$ on the ferromagnetic ball below. When the generated suction is equal to the gravity of the ferromagnetic ball, the ferromagnetic ball is suspended in the air, and the gap x between them is the suspension height of the train. At this time, the ferromagnetic ball is in an unstable equilibrium state. When the ferromagnetic ball is disturbed by external force f_d, it is very easy to lose its balance. In order to maintain the stability of the system, it is necessary to add a feedback link. Therefore, a control algorithm must be designed to ensure that the suspension meets the technical requirements of stable control.

When the train reaches the set suspension height, the electromagnetic suction is equal to the gravity of the train, and the train will be suspended. It can be seen that the control of single point suspension system is equivalent to the control basis of maglev train suspension. The dynamic equation is

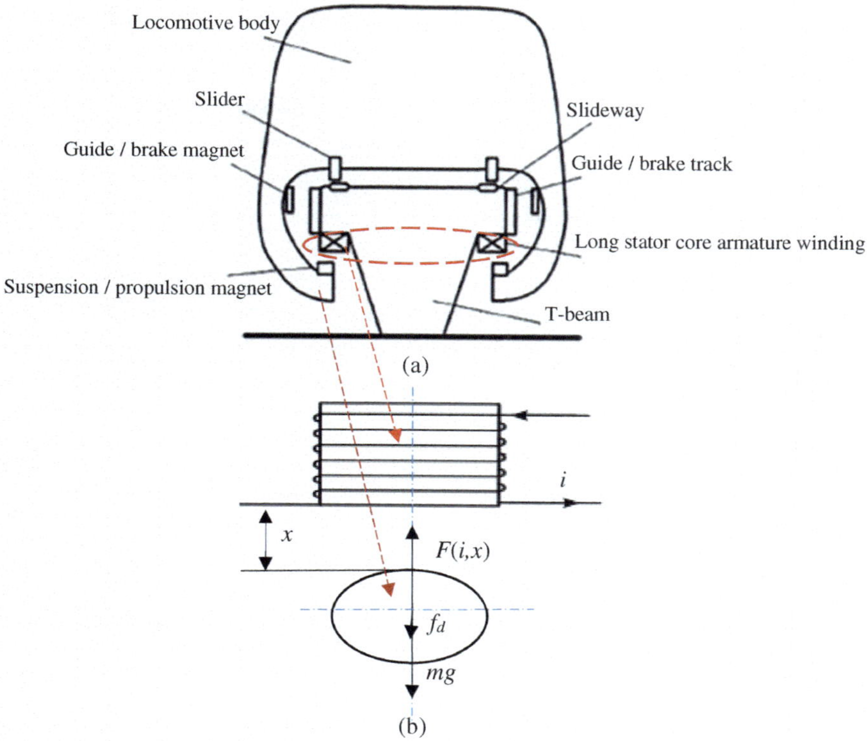

Fig. 7.1 EMS system principle and its basic dynamic model

$$m\frac{d^2x}{dt^2} = mg + f_d + F \tag{7.1}$$

where m is the mass of ferromagnetic ball; x is the suspension height of ferromagnetic ball; g is the earth's gravitational acceleration; f_d is the external interference force on the ferromagnetic ball; and F is $F(i, x)$, which is the suction force on the ferromagnetic ball. Among them,

$$F = K\left(\frac{i}{x}\right)^2 \tag{7.2}$$

where i is the current in the electromagnetic coil; K is the definition constant, $K = -\frac{\mu_0 AN^2}{4}$; μ_0 is the magnetoconductivity of air; A is the magnetic flux cross-sectional area of the electromagnet; and N is the number of turns of solenoid coil. Substituting formula (7.2) into formula (7.1) can obtain the following equation

$$m\frac{d^2x}{dt^2} = mg + f_d + K\left(\frac{i}{x}\right)^2 \tag{7.3}$$

Ignoring the influence of the change of the position of the suspended ferromagnetic ball on the inductance, the equation can be obtained as follows

$$U(t) = Ri(t) + L\frac{di(t)}{dt} \tag{7.4}$$

where U is the voltage at both ends of the electromagnet; R is the equivalent resistance of electromagnet; and L is the equivalent inductance of electromagnet winding.

When the ferromagnetic ball is in equilibrium, gravity and electromagnetic suction are equal. According to Newton's second law, the resultant force on the ferromagnetic ball is zero, that is

$$mg + F(i_0, x_0) = 0 \tag{7.5}$$

where $F(i_0, x_0)$ is the electromagnetic attraction of the electromagnet to the ferromagnetic ball when the suspension height of the ferromagnetic ball is x_0 and the equilibrium current is i_0.

After simultaneous Eqs. (7.3), (7.4) and (7.5), the following equations are obtained

$$\begin{cases} m\frac{d^2x}{dt^2} = mg + f_d + K\left(\frac{i}{x}\right)^2 \\ mg + F(i_0, x_0) = 0 \\ U(t) = Ri(t) + L\frac{di(t)}{dt} \end{cases} \tag{7.6}$$

Formula (7.6) is called the equation of motion and dynamics of a single point suspension system.

In order to ensure the reliability of this suspension and the stability of train operation, a linear motor with high power needs to be adopted, and the magnetic field maintains stable strength and suspension force through the accurate control of the current in the electromagnet, so as to maintain a gap of about 10 cm between the vehicle body and the guide rail. EMS usually adopts air gap sensor to implement gap feedback control of the system [11].

This is the obvious technical advantage of maglev train compared with wheel rail train.

However, EMS is a magnetic levitation with suction force. Both the track and the vehicle end need to generate magnetism through the power coil to attract each other, so that the vehicle is in a suspended state and the power consumption is high.

Relatively speaking, when the gap between the maglev vehicle and the track is small, the risk of its high-speed operation will be higher than that of the superconducting maglev vehicle suspended by repulsive force. Therefore, more people believe that the normal conductivity maglev train can only be more suitable for the operation of medium and low-speed maglev lines.

The Shanghai Maglev special line built in 2002 with German TR08 technology belongs to EMS.

EDS (Electro Direct-repulsion Suspension) train in Japan, or superconducting magnetic repulsion train. This is a maglev operation mode formed by the repulsion between the magnetic force generated by the vehicle electromagnetic coil through the superconducting effect and the track magnetic force.

The principle of superconducting magnetic repulsion (EDS) is a technical system structure composed of superconducting magnets (placed in liquid helium storage tank) installed at the bottom of the vehicle and a series of aluminum ring coils laid on both sides of the track.

When the train is running, the coil (superconducting magnet) on the train is connected with current to generate a strong magnetic field. The aboveground coil (aluminum ring) is tangent to it and generates a magnetic force opposite to the magnetic field direction of the superconducting magnet on the train, forming the repulsive force of the two magnetic fields. When the repulsion force is greater than the vehicle mass, the vehicle floats. Therefore, the superconducting magnetic repulsion train uses the relative motion between the superconducting magnet placed on the vehicle and the passive coil laid on the track to generate the suspension force to lift the vehicle body. (see Fig. 7.2)

Since the resistance of superconducting magnet is zero, it hardly consumes energy in operation. The biggest feature of superconducting magnet is that its magnetic field intensity can reach a very large degree. The strong repulsive force between the superconductor and the guide rail can make the vehicle float easily. When the vehicle moves downward, the reduction of the distance between the superconducting magnet and the suspension coil will increase the current in the coil, increase the suspension force and then make the vehicle automatically return to the original suspension position. This gap is related to the speed. Generally, the vehicle body can be suspended only at 100 km/h. Therefore, the vehicle must be equipped with mechanical auxiliary support devices, such as auxiliary support wheels and corresponding spring supports,

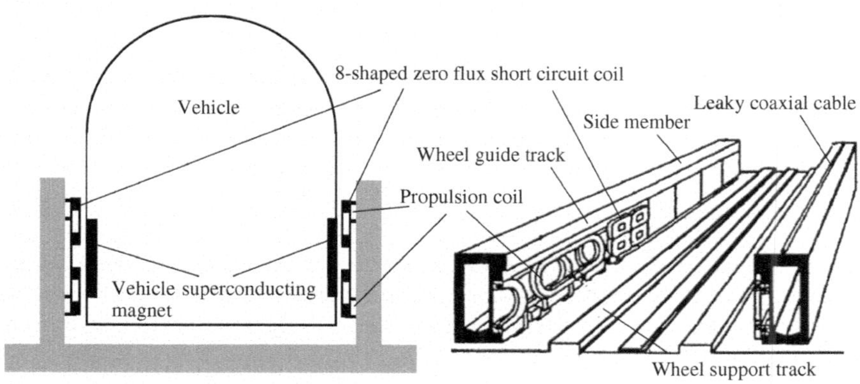

Fig. 7.2 Schematic diagram of EDS system structure

so as to ensure the safe and reliable landing of the train. At the same time, the control system shall also be able to realize the accurate control of train start and stop [12].

2 Guidance mode of maglev train

The guidance system with constant magnetic conduction and suction is similar to the suspension system. A group of electromagnets specially used for guidance are installed on the side of the vehicle. A certain gap shall be maintained between the vehicle body and the side of the guide rail. When the vehicle deviates left and right, the guide electromagnet on the vehicle interacts with the side of the guide rail to restore the vehicle to the normal position. The control system maintains the lateral clearance by controlling the current in the guide magnet, so as to control the running direction of the train.

The superconducting magnetic repulsion guidance system can be composed of the following three ways.

(1) Install mechanical guiding device on the vehicle to realize train guidance

This device usually adopts the lateral guide auxiliary wheel on the vehicle to interact with the side of the guide rail (rolling friction) to produce the restoring force, which is balanced with the lateral force generated when the train runs along the curve, so as to make the train run along the center line of the guide rail.

(2) Install a special guide superconducting magnet on the vehicle

This technology uses a special guide superconducting magnet to generate a magnetic repulsion force with the ground coil and metal belt on the side of the guide rail, which is balanced with the lateral force of the train to keep the train in the correct running direction. This guidance mode avoids mechanical friction. As long as the current in the lateral guidance coil of the ground track is controlled, the train can maintain a certain lateral clearance.

(3) Guided by magnetic force

The so-called magnetic guidance is realized by laying an "8" shaped closed coil in the "zero magnetic flux" guidance system. When the superconducting magnet set on the train is located on the symmetrical center line of the coil, the magnetic field in the coil is zero. When the train generates lateral displacement, the magnetic field in the "8" shaped coil is zero, and a reaction force is generated to balance the lateral force of the train and make the train return to the position of the line center line.

3 Propulsion mode

The key technology of Maglev propulsion system is to expand the rotating motor into a linear motor. Its basic composition and action principle are similar to that of ordinary rotating motor. After expansion, its transmission mode will change from rotating motion to linear motion.

Linear motor is divided into short stator asynchronous linear motor and long stator synchronous linear motor. The traction mode of short stator asynchronous linear motor is to install a full set of traction devices such as three-phase armature winding, traction transformer and converter on the vehicle, and install the induction rail on the track as the motor rotor. The vehicle generally obtains the power supply from the ground power supply system by contact current collection. This method has the characteristics of simple structure, easy maintenance and low cost. It is suitable for medium and low-speed urban transportation and suburban transportation, as well as short-range tourism line system. However, its main disadvantage is its low traction power and low efficiency, which is not conducive to high-speed operation. China's Changsha airport line, Beijing S1 line maglev train and Japan's HSST maglev train all adopt this propulsion mode.

Superconducting magnetic repulsion maglev adopts long stator synchronous linear motor. The super conductive magnet is installed on the vehicle, and a passive closed coil or non-magnetic metal plate is set along the track. The adoption of superconducting electromagnetic coil as maglev device provides convenient conditions for the excitation coil of linear synchronous motor to be in superconducting state. They can coexist in the same cooling system, or the same coil can play the role of suspension, guidance and propulsion at the same time.

German tr series high-speed maglev locomotives lay stator coils (called long stators) on the whole track. The suspension magnet on the vehicle also acts as the rotor of linear motor, and all traction power supply and transformation, converter control, switch control and other equipment are set on the ground. Considering the power loss, back electromotive force and other factors of the stator coil, the stator coil on the line shall be divided into multiple sections (called traction subareas), and each traction subarea shall be equipped with a complete traction power supply and transformation system. Only the ground traction system of the section through which the train passes is working. When the train crosses the section, the adjacent traction sections will be handed over automatically. In order to reduce the number of ground traction equipment, the length of traction subarea should be as long as possible (up to 30–50 km). In order to further reduce the loss of stator coil, a traction subarea is divided into several shorter stator sections (usually hundreds to more than 1000 m). Each stator section controls whether to connect traction current through the ground switching station. In such a traction subarea, only the stator section where the train is located is powered. In order to reduce the impulse of train switching between sections, the switching points of stator sections on the left and right sides of the track are staggered from each other, so as to ensure that at least one stator section on the left and right sides is powered on at the same time. The advantages of long stator linear motor are high traction power, higher efficiency than short stator and higher traction speed. Although its disadvantages are many ground equipment, complex system and high engineering cost, long stator linear motor has to be the inevitable choice of high-speed and ultra-high-speed maglev, because it is the need of technology.

4 **Performance evaluation**

The speed of normal conducting magnetic levitation (EMS) train can reach 400 to 500 km/h, and the speed of superconducting magnetic levitation (EDS) train can reach 500 to 600 km/h. Its high speed makes it superior to flying in the travel distance between 1000 and 1500 km. Without wheels and friction, it consumes 30% less electricity than the most advanced high-speed train. At the speed of 500 km/h, the energy consumption per seat/km is only 1/3 to 1/2 of that of aircraft, which also reduces the energy consumption by 30% compared with that of cars.

Compared with EMS, due to the superconducting effect of EDS, the end-coil resistance of the vehicle is zero, so the electric energy consumed for suspension is greatly reduced, which makes it possible for economic operation. In addition, the large suspension gap of EDS also increases the safety of the system.

7.3 System Architecture of Rail Transit Intelligent Operation and Maintenance

Only countries that can manufacture intelligent equipment can realize the intelligent operation and maintenance (operation and maintenance) of equipment. Furthermore, it can be understood that intelligent technology manufacturing is an important direction of China's manufacturing technology transformation and upgrading, and is closely related to the subsequent intelligent application technology.

The development direction and technical level of a country's industrial modernization will have a far-reaching impact on the overall situation of the rail transit system, which also determines the construction of the comprehensive intelligent system of rail transit.

China's investment in rail transit is increasing. Due to the characteristics of high speed, large traffic volume, low energy consumption and small land occupation, urban rail transit has become an effective means to solve the problems of large passenger flow, traffic congestion and slow speed in transportation, which also promotes the further development of China's rail vehicle industry in the direction of intelligent manufacturing.

At present, in order to ensure the stability, safety and reliability of high-speed railway and urban rail transit system, the manufacturing of intelligent technology and equipment of rail vehicles is the industrial foundation of building operation and maintenance intelligent technology and finally can truly ensure the substantive improvement of system operation safety, comfort and efficiency [13].

7.3.1 Development Ideas of Rail Transit Intelligent Technology and Equipment

China's rail vehicle system should be developed around the idea of "original innovation, green intelligence". At the same time, with the goal of building an advanced

R&D system, we will comprehensively promote the transformation of manufacturing technology based on modern technologies such as intelligent manufacturing and network informatization.

Rail transit intelligent technology equipment specifically includes real-time online intelligent monitoring of vehicle power and related components, data information transmission, collection and analysis, intelligent detection, life prediction of vehicles and key components, various auxiliary safety and management functions, as well as self-learning and self-correction of the system.

This involves the development and application of mechanical structure optimization design, the configuration of high-strength and lightweight materials, intelligent sensors and other technologies.

The following takes urban rail transit as an example to illustrate the development ideas of intelligent technology.

1 **Development status of intelligent high-speed rail at home and abroad**

In the new round of scientific and technological revolution in the field of rail transit, intellectualization has become a breakthrough for the world's manufacturing powerful nation to seize the highland.

Through its powerful information sharing function, the computer-controlled railway (CyberRail) system in Japan enables passengers to realize fast and convenient transfer between railway transportation and other transportation. CyberRail system can obtain passengers' personal travel needs in advance, obtain the optimal travel arrangement through intelligent algorithm, and push route arrangement and navigation information to passengers in real time. At the same time, it can help the railway company optimize the train timetable and realize the unification of passengers' personalized needs and railway company's economic indicators [14].

The "railway 4.0" development plan proposed by German Railway Corporation involves many aspects, such as passenger line design, line use efficiency, logistics solutions, maintenance arrangements and so on. The plan not only enables passengers to obtain better route arrangement and transfer experience, but also enables railway companies to realize efficient and fast operation planning, automatic diagnosis of facility faults, automatic generation of maintenance plan, etc. [15]

The digital railway (DIGITALSNCF) project proposed by the French national railway company is also a good engineering technology. The project can greatly improve the production efficiency and work quality of enterprises by sharing the data of trains, road networks and station buildings through industrial Internet and Internet of Things, and using intelligent search, decision-making and other methods. At the same time, the project can also significantly reduce the energy consumption in the process of enterprise production and operation. The project can also meet the needs of passengers for punctuality and ride comfort.

China's research on intelligent high-speed rail started relatively early. A special research center has been established, and corresponding policies are constantly launched to support the research and development of intelligent high-speed rail technology. For example, the national railway intelligent transport system engineering technology research center, established in 2000, has put forward the overall framework of China's railway intelligent transport system (RITS) [16].

In 2006, the outline of China's medium- and long-term national science and technology development plan (2006–2020) has proposed to develop information and intelligent technology of transportation system, safe and high-speed transportation technology, and improve transportation network capacity and transportation efficiency.

In 2015, the "made in China 2025" strategy also proposed to carry out integrated innovation and engineering application of the integration of new generation information technology and manufacturing equipment in key links of key manufacturing fields, and promote industrial intelligent application.

In 2019, the outline for building a transportation power issued by the State Council clearly pointed out that the intelligent detection and operation and maintenance technology of transportation equipment should be promoted and applied, and promoted the deep integration of new technologies such as big data, Internet and artificial intelligence with the transportation industry.

Of course, intelligent high-speed rail is not a simple superposition of intelligent technology and high-speed rail business, but a deep integration of advanced intelligent technology and high-speed rail equipment. After years of testing and practice, the existing intelligent technologies such as big data, Internet of Things, image recognition, three-dimensional imaging and robots have been applied in China's high-speed rail system to varying degrees [17].

2 Higher embodiment of Rail Transit Intelligent Technology

Take the high-speed rail as an example. In the future, passengers will not only experience safe and reliable intelligent travel, but also enjoy comfortable and convenient intelligent services during travel.

(1) Secure and reliable intelligent travel

Intelligent high-speed rail trains can integrate the functions of sensing, diagnosis, prediction and control algorithms to sense the running status of trains and the information of their operating environment.

The train is connected with other ground systems such as motor car depot, operation command center and ticket center through wireless transmission device and obtains environmental and vehicle status information such as wind level, rainfall, snow depth, earthquake, illegal intrusion objects along the line through on-board sensor network and other equipment.

During the operation of the train, the intelligent system will carry out optimization calculation with safety, stability and comfort as the control objectives, and combine with the braking technology of multiple units. It can make decisions and judgments according to the speed limit requirements of the line and realize the technical functions of automatic train driving, efficient and energy-saving operation, active accident prevention, rapid fault disposal, etc.

For example, the Beijing Zhangjiakou intelligent high-speed railway has realized manned automatic driving with a speed of 350 km per hour for the first time in the world by using CTCS-3 (Chinese Train Control System-3) + ATO (Automatic Train Operation) train control system, BeiDou satellite navigation system, etc. When the train is running, the driver will no longer operate the train, but only need to pay attention to the fault emergency disposal. This kind of automatic driving design reduces the driver's pressure by 40% and greatly improves the train operation efficiency. Through the on-board positioning device, BeiDou satellite navigation and other technologies, not only the accuracy error of the driving speed can be controlled within 2 km/h, but also the error of the final parking position can be controlled within 10 cm, and about 15% of the energy can be saved. In addition, the Beijing Zhangjiakou high-speed railway also has the functions of automatic train departure and parking at the station, automatic door opening, linkage control of door and platform door, door anti-trap and so on, so as to provide passengers with safer services.

(2) **The system can provide more comfortable and convenient intelligent services**

The intelligent system can bring a "safe, warm, convenient and green" ride experience to passengers throughout their journey. The intelligent service of high-speed rail has made development and innovation in passenger travel safety, passenger information, intelligent control of train equipment, etc.

Passenger travel safety is regarded as the top priority of intelligent services. Image recognition technology will play a significant role in passenger access safety, baggage safety management, emergency handling and so on. After a passenger purchases a ticket, his/her personal information, riding information, image information, etc., will be collected by the ground control center through the 5 g network and transmitted by the train-ground wireless communication system to the entry and exit station and waiting system. Passengers can easily get in and out of the station and get the automatic guidance of the train carriage and its seat number through face recognition technology. People with disabilities can quickly find the location of train carriages, seats, toilets, etc., according to voice, sound and light prompts. Baggage safety is also a concern of passengers when they travel, and can also be well guaranteed. The digital camera equipped in the carriage of intelligent high-speed rail trains collects the passenger information and the characteristics of the luggage, such as shape, color and quantity, and stores these characteristics in the image control system. Once the baggage is found to be forgotten on the train, the image control system will scan the baggage first, draw the owner range of the lost baggage through the comparison of the characteristics of the front and rear images, quickly query the owner information of the lost baggage through the ticketing system and return the baggage in time. The intelligent service also carries out safety monitoring design for accidents such as fighting, fainting and fainting in the carriage. The digital camera collects the activity information of human arms, legs, feet and other parts. Through in-depth learning, it generates multiple sets of human body algorithm models for different application scenarios to identify the nature of accidents, so as to facilitate the timely implementation of the safety management of emergencies in the compartment. In addition, the intelligent service scientifically manages the driving authority of the train at different levels and adopts fingerprint identification or palmprint identification in the cab and special areas to ensure the safety of train operation.

In addition, the purpose of information service is to bring passengers a happy journey. The news information, train operation information, video programs, advertisements, etc., sent by the ground control center are transmitted to the high-speed rail train through the train-ground wireless communication system, and passengers can enjoy rich and colorful multimedia programs through the TV device in the carriage. By operating the interactive intelligent terminal equipped on the intelligent high-speed rail, passengers can easily access train operation information, station information, news, video, advertising, etc. Passengers can also use their mobile phones to surf the Internet through Wi-Fi in the carriage, and then select food, query a variety of social information, and buy goods on the App.

Intelligent high-speed rail also uses intelligent technology to control and manage multiple equipment on the vehicle, including windows, lighting, air conditioning, etc. Automatic dimming system is adopted for windows of high-speed trains. The dimming system is located in the interlayer of the train window, which can control the light transmittance of the window according to the intensity of the external light, and the light adjustment range can be up to 28%. In addition, the light transmission and shading areas can also be adjusted by passengers according to their own needs. At present, this technology has been installed and debugged on Beijing Xiong'an high-speed railway. The lighting system adopts the intelligent RGB (red, green and blue) dimming technology, which can adjust the color temperature and brightness of the lamps according to the operation status and the internal and external environment of the vehicle, and bring new visual effects to the passengers [18].

The intelligent air conditioning system can automatically adjust the temperature and humidity in the compartment according to the external climate conditions. In severe haze weather, the intelligent air purification device will exert its power to suppress toxic and harmful substances in the air and keep the air in the compartment fresh. Through the intelligent train management system, the internal environment of the vehicle becomes more warm and comfortable.

(3) **Intelligent operation and maintenance management**

Effective operation and maintenance is an important guarantee for the safe and efficient operation of high-speed trains. The intelligent operation and maintenance system for high-speed rail trains aims to further improve maintenance efficiency, reduce operation and maintenance costs, extend train service cycle, ensure safe and stable operation, and cooperate with operating enterprises to complete forward-looking planning. It adopts a service architecture with advanced perception, cloud edge integration, intelligent decision-making, remote collaboration and other functions at the core of fault prediction and health management (PHM).

(1) **Advanced perception technology**

The intelligent operation and maintenance system adopts a more comprehensive sensing coverage, a higher precision sensing method and a more intelligent sensing fusion technology to promote the application of multi-parameter variable composite sensing and wireless sensor network in train status monitoring. Therefore, it can break through the limitations of vehicle structure space and complex monitoring application environment and realize multi-dimensional comprehensive perception of train status. For example, the linear array camera and the area array camera form a stereo vision sensor: The two-dimensional linear array camera realizes rapid positioning, and the laser and the area array camera complete the parameter measurement and health scanning of wheel sets, Pantographs, catenary, underbody equipment, etc.

In addition, different sensing devices are used in different areas of the vehicle according to the train operation environment: Millimeter radar is used to detect whether there are obstacles in the wide area in front of the train track line. The laser radar is used to realize the intrusion detection of train operation limit in specific area on the ground and the monitoring of large-scale slope landslide. The online monitoring of rail health status is completed by using acoustic analysis technology.

(2) **Cloud edge integration**

The so-called cloud edge integration means that the edge end (on-board PHM) supporting real-time data processing cooperates with the cloud end (ground PHM) supporting large-scale data analysis, prediction and decision-making, and integrates train status, line characteristics, trackside detection, environment and other data to form an intelligent high-speed railway PHM system integrating distributed train status monitoring, analysis, centralized decision optimization and knowledge mining. Object (or system) health state insight and decision reference can also provide services for different departments of high-speed rail transit system through App to optimize collaboration and improve efficiency.

Take the bearing PHM of the train axle box as an example. The train axle box is the connecting device between the axle and the bogie or vehicle body, and the bearing equipment is installed inside. The bearing is close to the wheel rail contact point, there is very large vibration noise in the operating environment, and the early fault features are very weak. Therefore, it is necessary to develop advanced signal processing and feature recognition algorithms to accurately predict the fault. These algorithms require a lot of computation. If the original data are sent back to the remote center for analysis, there will be a huge cost of communication and computing resources. The cloud edge integration function of the intelligent operation and maintenance system combines a large number of tests with edge computing technology. The on-board hardware terminal completes signal processing and feature extraction and analysis, and then transmits the calculated health features to the data center, and then carries out fault identification through machine learning algorithms such as pattern recognition. In this way, the original transmission content per second has changed from hundreds of megabits (Mb) of original data to thousands of bits (kb) of characteristic information, which not only meets the real-time requirements of data analysis, but also reduces the transmission and computing pressure of the data center.

(3) **Intelligent decision**

Through data analysis and fault knowledge accumulation, the intelligent operation and maintenance system realizes automatic fault diagnosis and identification and component health analysis. The intelligent operation and maintenance system breaks the graded maintenance system, gives intelligent adjustment suggestions for different systems or equipment maintenance cycles according to the maintenance decision-making model of components and systems based on the dimensions of mileage, environment and operation status, realizes the prediction of maintenance tasks and forms an intelligent dynamic adjustment maintenance cycle, and finally forms an intelligent maintenance scheme that comprehensively improves the system operation safety, operation efficiency and reduces the maintenance cost. At the same time, it connects the original independent automation systems and information systems, automatically triggers the generation of task work orders and promotes the follow-up maintenance business. At the same time, the maintenance information is fed back to the fault analysis module to guide and optimize the subsequent fault handling process. Carry out intelligent management of the whole process of planning, process, accessories, quality management, etc., and carry out closed-loop tracking and feedback on the predictive maintenance model in combination with the work order execution, so as to continuously verify and optimize the model.

(4) **Remote collaboration**

The so-called remote collaboration means that when the train is running, the intelligent system can automatically screen the real-time fault data, analyze the relevant state parameters of the screening results, and finally identify the cause of the fault. According to the maintenance and decision-making knowledge base, the ground support system can timely provide technical support for emergency maintenance on the way and guide the on-board machinist to carry out fault emergency disposal and maintenance. In the maintenance workshop, the inspection robot can intelligently sense the working status of the equipment in real time and timely notify the operators for processing, so as to comprehensively improve the data acquisition accuracy and maintenance efficiency. The maintenance robot can go deep into the positions that are difficult for maintenance personnel to reach, quickly scan the detection parts through the high-definition vision system and send the collected high-definition image data back to the data center. The data center will compare and analyze the collected images with the corresponding images of the "healthy" high-speed rail system, so as to accurately determine the faults of train components such as bolts, wheel sets and brake disks. Therefore, the workload of manual inspection can be greatly reduced. In the monitoring center of manufacturing enterprises, the integrated monitoring service system using advanced information and visualization technology can timely judge the current operation status of products and then provide operation suggestions to the operating units. The coordinated operation of maintenance site and remote experts can significantly improve the system emergency handling capacity and maintenance efficiency.

The intelligent operation and maintenance system connects the parts demand, inventory management, sales delivery, parts logistics, business analysis and monitoring, personnel reserve, technical data and other maintenance subsystems of high-speed rail trains, realizing the unified management and scheduling of national maintenance resources, further enhancing the linkage of all links and making operation and maintenance services more efficient [19].

After years of development, China's high-speed rail network has led the world, forming a complete set of technical systems and equipment with independent intellectual property rights.

7.3.2 Vision of Rail Transit Intelligent Technology

Future oriented rail transit technology development trends and innovative technologies, including: "Rail transit vehicle safety design", "Rail transit system active safety assurance", "Rail Transit Intelligent Sensing and artificial intelligence technology", "Unbalanced classification and identification of high-speed train braking system fault diagnosis", "Rail transit linear motor traction technology", "Third generation semiconductor silicon carbide power device technology and application", "The latest development of the next generation high-speed rail transmission technology and MMC", "Fully automatic train driving technology", "Exploration and application of new technologies for Metro intelligent power supply", "Application of electronic arc extinguishing technology in DC switch", "Health monitoring of IGBT module during thermal cycle", "GPU accelerated intelligent rail transit", "Intelligent rail transit equipment", etc. These new concepts will eventually be transformed into practical and feasible technologies to be popularized [20].

(1) **Application prospect of intelligent system for urban rail vehicles**

In order to deeply understand and explore the development direction and application prospect of rail transit vehicles, and constantly move closer to the ultimate goal of modernization and intelligence, the characteristics of rail transit vehicles can be analyzed and defined according to several types of development directions of rail transit vehicles.

As far as production and manufacturing systems are concerned, they can be divided into traditional and fine types. In the process of manufacturing, traditional manufacturing system has independent data operation equipment and its operation unit, and it has realized electronic technology data function and automatic network management (including basic finance and office). Compared with the traditional manufacturing system, the fine manufacturing system has an interconnected network data information system and architecture. The fine manufacturing system can realize three-dimensional technology and mathematical modeling and has related software services and maintenance network platform system. The fine manufacturing system can realize smooth connection and connection in the production process. In the process of manufacturing, the system already has a relatively complete Internet Interworking system between network data information and equipment, can smoothly realize three-dimensional technology, complete the task of virtual simulation with life cycle, build a real-time interworking platform between equipment and network information, and its business process management in all aspects is also relatively perfect. In the manufacturing process of the digital system, the network digital information of the equipment can realize all-round data collection and application, and the production and manufacturing management process can realize visual safety monitoring. In the manufacturing process of the intelligent system, its equipment has perfect functions of self-perception, self-analysis, self-learning and decision-making. This is because the intelligent system has the ability to intelligently collect, sort, calculate, process and analyze big data, and its cloud network can smoothly connect with the industrial network. The coordination control management system platform has modern intelligent capabilities such as self-control management, learning regulation and Multi-Chain Cross management.

The future development and application trends of the intelligent system of urban rail vehicles mainly include the acceleration of enterprise restructuring and merger, the increasing globalization of industrial chain configuration and the proportion of technology "Refocusing" in the sales of components and after-sales services will increase significantly. The purpose of enterprise reorganization and merger is to reduce risks and costs. Enterprise alliance can not only promote technological innovation, but also improve production capacity and efficiency. The significance of the globalization of industrial chain allocation lies in the ability to cope with the complex and changeable market environment. In fact, different regions need different configurations. The "Refocusing" of technology means that the key technology should occupy the core position in intelligent manufacturing. Enterprises need to focus on the development of core technologies to improve the flexibility of manufacturing. The sales and after-sales service of component parts are the downstream part of the whole industrial chain. However, such value-added activities are also crucial. In the face of fierce competition in the market environment, the importance of the service industry is also the trend of future development.

(2) **Development trend of Urban Rail Transit**

Relying on "Internet + transportation", urban rail transit is also gradually developing toward digitalization, networking, operation automation and intelligent operation. With the rapid development of Internet of Things, cloud computing, big data, artificial intelligence and other technologies, the digital and intelligent management of urban rail transit network can be realized by building an Internet of Things platform to coordinate the management of station equipment and facilities and build various station application scenarios. Relying on new technologies and equipment, train operation has changed to scenario-based full-automatic operation. With the maturity and application of intelligent technology and industry, the passenger services of the station, such as ticketing service, information service, passenger transport organization and life service, have gradually changed to flexible, diverse, convenient and intelligent. The internal management of the station has changed to be scientific, efficient and refined.

In the future, the operation and management of the station will tend to be "wisdom", which is mainly reflected in two aspects: intelligent service and intelligent management. Among them, smart service refers to providing "smart" comprehensive and personalized services for passengers' travel with intelligent scientific and technological means. Intelligent management is an efficient and safe management that uses intelligent equipment and facilities to realize the "autonomy" of stations. In recent years, several major cities in China (such as Beijing, Shanghai, Guangzhou, and Shenzhen) have made great progress in automatic train operation, smart station construction, Metro cloud application relying on Internet of Things technology, etc. These technological developments have driven the innovation of other industries, so they are leading the urban rail transit industry into a new stage.

Based on the big data platform, the urban rail transit system has been able to accurately serve passengers through subway App, intelligent guidance, intelligent customer service and other technical means. Online, it provides passengers with the required conductor information and the transfer solution of the whole transportation network, so that passengers can first draw up the riding scheme and determine their own optimal transfer path on the online client. In other words, the system can also accurately formulate the transportation solutions for each passenger. Offline, the system can provide passengers with transfer and various service information relying on intelligent guidance and intelligent customer service in the station. At the same time, the "multi-function box" around the station can be combined to meet the precise needs of all kinds of passengers.

References

1. Lin H, Wang L, Wei Y (2019) A study of the EU Shift2 rail plan. Foreign Railway Veh 56(1):11–16 (in Chinese)
2. Ahiko Y (2017) Development of N700S validation test vehicle. Railway Age (1):10
3. Su X (2019) N700S confirmation test vehicle of high-speed train. Mod Urban Transit 4:79–83 (in Chinese)
4. Han W, Dong X (2017) Analysis on the development trend of foreign high-speed EMUs. Locomotive Rolling Stock Technol 4:8–10 (in Chinese)
5. Lin G, Lian J (1998) Development of Maglev high speed railway in Japan and technical and system characteristics of Yamanashi test line. Electr Drive Locomotive (4):7–10 (in Chinese)
6. Miao B, Zhang W, Chi M et al (2019) Analysis and prospect of key technical features of next generation high speed trains. J China Railway Soc 3:58–70 (in Chinese)
7. Yan L (2002) Consideration on the development strategy of high-speed Maglev in China. Eng Sci 4(12):40–46 (in Chinese)
8. Zhang R, Yan L, Xu S et al (2004) A new high speed maglev train—Swiss vacuum pipeline high speed maglev train scheme. Converter Technol Electr Traction (1):44–46 (in Chinese)
9. Shen Z (2005) On developing high-speed evacuated tube transportation in China. J Southwest Jiaotong Univ 40(2):133–137 (in Chinese)
10. Feng Z, Fang X, Li H et al (2018) Technological development of high speed maglev system based on low vacuum pipeline. Strategic Study of CAE 20(6):105–111 (in Chinese)

11. Yang J, Shi H, Hu H et al (2020) ADRC algorithm for single-point suspension system. J Ordnance Equipment Eng 41(4):167–171 (in Chinese)
12. Wang Y, Chen S, Su Z (2001) Comparison and analysis of EMS and EDS maglev train systems. Rolling Stock 39(10):17–20 (in Chinese)
13. Yu Z, Ren K, Xu W (2020) Development trend of high speed rail transit. Equipment Manuf Technol 3:230–240 (in Chinese)
14. Shi T, Zhang H, Jia L (2003) Railway intelligent transport system in Japan—CyberRail framework and research status. China Railway Sci 24(6):82–88 (In Chinese)
15. Nie N, Guan K, Zhong Z (2017) German railway 4.0 strategy. China Railway (5):86–90 (in Chinese)
16. Wang T (2019) Study on the development strategy of China intelligent high speed railway. China Railway 1:9–14 (in Chinese)
17. Liang J (2020) A new era of intelligent railway equipment. Science 72(2):17–22 (in Chinese)
18. Zhang X, Ying J, Cheng Y et al (2008) Intelligent color automatic adjustment method for TV camera. Chinese Invention Patent CN200810201622.5. October 2008
19. Sun Z (2017) "Fuxing" bullet trains open up a new era in Chinese Railway construction. Science 69(6):22–24 (in Chinese)
20. Sun Z (2016) From pursuer to leader in the advanced technology of high-speed rail. Science 68(6):27–31 (in Chinese)

The manufacturer's authorised representative in the EU is Springer Nature Customer Service Centre GmbH, Europaplatz 3, 69115 Heidelberg, Germany. If you have any concerns regarding our products, please contact ProductSafety@springernature.com

Printed and bound by CPI Group (UK) Ltd, Croydon, CR0 4YY
30/09/2026
02225315-0007